Process Development Division 2015

Core Programming Area at the 2015 AIChE Annual Meeting

Salt Lake City, Utah, USA
8-13 November 2015

ISBN: 978-1-5108-1871-2

Printed from e-media with permission by:

Curran Associates, Inc.
57 Morehouse Lane
Red Hook, NY 12571

Printed by Curran Associates, Inc. (2016)

For permission requests, please contact AIChE
at the address below.

AIChE
120 Wall Street, FL 23
New York, NY 10005-4020

Phone: (800) 242-4363
Fax: (203) 775-5177

www.aiche.org

Additional copies of this publication are available from:

Curran Associates, Inc.
57 Morehouse Lane
Red Hook, NY 12571 USA
Phone: 845-758-0400
Fax: 845-758-2634
Email: curran@proceedings.com
Web: www.proceedings.com

TABLE OF CONTENTS

429453 A Multivariate Herschel-Bulkley Rheological Model for Nanoparticle-Based Drilling Fluids

Abstract:

Injection fluids used during oil and gas drilling require specific rheological properties, and a bentonite-based drilling fluid enhanced with iron nanoparticles has been analysed to investigate its suitability for high-performance drilling applications. Accordingly, it is necessary to model the shear stress and viscosity in terms of the fluid shear rate, temperature and concentration of iron nanoparticles (Dandekar, 2013). By developing models relating these variables it is possible to examine how the viscoplastic fluid behaves under different drilling conditions and if it can reliably ensure process performance. Under steady shear, flow behaviour will vary with concentration and temperature (Andertova & Rieger, 2009), and careful tuning of rheological characteristics is critical to the design of the process. The currently prevalent tri-parametric model relating shear stress and shear rate is the Herschel-Bulkley model (Kelessidis et al., 2006; Kelessidis & Maglione, 2008; Pouyafar & Sadough, 2013) which combines the effects of power-law and Bingham plastic behaviour in a fluid. The parameters estimated from experimental campaigns are temperature- and concentration- dependent, and literature models are usually condition-dependent. Therefore, it is important to develop a unified model including all three independent variables (shear rate, temperature and iron nanoparticle concentration) in order to describe in detail the rheological behaviour of the injection fluid, which is of critical importance during drilling operations.

The literature has been extensively surveyed to identify the most widespread and suitable models of drilling fluids (Nguyen & Boger, 1992; Balmforth et al., 2014). A Grace M3600 Fann type rotational viscometer has been used to measure shear stress and viscosity at three temperatures and several iron nanoparticle concentrations (Yan & James, 1995). Multivariate nonlinear least squares regression (Berge, 1993) has been employed in order to determine the most accurate model (Puxty et al., 2005) for shear stress and viscosity as functions of shear rate, temperature and additive concentration; the multi-parametric estimation problem has been subsequently solved using the novel experimental data, in order to compute the optimal set of parameters which minimise the sum of squared errors (SSE) and maximise the respective coefficient of determination, R^2(Graybill & Iyer, 1994). Ensuring that a global optimum is achieved is of particular importance, because the strongly nonlinear multivariate expression may induce trapping in various local minima; the multivariate nonlinear regression has thus been performed using a systematic strategy and several starting parameter sets to confirm optimality. The multivariate and multi-parametric nonlinear estimation models have been plotted against the experimental data, the standard error for shear stress and viscosity at each combination of independent variables has been computed, and error bars have been obtained to illustrate uncertainty.

Nonlinear model plots have been analysed to show that injection fluids are shear thinning fluids with yield stress which follow the Herschel-Bulkley fluid model for shear stress and the power law model for viscosity (Abu-Jdayil & Ghannam, 2014). The multiplicative rule has been found to provide reliable estimates, combining the Herschel-Bulkley form with an Arrhenius model for temperature and a nonlinear model for concentration dependence. Higher temperatures cause increased standard error for shear stress and viscosity, indicating that temperature affects the structure and rheology of the fluid significantly, increasing the expected prediction uncertainty. Further investigation of concentration effects for various additives has also been conducted, indicating that the model already provides reliable fluid behaviour prediction via quantitative estimation of shear stress and viscosity, in a wide spectrum of experimental conditions relevant to drilling.

LITERATURE REFERENCES

1. Abu-Jdayil, B., Ghannam, M., *Energ. Source Part A*, 2014, **36**(10), 1037.
2. Andertova, J., Rieger, F., *Ceramics-Silikáty*, 2009, **53**(4), 283.
3. Balmforth, N.J., Frigaard, I.A., Overlez, G., *Annu. Rev. Fluid Mech.* 2014, **46**, 121.
4. Berge, J.M.F., *Least Squares Optimization in Multivariate Analysis*, DSWO Press, 1993.
5. Dandekar, A.Y., *Petroleum Reservoir Rock and Fluid Properties*, CRC Press, 2013.
6. Graybill, F.A., Iyer, H., *Regression Analysis: Concepts and Applications*, Duxbury Press, 1994.
7. Kelessidis, V.C., Maglione, R., Tsamantaki, C. et al., *J. Petrol. Sci. Eng.*, 2006, **53**(3-4), 203.
8. Kelessidis, V.C., Maglione, R., *Colloid Surface A*, 2008, **318**(1-3), 217.
9. Nguyen, Q.D., Boger, D.V., *Annu. Rev. Fluid Mech.*, 1992, **24**, 47.
10. Pouyafar, V., Sadough, S.A., *Metall. Mater. Trans. B.*, 2013, **44B**(5), 1304.
11. Puxty, G., Maeder, M., Hungerbühler, K., *Chemometr. Intell. Lab.*, 2006, **81**(2), 149.
12. Sheng, J.J., *Modern Chemical Enhanced Oil Recovery – Theory and Practice*, Elsevier, 2011.
13. Yan, J., James, A.E., *J. Non-Newton. Fluid*, 1997, **70**(3), 237.

430072 Shear Thickening Fluid (STF) – Nanocomposites for Improved Puncture and Hypervelocity Impact Resistance

Abstract:

Concentrated colloidal dispersions exhibit unique flow properties that depend on the applied rate of deformation, among these being the phenomena of shear thinning and shear thickening. Shear thinning occurs at low shear rates and results in a reduction of the dispersion viscosity. Contrarily, shear thickening manifests itself at high shear rates where the dispersion exhibits a reversible increase in viscosity. For the most concentrated dispersions, a sudden reversible transition from a flowing state to a rigid, solid-like state. In many industrial processes, shear thickening is often undesirable as it can damage equipment and limit flow rates for pumping. However, when integrated with ballistic fabrics as STF-Armor™, this reversible fluid-to-solid transition makes these dispersions attractive for use as field-responsive protective materials such as soft body armor and correctional armor applications.

In low-earth orbit, astronauts performing extra-vehicular activities are exposed to the dangers of micrometeoroid and orbital debris (MMOD). While generally less than a centimeter in size, these MMOD particles can travel at velocities on the order of tens of kilometers per second, rending them highly energetic and a threat to puncture the thermal micrometeoroid garment (TMG) lining the pressurized air bladder. An equally dangerous threat to astronauts occurs as the result of MMOD impacts into the exterior of space vehicle. These impacts result in the formation of craters with sharp raised lips. These sharp surfaces are a threat to puncture the TMG and have resulted in the premature termination of spacewalks outside the International Space Station due to suit tears. While the TMG is designed to withstand direct MMOD impacts, the record of suit tears during spacewalks suggests the need to enhance the resistance to cutting and puncture threats which involved drastically different energy and time scales than direct MMOD impacts. As such, STF-Armor™ is an attractive candidate composite material to incorporate within the TMG.

We replaced the traditional neoprene-coated nylon absorber layers in the TMG with STF-Armor™ and studied their resistance to quasi-static puncture and hypervelocity impact threats. We used hypodermic needles to simulate the cutting and puncture threats associated with the sharp cratered surfaces on the International Space Station. Under quasi-loading, TMG lay-ups containing STF-Armor™ increased the peak force required to puncture the lay-up by 72% while the flexibility of the lay-up was unaltered and its areal density was reduced by 13%. The hypervelocity impact resistance of the lay-ups containing STF-Armor™ was investigated using spherical aluminum projectiles at velocities ranging from 4.5-7.5 km/s. The ballistic limit of these lay-ups containing STF-Armor™ is shown to provide meaningful hypervelocity impact protection. On-going experiments concerning the effects of the real space environment on the puncture and hypervelocity impact resistance of lay-ups containing STF-Armor™ and an International Space Station Flight Opportunity will be discussed.

435594 Integration of Advanced Physical Property Models in the Multi-Scale Simulation of Refinery Heat Exchangers Undergoing Crude Oil Fouling

Abstract:

Improving energy efficiency in oil refineries is becoming increasingly important, especially in the crude distillation unit where energy consumption is excessively high. In that unit, energy from the hot distillation products is recovered in the pre-heat train, an extensive network of heat exchangers, to heat the crude oil from storage temperature before entering the distillation column. This process is however hindered by fouling, thereby resulting in gradually lower energy efficiencies as the deposits build up on the heat transfer surfaces. In this context, modelling of heat transfer equipment and fouling has emerged as a promising tool to assist in understanding the complex phenomena involved in fouling, monitoring of heat exchangers, prediction of fouling behaviour and proposition of mitigation options.

Recent work in this area has led to the development of a dynamic, distributed model of a shell and tube heat exchanger [1], implemented in Hexxcell Studio™ [2], which has been successful in predicting fouling in heat exchanger units. The model assesses deposition and accounts for the variation of the physical properties of the fluids involved as function of local conditions.

Thermo-physical properties are involved directly in heat and mass balances and heat transfer rates, and thus an accurate prediction of those properties is important for monitoring, performance assessment and design purposes. Much work over the past two decades has led to the development of the Statistical Association Fluid Theory (SAFT) into an equation-of-state (EOS) form. Due its physical basis, this EOS has found good use in describing the fluid phase behaviour and thermodynamic properties of complex mixtures [3].

In this work, part of the UNIHEAT project [4], advanced SAFT models are incorporated into the above heat exchanger model. A group contribution version of the SAFT-γ theory is used employing a coarse grained (CG) approach [5]. The thermodynamic models are used to calculate properties such as heat capacity and density as function of local conditions (temperature, pressure) and composition. Crude oil is modelled as a mixture of a finite number of pseudo-components with composition characterised based on their true boiling point (TBP) curve, using Multiflash [6]. Using critical properties and acentric factors of the pseudo-component fractions, the required Mie force-field parameters were obtained from a corresponding states treatment. Once the mixture representing the crude oil is fully defined and the parameters for each pseudo-component are determined, the thermo-physical properties can be estimated as function of local operating conditions. This multi-scale approach is demonstrated for two crude oils in a typical industrial heat exchanger at refinery conditions. Finally, results are compared to those obtained using other physical properties modelling methods, such as cubic EoS and correlations from oil industry.

Acknowledgments

This research was performed under the UNIHEAT project. The authors wish to acknowledge the Skolkovo Foundation and BP for financial support. The support of Hexxcell Ltd, through provision of Hexxcell Studio™, is also acknowledged.

[1] Coletti F, Macchietto S. A Dynamic, Distributed Model of Shell-and-Tube Heat Exchangers Undergoing Crude Oil Fouling. Ind Eng Chem Res. 2011;50(8):4515–4533.

[2] Hexxcell Ltd., 2015. Hexxcell Studio™. http://www.hexxcell.com.

[3] Müller, E. A. and Jackson, G. Force-Field Parameters from the SAFT-γ Equation of State for Use in Coarse-Grained Molecular Simulations. *Annual review of chemical and biomolecular engineering*, *5*, 405-427.

[4] Macchietto, S., 2015. Energy Efficient Heat Exchange in Fouling Conditions: the UNIHEAT Project. Proceedings, Heat Exchanger Fouling and Cleaning XI, Enfield (Ireland), June 7-11.

[5] Mejía, A., C. Herdes, and E.A. Müller, Force Fields for Coarse-Grained Molecular Simulations from a Corresponding States Correlation. Industrial & Engineering Chemistry Research, 2014. 53(10): p. 4131-4141.

[6] Infochem, Multiflash. http://www.kbcat.com/technology/infochem-software.

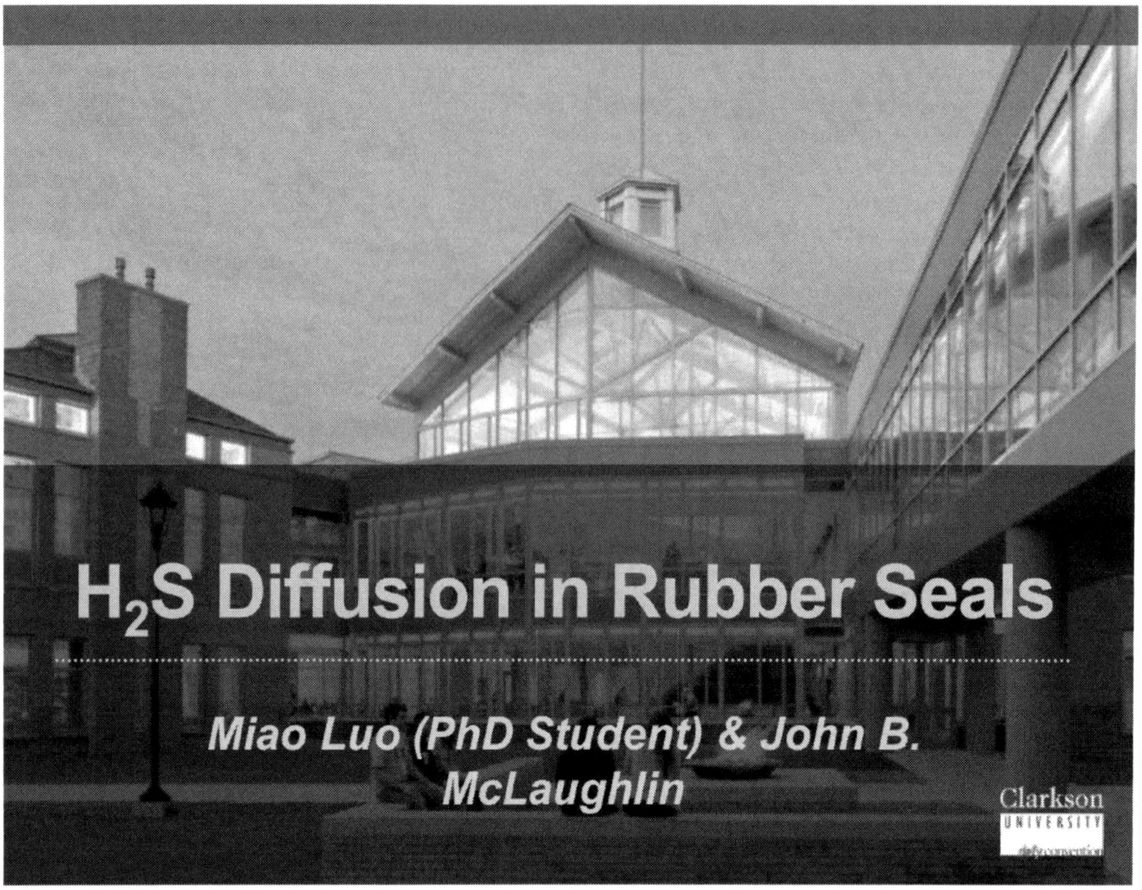

Outline for presentation

- Literature review of previous work on this field
- Molecular dynamic methods we use in simulations
- Typical results and discussion

12/17/2015 2

Clarkson UNIVERSITY

Degradation of Hydrogenated Nitrile-butadiene Rubber in Aqueous Solutions of H2S or HCl

- The properties of HNBR exposed to aqueous solutions of H_2S deteriorated significantly compared to HNBR exposed to aqueous solutions of HCl and an unexposed sample.
- The mechanism for this deterioration is the formation of C(=O)—NH_2, C—S—C and C=S detected by ^{13}C nuclear magnetic resonance(^{13}C NMR), X-ray photoelectron and infrared spectroscopies.
- The formation of C—S—C was due to a radical reaction initiated by the mercapto radical and that of C=S was due to a nucleophilic reaction initiated by mercapto cations.

Cong, C. B., Cui, C. C., Meng, X. Y., Lu, S. J., & Zhou, Q. (2013). Degradation of hydrogenated nitrile-butadiene rubber in aqueous solutions of H2S or HCl.*Chemical Research in Chinese Universities*, *29*(4), 806-810.

Clarkson University

12/17/2015 3

Degradation of Hydrogenated Nitrile-butadiene Rubber in Aqueous Solutions of H2S or HCl

sample	Hardness(shore A)	Tensile strength/Mpa	Ultimate elongation(%)
HNBR-1	68	22.8	657
HNBR-2	66	19.3	571
HNBR-3	63	18.6	270

- HNBR-1 is the unexposed sample
- HNBR-2 was exposed to 10% hydrochloric acid and heated at 140 °C for 24 h
- HNBR-3 was exposed according to the NACE TM0187-2003 standard. According to the author, the temperature for this experiment is around 100 °C

Clarkson University

12/17/2015 4

Degradation of Hydrogenated Nitrile-butadiene Rubber in Aqueous Solutions of H2S or HCl

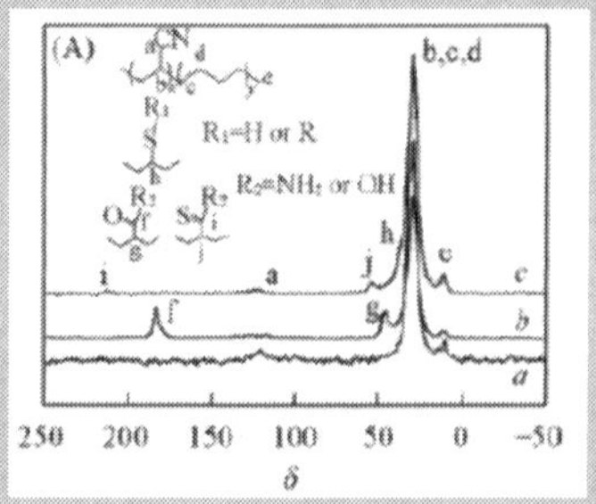

^{13}C NMR

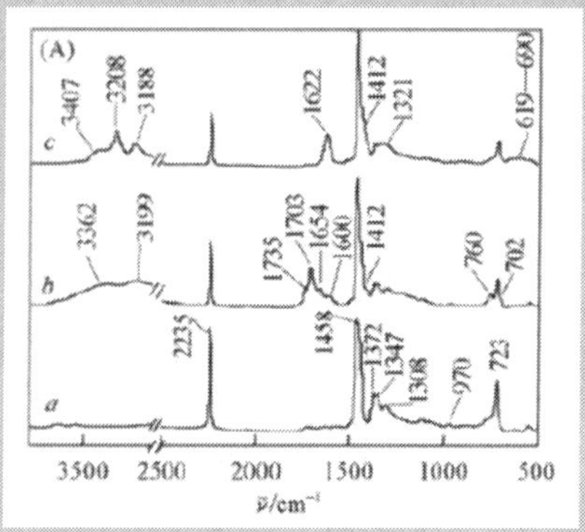

infrared spectroscopy

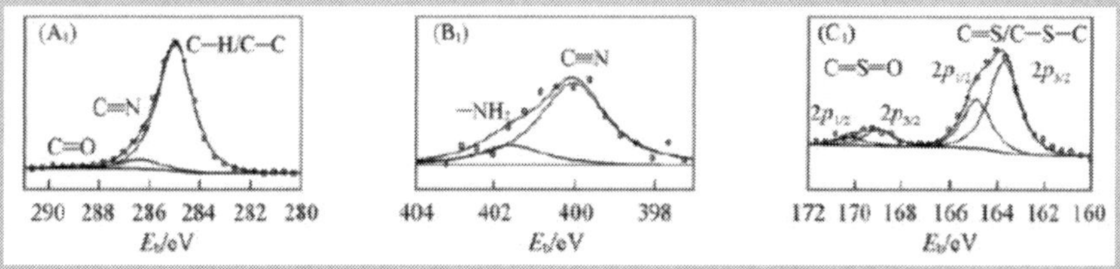

X-ray photoelectron spectroscopy

Degradation of Hydrogenated Nitrile-butadiene Rubber in Aqueous Solutions of H2S or HCl

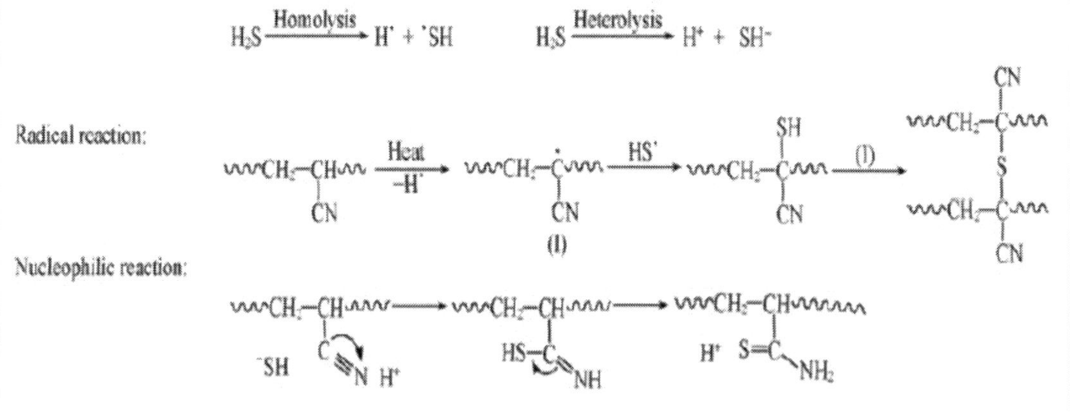

Scheme 1 Reaction mechanism of HNBR in H_2S aqueous solution

Diffusion Coefficients of Small Gas Molecules in Amorphous cis-1,4-polybutadiene Estimated by Molecular Dynamics Simulations

- Small gas molecules (Ar, O_2, N_2, CO_2, and CH_4) diffusivities in amorphous *cis*-1,4-polybutadiene in the temperature range 250 to 400K were studied by performing MD simulations.
- Diffusion coefficients were calculated from long NVT MD simulations for up to 3 ns at temperatures ranging from 250 up to 400 K.
- Calculated diffusion coefficients compare well with experimental data as well as previous published work, though a systematic overestimation is found due to the finite-size effect of the model.

Meunier, M. (2005). Diffusion coefficients of small gas molecules in amorphous cis-1, 4-polybutadiene estimated by molecular dynamics simulations. *The Journal of chemical physics*, *123*(13), 134906-134906

Diffusion Coefficients of Small Gas Molecules in Amorphous cis-1,4-polybutadiene Estimated by Molecular Dynamics Simulations

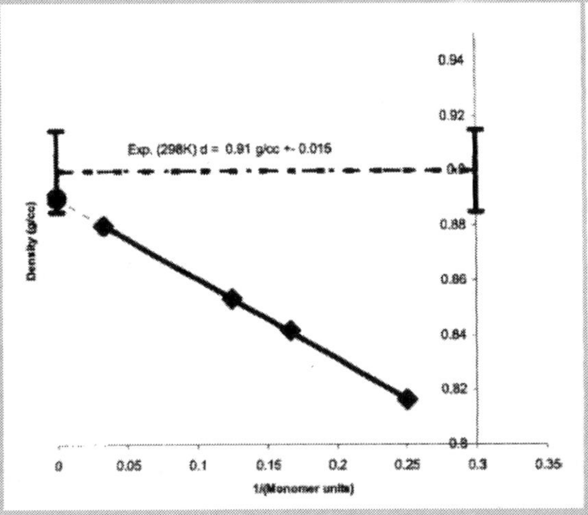

Density versus 1/monomer units

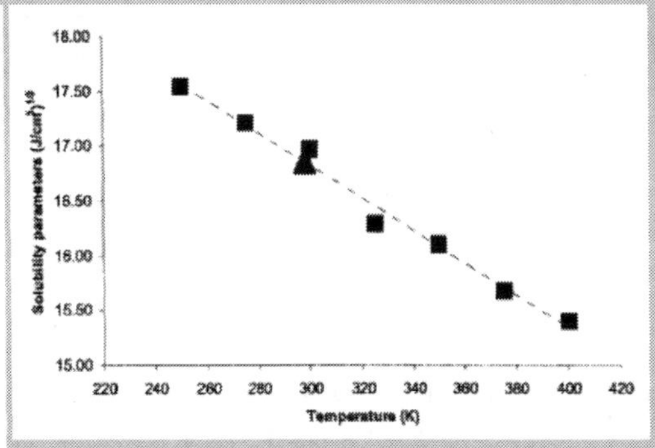

Solubility parameter versus temperature

Molecular Dynamic Simulations Method

The increase of mean square displacement (MSD) with time is related to the diffusion coefficient D as follows:

$$D = \frac{1}{6N_a} \lim_{t\to\infty} \frac{d}{dt} \sum_{i}^{N_a} \langle [r(t) - r(0)]^2 \rangle$$

where N_α is the number of diffusive atoms in the system. The MSD in Forcite is already averaged over all atoms in targeted molecules.

To determine the diffusion coefficient we fit a straight line, $y = a*x+b$, to the data in the diffusive regime and extract the slope a. The units of a are Å^2/ps. According to the above definition, D is obtained as follows:

$D = a/6$

M. Meunier "Diffusion coefficients of small gas molecules in amorphous cis-1,4-polybutadiene estimate by molecular dynamics simulations", J. Chem. Phys., 123, 134906 (2005).

Clarkson University

12/17/2015 9

Diffusion of Small Gas Molecules in Amorphous cis-1,4-polybutadiene

	Ar	O_2	N_2	CO_2	CH_4
Expt(298K)	4.06	1.5	2.96	1.05	
Calc(300K)	7.03	9.5	8.8	5.3	7.5
Calc(ref)					2.25

- D is multlplied by 10^{-6} cm^2/s
- Inhomogeneity in the way that the chains are placed in the computational box may result in gas molecules becoming trapped in low density regions, and this will increase the computed value of the diffusion coefficient.
- The flexibility of polymer chains may be too small because the chains are short, and this may also cause an apparent increase in the diffusivity of the gas molecules.

M. Meunier "Diffusion coefficients of small gas molecules in amorphous cis-1,4-polybutadiene estimate by molecular dynamics simulations", J. Chem. Phys., 123, 134906 (2005).

Clarkson University

12/17/2015 10

Molecular Dynamic Simulations Method

- Build the amorphous cell of the molecules.
- Perform geometry optimization on the system.
- Use the NPT ensemble to obtain the correct density.
- Use the NVT ensemble to ensure that the system is stable at the given temperature.
- To calculate the diffusion coefficient, it is necessary to perform a dynamic simulation using the NVE ensemble to exclude the influence on the thermostat.
- The COMPASS force fields are used in all of the molecular dynamics simulations.

Sun, Huai, 1998, COMPASS: an ab initio force-field optimized for condensed-phase applications overview with details on alkane and benzene compounds

Sample Used in our simulations

FKM: The FKM used in our simulation is composed of PVDF(vinylidene fluoride), HFP(hexafluoropropylene), TFE(tetrafluoroethylene). The mole ratio is VDF:HFP:TFE=2:1:1, which causes the fluorine content to be 70%. The molecular weight is 14613.9.

HNBR: The HNBR used in our simulations contains about 17% acrylonitrile and 5% butadiene. The molecular weight is 5550.99

All the HNBR and FKM indicated in the slides are the same.

Grafted HNBR: grafted HNBR is obtained by attaching the Capstone™62-MA to the chains of HNBR with different mole ratios.

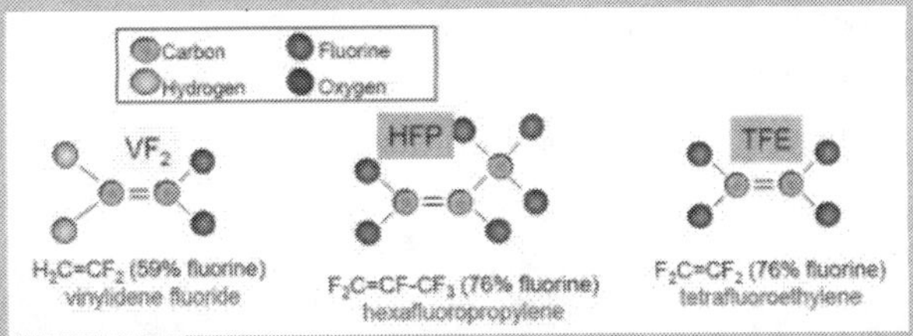

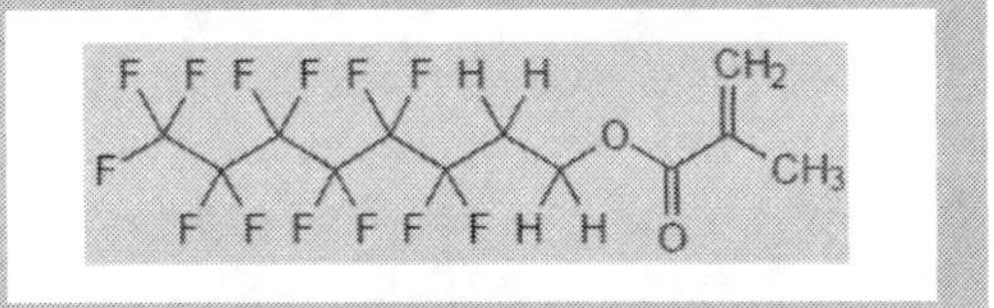

Capstone™62-MA

Simulation Results for the Diffusivities of H2S

	FKM	HNBR	Grafted HNBR
180 F	$1.168*10^{-3}$	0.056	$2.47*10^{-3}$
250 F	$1.62*10^{-3}$	0.156	$4.45*10^{-3}$
350 F	$2.92*10^{-3}$	0.226	0.011

The diffusion coefficients in the table have the unit 10^{-4} cm^2/s.
The method used to calculate these diffusion coefficients was shown on slide 9.

12/17/2015 13

Simulation Results Plots for H2S Diffusivities

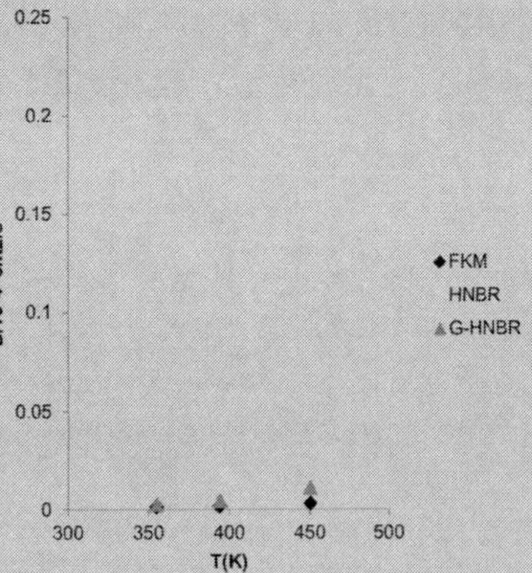

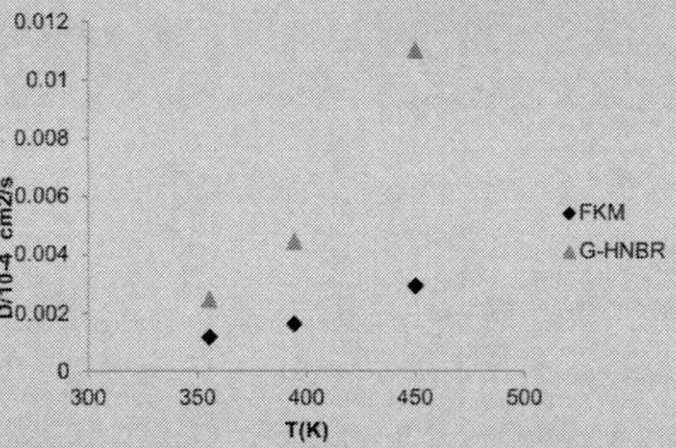

These two plots shows how the value of the diffusion coefficient varies with temperature. The data in these plots are taken from slide 13.

12/17/2015 14

Simulation Results

Mole ratio of capstone	Density (100 mers)	Density (200 mers)	SP (100 mers)	SP(200 mers)
0%	0.869			
2%	0.918	0.920	16.297	16.44
4%	0.966	0.965	16.081	15.97
6%	1.007	1.008	15.844	15.54
8%	1.045	1.048	15.318	14.82

Mole ratio of capstone x% in the chart means x capstone molecules are grafted to the HNBR chains with 100 mers. X% is the mole ratio of capstone molecules to the number of mers of HNBR chains.
The density has the unit g/cm^3.
SP= Solubility Parameter which has the unit $(J/Mpa)^{1/2}$

Effect of Capstone Content on Cohesive Energy Density and Mass Density

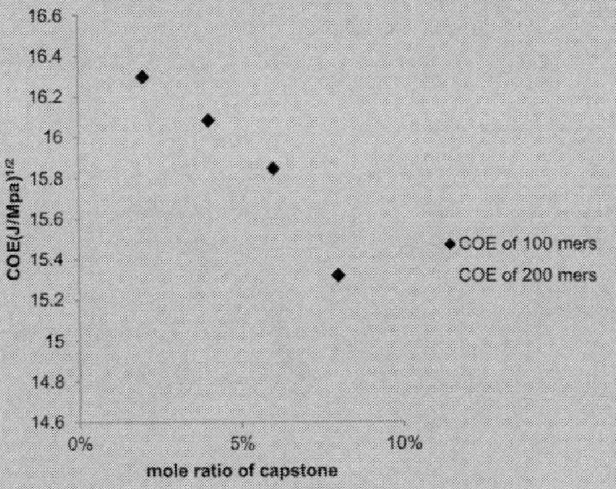

The cohesive energy density is the amount of energy needed to completely remove unit volume of molecules from their neighbors to infinite separation (an ideal gas). This is equal to the heat of vaporization of the compound divided by its molar volume in the condensed phase

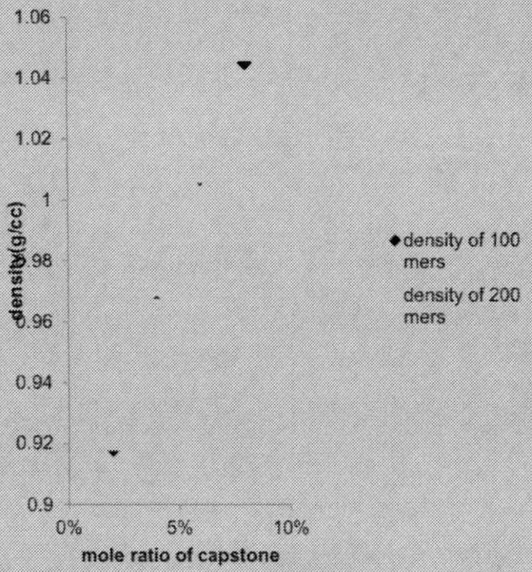

From the graph we can see that the chain length in the simulation has almost no influence on the density and little influence on the cohesive energy density.

Diffusivity of H2S in grafted HNBR

Mole ratio of capstone	Diffusion Coefficient Value(cm^2/s)
0%	$14.8*10^{-7}$
2%	$9.933*10^{-7}$
4%	$6.103*10^{-7}$
6%	$4.704*10^{-7}$
8%	$3.126*10^{-7}$

Mole ratio of capstone x% in the chart means x capstone molecules grafted to the HNBR chains with 100 mers. X% is the mole ratio of capstone molecules to the number of mers of HNBR chains.

12/17/2015 17

Clarkson UNIVERSITY

Diffusivity of H2S in grafted HNBR

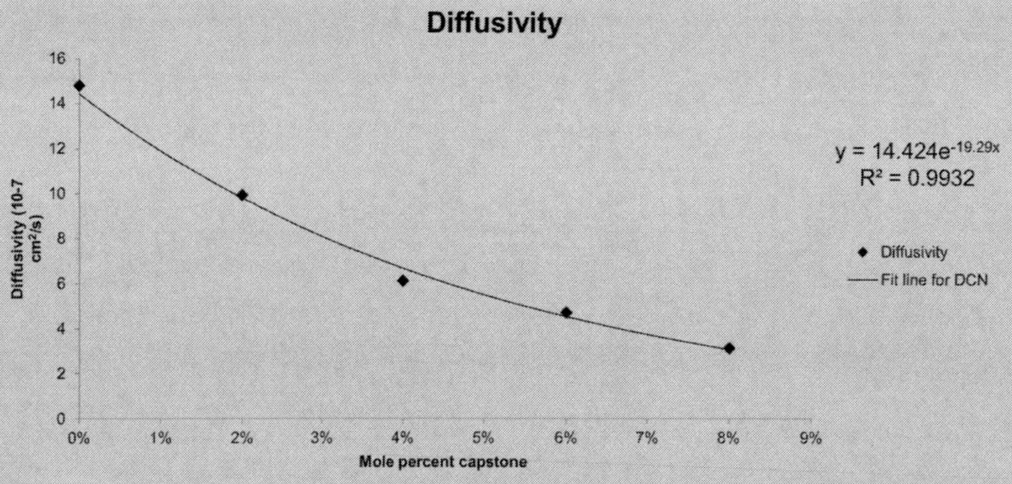

12/17/2015 18

Clarkson UNIVERSITY

Glass Transition Temperature of Grafted HNBR

temperature	$D(T)*10^9$
300	13.3
287.5	8.83
275	5.5
262.5	5.1
250	3.15
237.5	1.114
225	0.775

D is the self diffusion coefficient calculated by Materials Studio, with unit cm^2/s

Clarkson UNIVERSITY

12/17/2015 19

Glass Transition Temperature of Grafted HNBR

Following Nealey, we fit the data by the modified WLF equation:

$$log\frac{D(T)}{D(T_g)} = \frac{C_1(T - T_g)}{C_2 + T - T_g}$$

We fit the equation and obtain:

$$log\frac{D(T)}{1.737 * 10^{-6}} = \frac{3.75\,(T - 206.5)}{94.8 + T - 206.5}$$

with R^2=0.9732

Therefore, we estimate that the glass transition temperature of the grafted HNBR is approximately 206 K. This is somewhat smaller than the normal range of 263.95-235.35 for pure HNBR, and this improves the low temperature performance.

Torres, J. A., Nealey, P. F., & De Pablo, J. J. (2000). Molecular simulation of ultrathin polymeric films near the glass transition. *Physical Review Letters, 85*(15), 3221.

Severe, G., & White, J. L. (2000). Physical properties and blend miscibility of hydrogenated acrylonitrile-butadiene rubber. *Journal of applied polymer science,78*(8), 1521-1529.

Clarkson UNIVERSITY

12/17/2015 20

Synthesis of polymers grafted with short fluoroalkyl groups using diene copolymer precursors

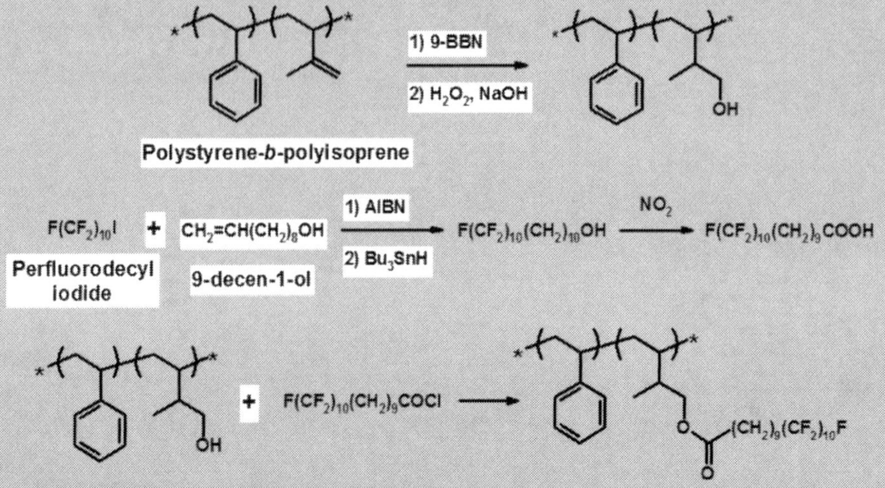

S. Krishnan, N. Wang, C. K. Ober, et al., Comparison of the fouling release properties of hydrophobic fluorinated and hydrophilic PEGylated block copolymer surfaces, Biomacromolecules ***2006****, 7, 1449-1462*

Clarkson University

12/17/2015 21

Synthesis of polymers grafted with short fluoroalkyl groups using diene copolymer precursors

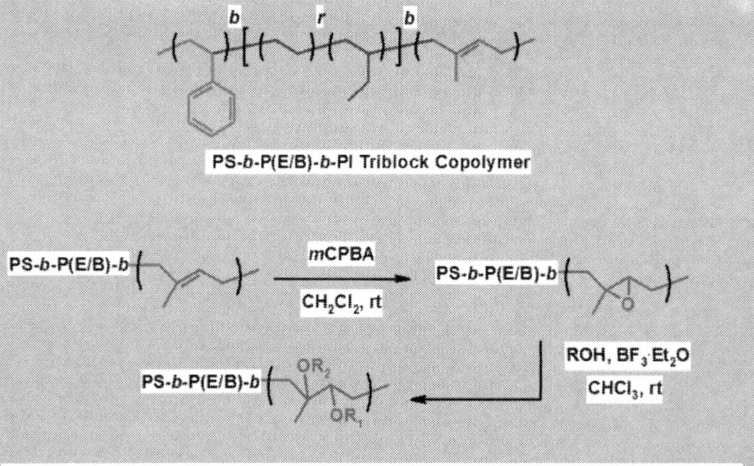

R_1, R_2 = H or R, where R = $-CH_2CH_2(OCH_2CH_2)_x(CF_2CF_2)_yF$

C. J. Weinman, J. A. Finaly, D. Park, M. Y. Paik, S. Krishnan et al., ABC triblock surface active block copolymer with grafted ethoxylated fluoroalkyl amphiphilic side chains for marine antifouling/fouling-release applications, Langmuir ***2009****, 25, 12266-12274*

Clarkson University

12/17/2015 22

Conclusion

- The density of capstone grafted HNBR increases linearly with an increase of mole ratio of capstone.
- The Diffusivity of H2S decreases exponentially with an increase of mole ratio of capstone.
- Grafting capstone chains to HNBR will decrease the glass transition temperature of HNBR which means better low temperature performance.

12/17/2015 23

Clarkson UNIVERSITY

NJIT

Energy storage materials with nano-encapsulated inclusions of an easy to melt metal

Ani Abraham

Mirko Schoenitz

Edward L. Dreizin

Department of Chemical, Biological and Pharmaceutical Engineering
New Jersey Institute of Technology Newark, NJ 07102 USA

NJIT

Introduction

- Phase change materials (PCM) or energy storage materials (ESM) are extensively studied recently.
  - Absorbs energy during melting and releases energy during solidification
  - Temperature is constant during the phase change process
- Common shortcoming of many ESM is their **low thermal conductivity**, which results in poor heat exchange.
  - Organic compounds: 0.15 – 0.3 W/(m °C)
  - Salt hydrates: 0.4 – 0.7 W/(m °C)

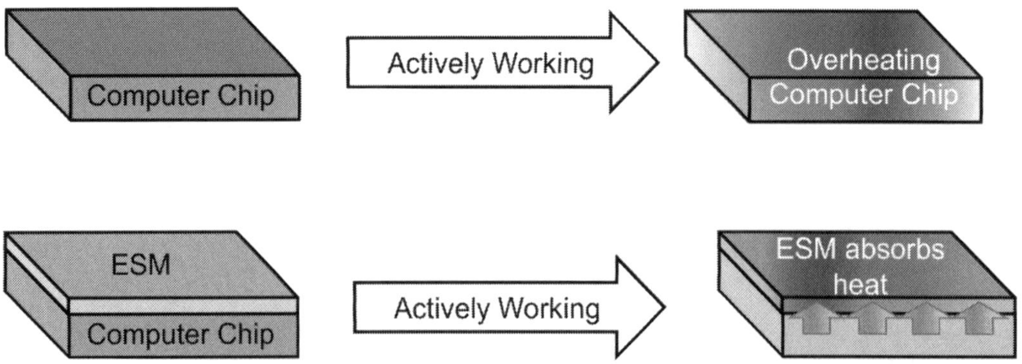

NJIT

Introduction

- Enhanced thermal conductivity of the ESM is critical for
  - Thermal management of electronics
  - Storage of solar thermal energy
- Metal-based ESM are attractive due to **high thermal conductivity** and **strength** (Ge et al., 2013)
  - **Containment** of liquid metal is challenging
  - **Bulk containers** are often used, limiting the type of structures for ESM
- Use of nanoparticles of Bi metal in Ag matrix and polymer resin was demonstrated recently (Liu et al., 2013 & 2015)
  - Ag matrix: **coalescence or migration** of liquid metal; alloy formation can occur;
  - Polymer resin matrix: **jeopardize mechanical strength** of ESM

References

Ge et al.,, *Renewable Sustainable Energy Rev.*, 21 (2013) 331-346.

Liu et al., *Nanoscale* 5 (2013) 7234-7237.

Liu et al., *ACS Nano* 9 (2015) 1341-1351.

NJIT

Technical Approach

- Prepare **reactive nano-thermite** using mechanical milling
- Heat treat the nanocomposite to **reduce the metal oxide inclusions**
- Obtain **nano-inclusions of a low-melting metal encapsulated with a protective ceramic layer in a metal matrix**
  - High thermal conductivity, good stability, and strength

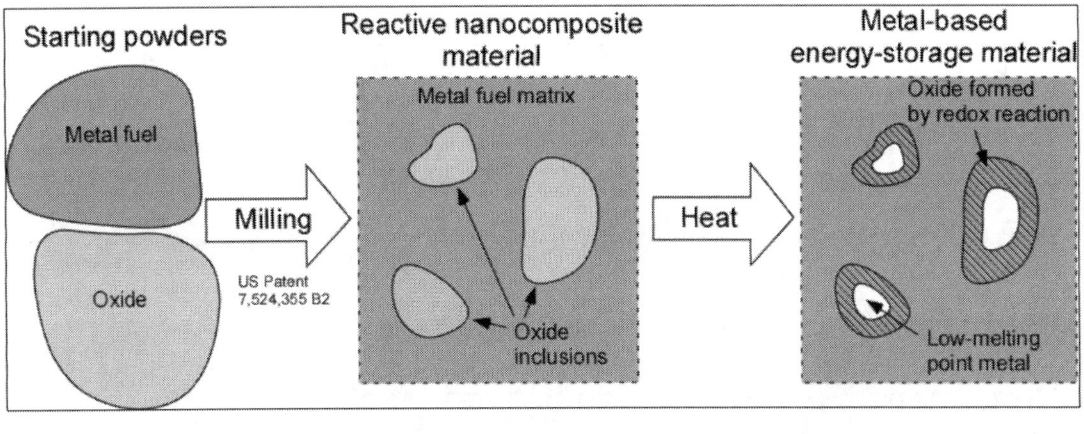

Material Synthesis

Ball Milling Parameters:

- Room Temperature
- Batch Size: 5g
- BPR: 10
- Milling Time: 45 - 60 min

Pellet Size

- DSC: 1/8" diameter pellet
- Mechanical Testing: 1/4" diameter pellet

Prepared Materials	Milling Time (min)	Vol % of Bi in ESM	Energy Storage (KJ/m³ of ESM)	Characterization Methods
$12Al/Bi_2O_3$	45	25	126,269	DSC, Mechanical Testing
6.5Al/Bi	60	25	126,269	DSC, Mechanical Testing
$6Al/Bi_2O_3$	45	38	193,960	Mechanical Testing
$4Al/Al_2O_3/Bi$	45	38	193,960	Mechanical Testing

Thermal Analysis

- DSC
  - 1/8" diam. pellets
  - Annealed:
    - 550 °C for 6 hrs
    - 300 °C for 64 hrs
- Nanocomposite thermite ($12Al/Bi_2O_3$)
  - Redox reaction
  - Exothermic peaks
- Metal composite (6.5Al/Bi - reference)
  - Endothermic melting peak
- Performance: thermal cycling between melting pt of Bi
  - Expect similar energy storage capabilities for both materials

Netzsch

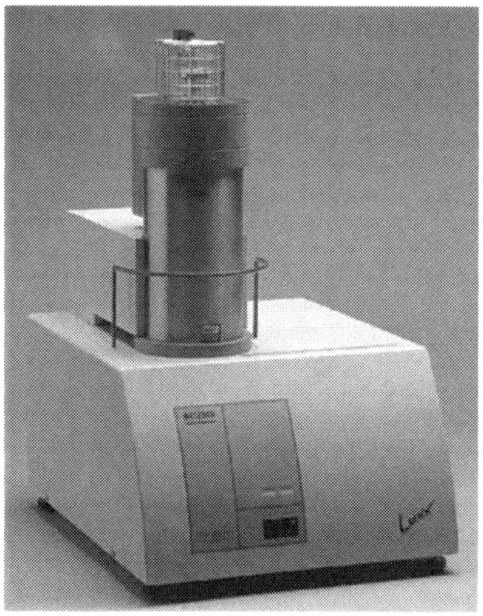

Fabrication of ESM structure

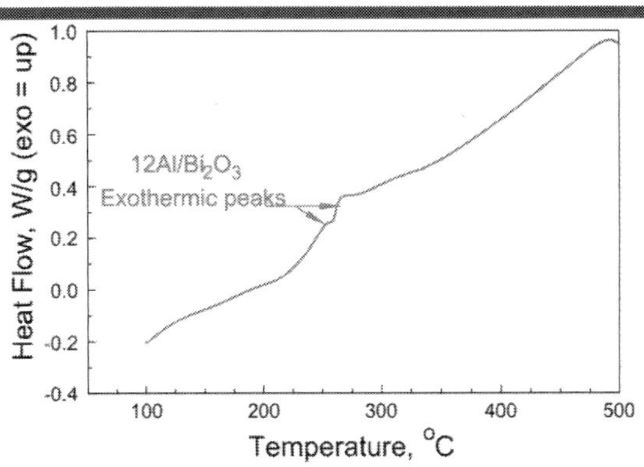

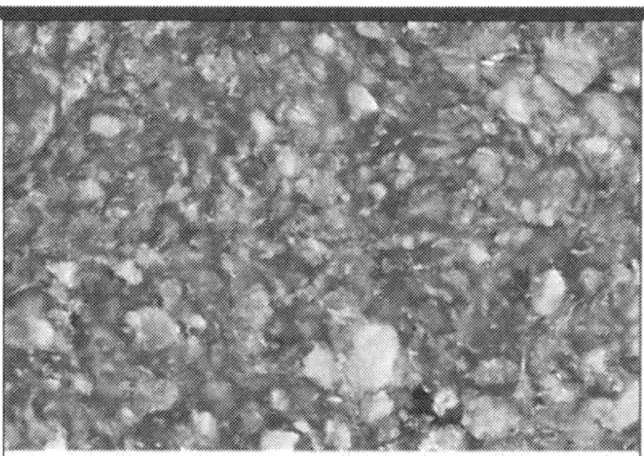

1 μm 12Al/Bi$_2$O$_3$ – Milled

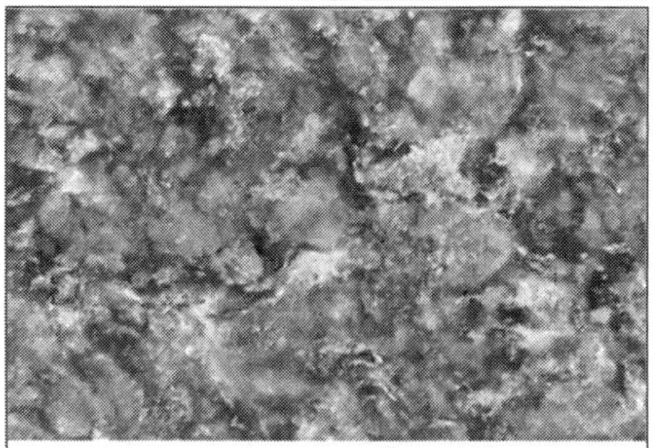

1 μm 12Al/Bi$_2$O$_3$ – Reacted – 300 °C

Precursor: Nanocomposite thermite

- 12Al/Bi$_2$O$_3$
  - micron- and nano-sized inclusions of Bi$_2$O$_3$
- Redox Rxn: Exothermic peaks
  $2Al + Bi_2O_3 \rightarrow Al_2O_3 + 2Bi$
- Annealed: 300 °C for 64 hrs
- ESM structure:
  - Fine Bi inclusions uniformly distributed
  - No coalescence or migration of inclusions

Reference composite: no encapsulation

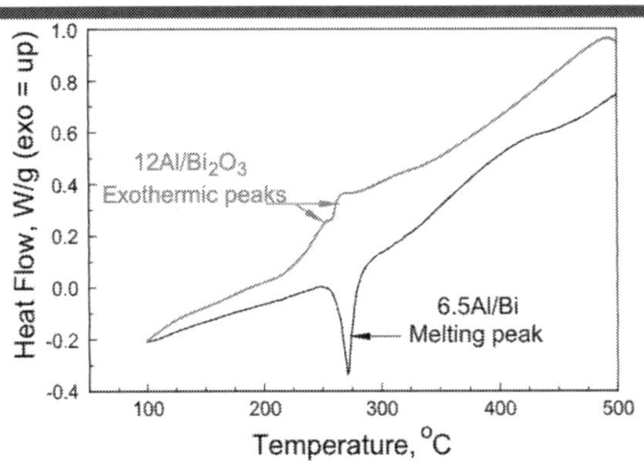

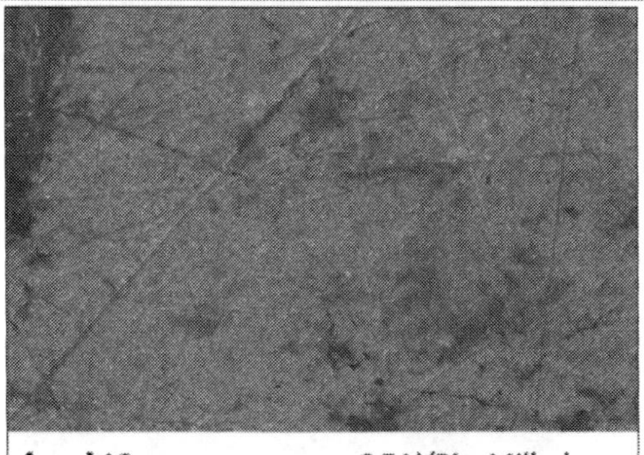

10 μm 6.5Al/Bi – Milled

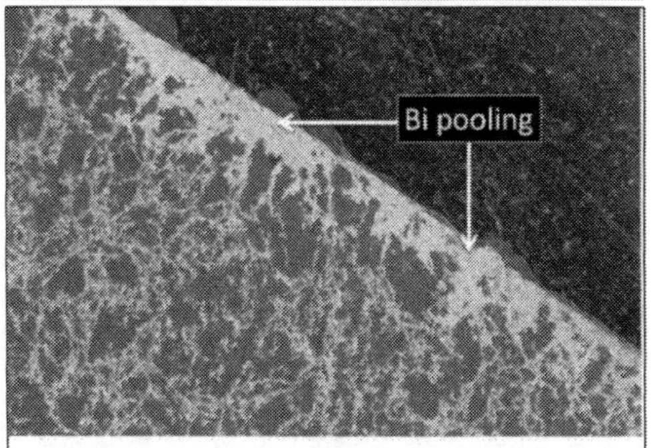

20 μm 6.5Al/Bi – Reacted – 300 °C

Metal composite:

- 6.5Al/Bi
  - nano-scaled inclusions of Bi
- Melting peak detected
- Annealed: 300 °C for 64 hrs
- Structure:
  - Bi redistributed

ESM Performance: Thermal Cycling

NJIT

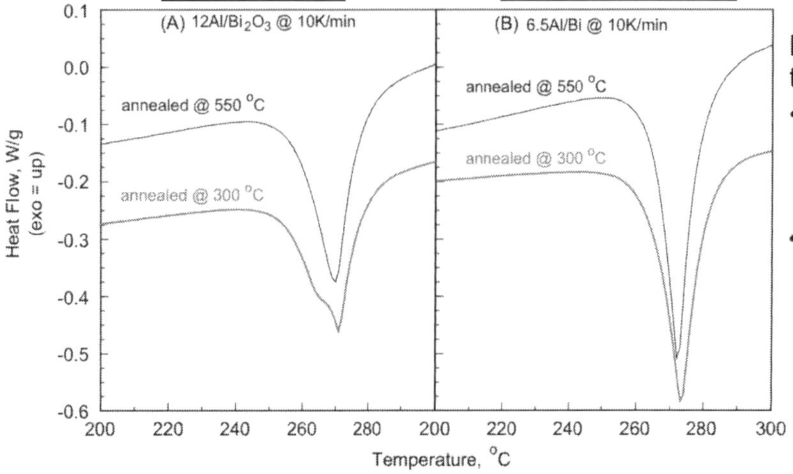

Lower melting point than pure Bi (271.4 °C)

- Melting point depression due to nano-sized Bi
- **Split melting peaks** for **encapsulated** sample annealed @ 300 °C

Precursor Material	Expected ESM Structure	Annealing conditions	Peak onset, °C	Energy Stored, J/g	
				Actual	Expected
$12Al/Bi_2O_3$ – Encapsulated	$10Al/Al_2O_3/2Bi$	550 °C for 6 hrs	256.9	26.40	27.5
		300 °C for 64 hrs	253.4	22.39	27.5
6.5Al/Bi – No Encapsulation	6.5Al/Bi	550 °C for 6 hrs	264.4	31.74	28.3
		300 °C for 64 hrs	264.2	29.40	28.3

ESM Performance: Al dissolution into molten Bi

NJIT

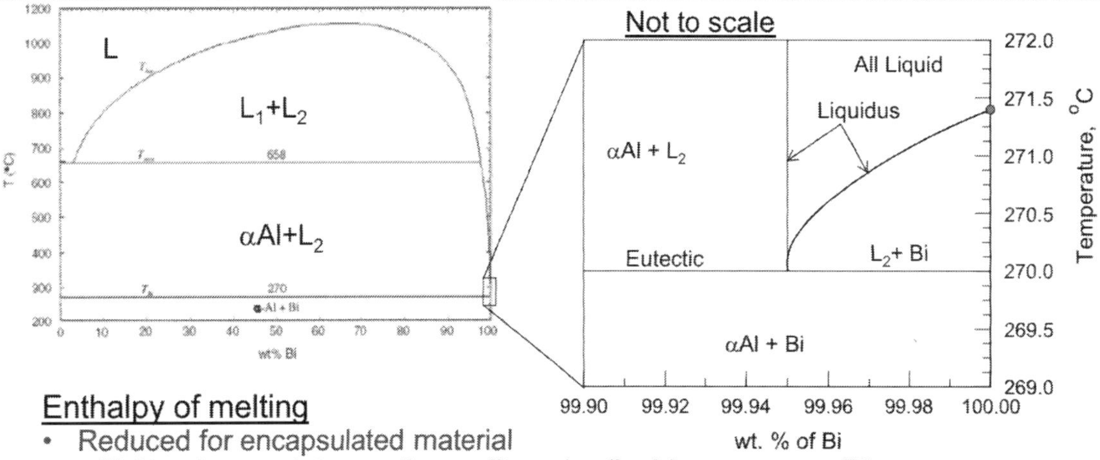

Enthalpy of melting

- Reduced for encapsulated material
  - ❑ Nearly complete reaction; split peaks: liquidus near pure Bi
- Increased for reference material with no encapsulation
  - ❑ Material starts deteriorating, forming Bi-rich L_2

Precursor Material	Expected ESM Structure	Annealing conditions	Peak onset, °C	Energy Stored, J/g	
				Actual	Expected
$12Al/Bi_2O_3$ - Encapsulated	$10Al/Al_2O_3/2Bi$	550 °C for 6 hrs	256.9	26.40	27.5
		300 °C for 64 hrs	253.4	22.39	27.5
6.5Al/Bi – No Encapsulation	6.5Al/Bi	550 °C for 6 hrs	264.4	31.74	28.3
		300 °C for 64 hrs	264.2	29.40	28.3

NJIT

Mechanical Testing

- Compression test
  - ¼" size pellets
  - Isothermal: 250 °C and 300 °C
  - Expect the encapsulated materials to be stronger
- Creep test
  - ¼" size pellets
  - Heated from 200 °C to 300 °C
  - Approaching Bi melting point:
    - Encapsulated material should hold its structure, while reference material may deform

Instron

NJIT

Mechanical testing: annealed pellet @ 300 °C

Compressive stress-strain @ 250 & 300°C

- Elastic and plastic deformation
- Above Bi mp → decrease in yield strength
- Encapsulated ESM is stronger than reference with no encapsulation
  - At 5% strain, ESM is twice as strong
- Encapsulated material is more mechanically sound

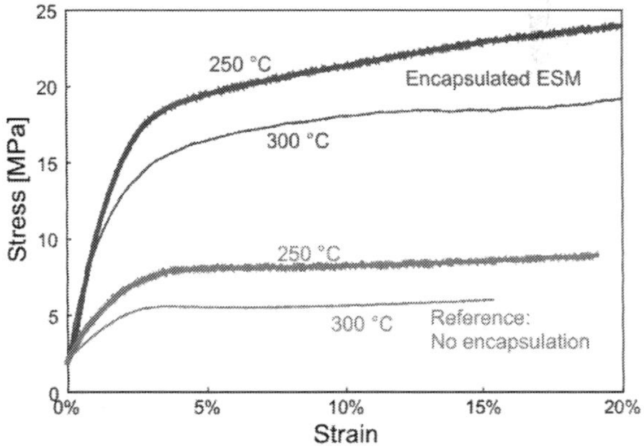

Constant load (3 kN); heated to 300 °C

- Deformation detected in the reference
- **ESM does not deform**
  - Encapsulated Bi

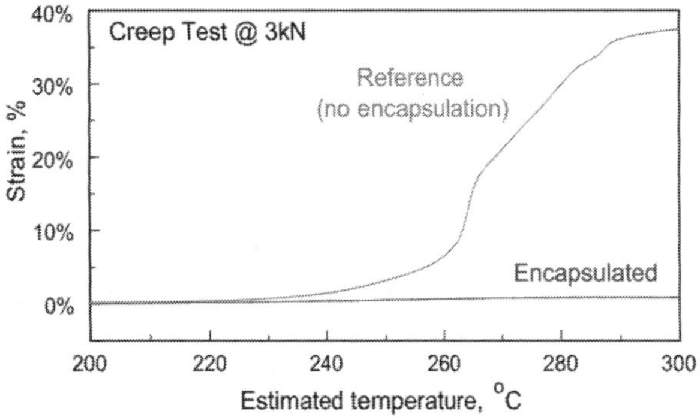

Cross-sectional SEM of Mechanical Tested Samples

1 μm 12Al-Bi_2O_3 – Compressed – 250 °C

1 μm 6Al/Bi_2O_3 – Creep Tested

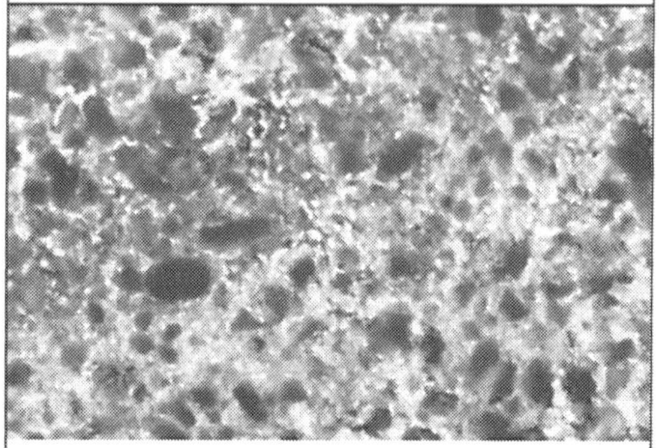

1 μm 4Al-Al_2O_3-2Bi – Creep Tested

Encapsulated Structure

- Equiaxial Bi inclusions dispersed uniformly
- Well-defined boundaries surrounded by a darker phase
- Al_2O_3 phases are hard to detect

Reference: no encapsulation

- Crystalline Al_2O_3 particles are visible
  - Redistributed Bi

Conclusions

- Nanocomposite (Al/Bi_2O_3) thermite used as a precursor material for fabrication of ESM
- Nano-inclusions of Bi are encapsulated with protective Al_2O_3 in an aluminum matrix
- Capability of energy storage by phase change is shown
- Encapsulated Bi is not observed to migrate following thermal cycling for ESM.
  - In reference composite materials, molten Bi pools towards the surface, tends to separate from Al.
- Difference in mobility between encapsulated and non-encapsulated Bi inclusions causes dramatic differences in the mechanical properties of the resulting composites.
  - ESM with encapsulated Bi inclusions are much stronger and capable of retaining their structure

NJIT

Thank You

Process Safety

ÉCOLE POLYTECHNIQUE FÉDÉRALE DE LAUSANNE

Molecular Structure Based Predictive Models for Thermal Stability of Chemicals

Nadia Baati, Annik Nanchen, Francis Stoessel, Thierry Meyer

École Polytechnique Fédérale de Lausanne, Switzerland
Swissi Process Safety, Switzerland

2015 AIChE Annual Meeting
November 2015 - Salt Lake City UT- USA

Simulations & Process Development

Simulations:

- Anticipation
- Greater influence
- Savings
  - Time
  - Material
  - Financial

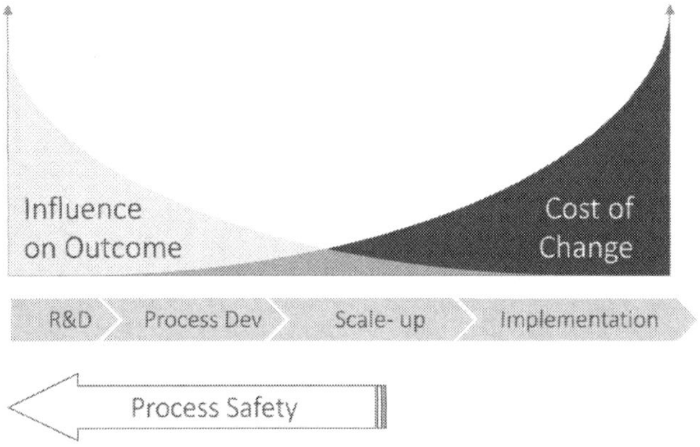

Industrial Accidents Causes

Since 1990:

> 700 industrial accidents

➢ Releases
➢ Fires
➢ Explosions

Causing Weaknesses

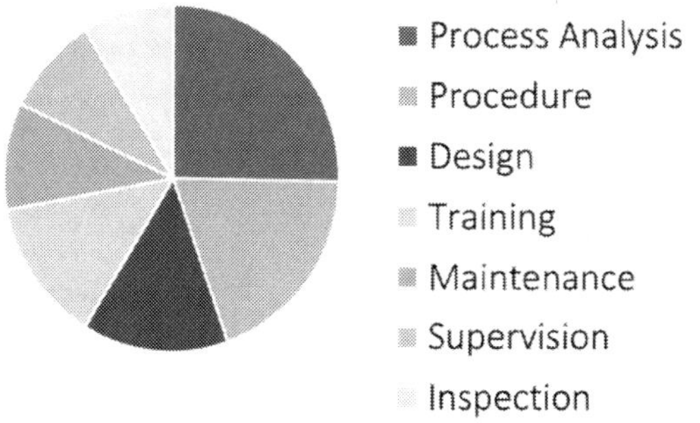

Jacobson et al, 2009

Thermal Risk Assessment

Several questions need to be answered

➢ Can this chemical decompose?
➢ In which temperature range?
➢ How much heat is released?
➢ At which rate?

➢ Differential Scanning Calorimetry
➢ What if a compound of interest is:
  ... Impossible to isolate
  ... Known to be highly toxic
  ... Physically unavailable ?

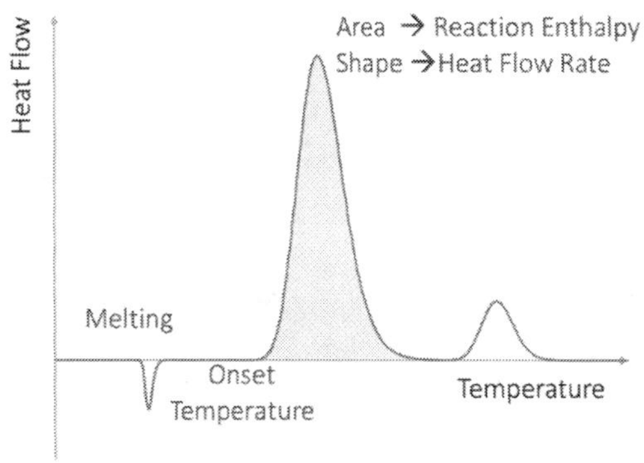

Motivations and Objectives

- ➢ Benefits of Simulations
- ➢ Flaws in Design and Analysis
- ➢ Practically Impossible Experiments

Predictive Simulations

- ➢ Forecasting DSC results
- ➢ Improve testing strategy
- ➢ Complement for design

8

Molecular-Based Simulations Methods

1. Quantitative Structure-Property Relationships (QSPR)

Structural aspects covered:

- ➢ Constitution
- ➢ Topology
- ➢ Geometry
- ➢ Electronic
- ➢ Quantum derived

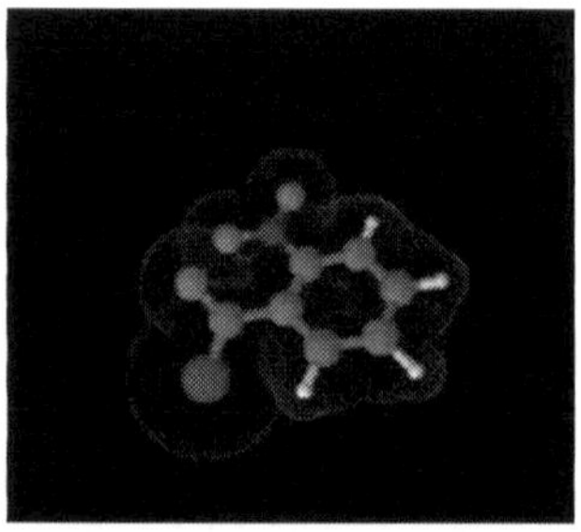

Codessa Pro
ChemDraw

9

Molecular-Based Simulations Methods

2. Group Contribution Methods (GCM)

Various Frameworks

- Marrero-Gani

Functional Groups

- Several levels

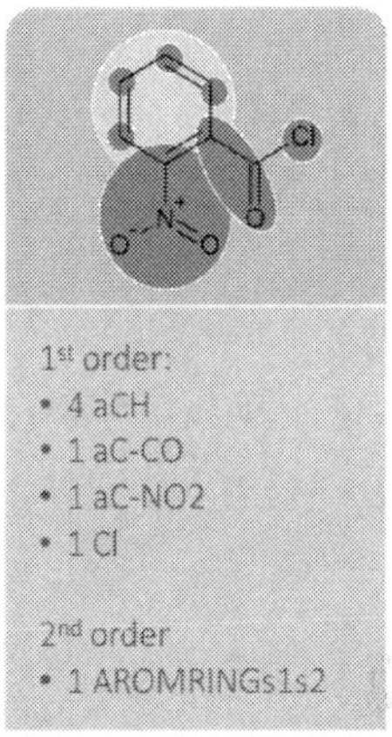

ICAS Software

Modeling Procedure

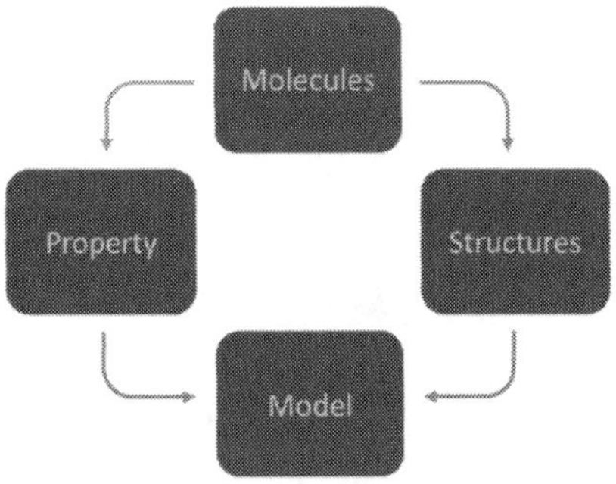

Modeling Procedure

Properties

- Amplitude (A)
- Peak Position (Tp)
- Peak Width (σ)
- Peak Asymetry (AS)
- Reaction Enthalpy (ΔHr)

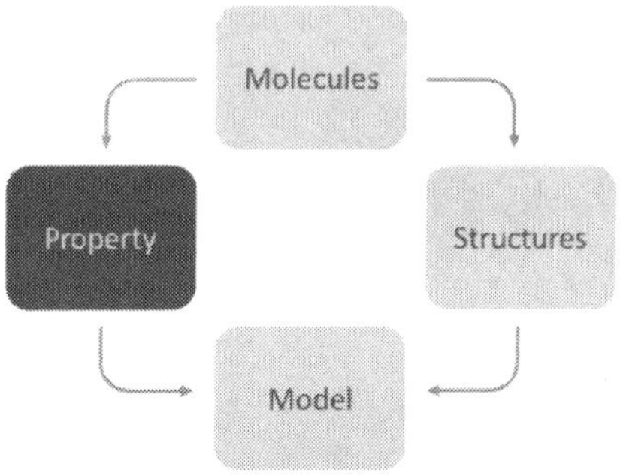

AKTS Software

Modeling Procedure

Properties

- Amplitude (A)
- Peak Position (Tp)
- Peak Width (σ)
- Peak Asymetry (AS)
- Reaction Enthalpy (ΔHr)

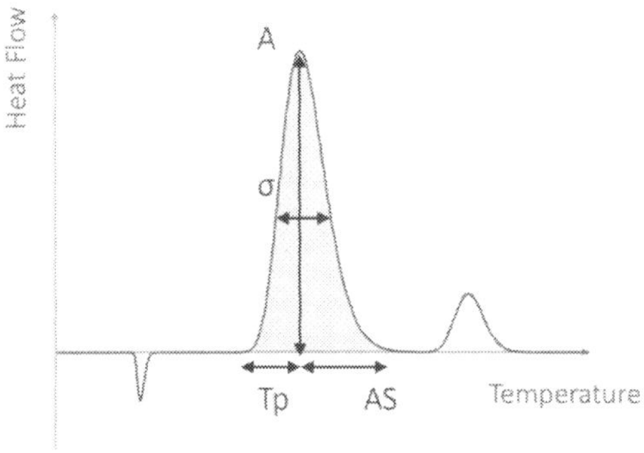

AKTS Software

Modeling Procedure

Structures

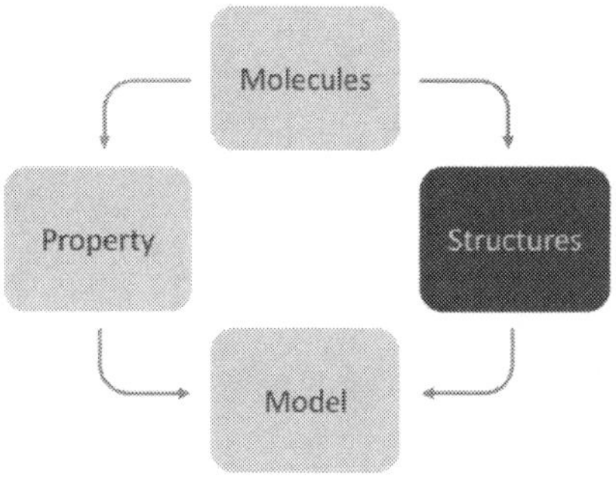

- Generate Structure
- Represent in selected framework
- Compute descriptors / groups

15

Modeling Procedure

Models

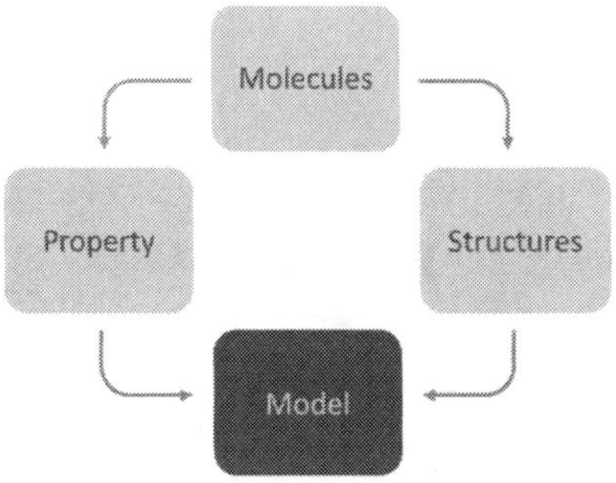

- Linear Regressions
- Training - Validation
- Local Subsets (Clusters)

16

Modeling Procedure

Models

- ➢ Linear Regressions
- ➢ Training - Validation
- ➢ Local Subsets (Clusters)
  - ➢ Classification

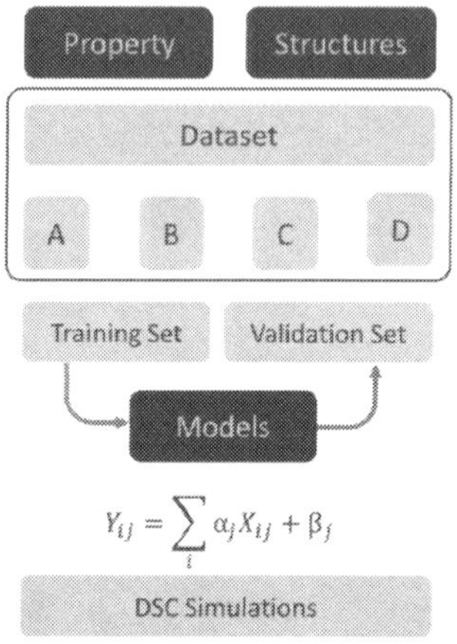

Classification

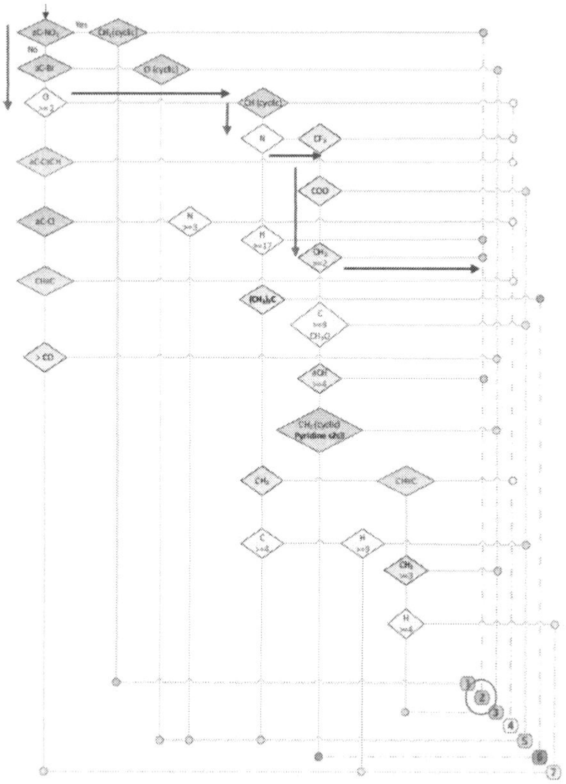

- ➢ Local Models
  - ➢ Chemical similarities
  - ➢ Particular structure
- ➢ Systematic Method
  - ➢ DSC similarities
  - ➢ Corresponding structures
- ➢ Example:

3-Hydroxy-1,2,3-benzotriazin-4(3H)-one:

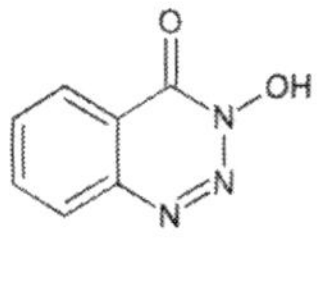

Results

GC Method, Local Model, Peak's Amplitude

Class 2

	Amplitude
Training Set R^2	0.95
Validation Set R_{EV}^2	0.89
Training Set Size	37 observations
Cross- Validation Set Size	5 observations

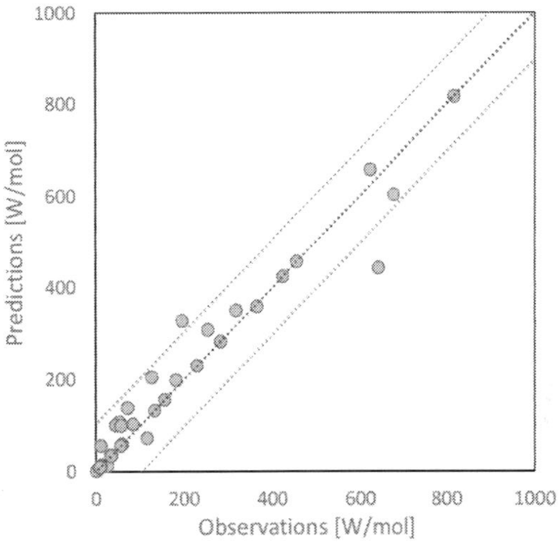

DSC Reconstruction

3-Hydroxy-1,2,3-benzotriazin-4(3H)-one:

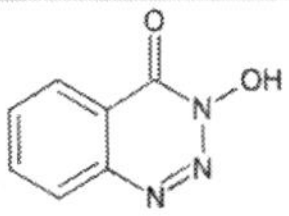

Group	Frequency	Coefficient	αi . Gi
aCH	4	76	303
aC fused with non-aromatic ring	2	192	384
OH	1	-162	-162
aC-NCO	1	-39	-39
$aC\text{-}NO_2$	1	57	57
AROM.FUSED[2]	1	-392	-392
H	5	3	15
N	3	167	501
O	2	64	129
C	7	-31	-214
$^0\chi$	3.54	22	79

Measured:

$$A = 676\ \mathrm{W/mol}$$

From GCM local:

$$A = \sum_i \alpha_i \cdot G_i$$

$$A = 660\ \mathrm{W/mol}$$

DSC Reconstruction

3-Hydroxy-1,2,3-benzotriazin-4(3H)-one:

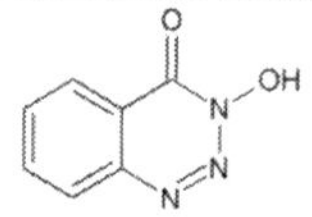

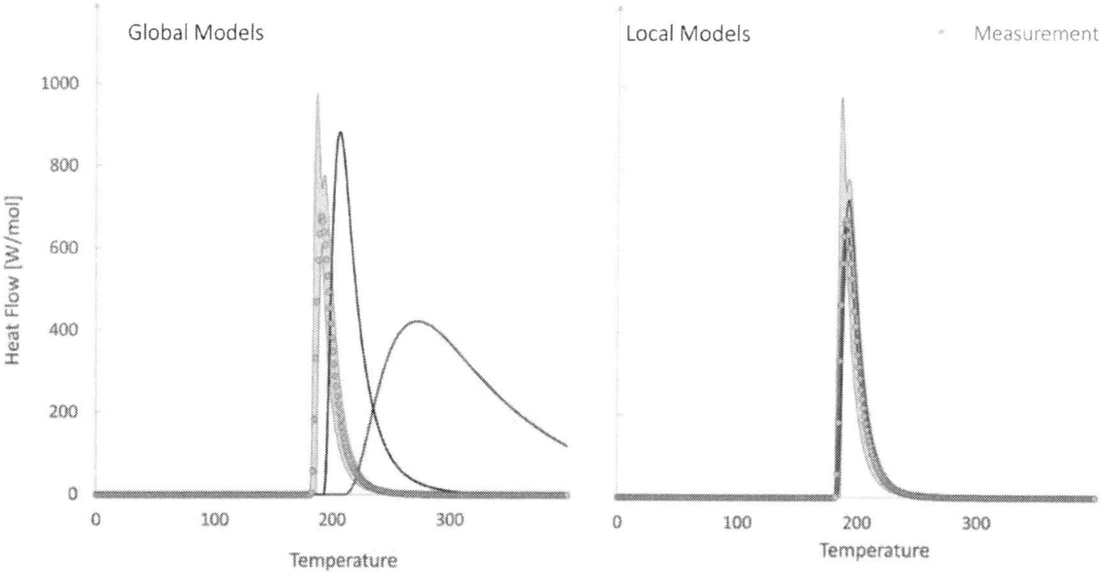

22

DSC Reconstruction

5-(2-Bromophenyl)-1H-tetrazole:

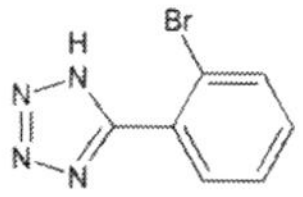

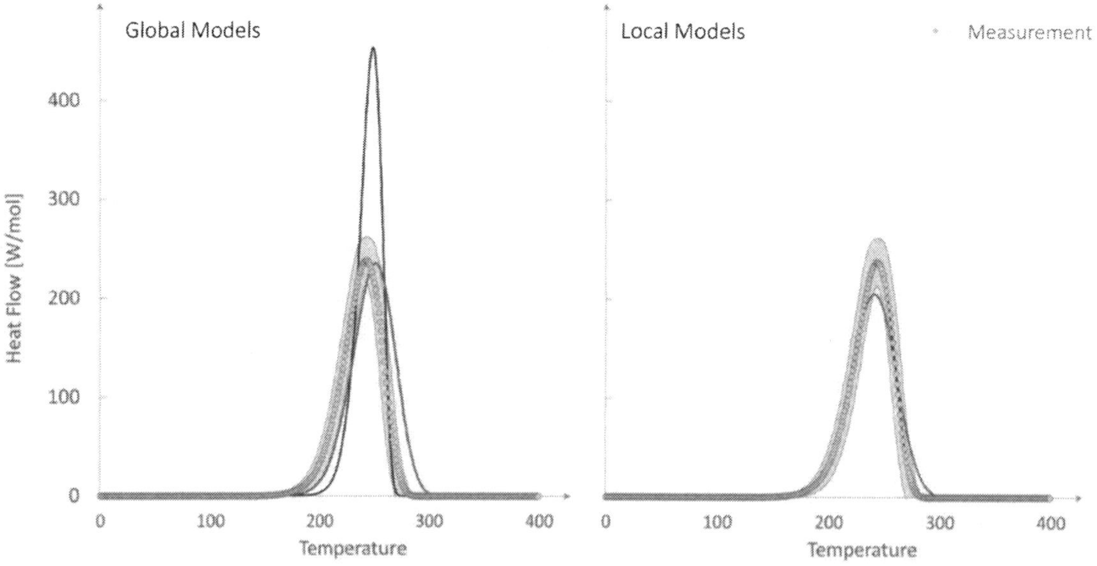

23

Comparison

	QSPR	GCM
➢ Structural Information	+	-
➢ Accuracy	=	=
➢ Computing Required		
• for descriptors/groups	-	+
• for models	=	=
• for application	-	+
➢ Physical Interpretation	-	+
Overall		
• for a molecule from DB	=	=
• for a new molecule	X	✓

24

Conclusions

➢ Methodology successful and adaptive

➢ Global models are limited

➢ Local models show higher performances

  ➢ Classification Tree

25

Applications

- Estimation of Missing Data
- Screening Tool
  - Test & evaluate several alternatives
  - Identify the most/least hazardous ones
- Process Design
  - Minimal intrinsic hazard
  - Inherently Safer Design
- Future Applications
  - Kinetic Parameters Estimation
  - Criticality Classification

26

Applications

- Estimation of Missing Data
- Screening Tool
  - Test & evaluate several alternatives
  - Identify the most/least hazardous ones
- Process Design
  - Minimal intrinsic hazard
  - Inherently Safer Design
- Future Applications
  - Kinetic Parameters Estimation
  - Criticality Classification

Not a replacement for experimental testing and proper characterization !

27

Acknowledgments

This project is supported by the

Swiss Commission for Technology and Innovation

(project n°:14711.1 PFIW-IW)

AKTS

Advanced Kinetic and Thermal Analysis Software

Novartis Pharma

Safety Laboratory

CAPEC

Computer Aided Process Engineering Center

Technical University of Denmark

28

TÜV SÜD

Process Safety

ÉCOLE POLYTECHNIQUE FÉDÉRALE DE LAUSANNE

Thank you for your attention

Nadia Baati

nadia.baati@epfl.ch

Group of Chemical and Physical Safety

EPFL, Switzerland

429798 Flame Retardant Polypropylene Matrix Nanocomposites

Abstract:

Polypropylene (PP) matrix nanocomposites reinforced with a variety of nanofillers including magnetic nanoparticles (NPs), carbon nanotubes, magnesium hydroxide, aluminum hydroxide, and silicon dioxide were prepared through in- or ex- situ solution mixing method for flame retardation purpose. In order to improve the filler dispersion quality, polymer-filler interfacial bonding and the flame retardant efficiency, the filler surface pre-treatment/functionalization and the aid of surfactant were used during the preparation of these different PP matrix nanocomposites. The filler surface modification and dispersion were investigated through transmission electron microscopy (TEM) and scanning electron microscopy (SEM). The chemical structures were studied by Fourier transform infrared spectroscopy (FT-IR), X-ray diffraction analysis (XRD), X-ray photoelectron spectroscopy (XPS), and et al. The physicochemical properties of thus prepared nanocomposites including thermal stability, flammability, melting rheological behaviors were studied using thermogravimetric analysis, micro-scale combustion calorimetry, UL-94 vertical burning, limiting oxygen index, and rheometry. The thermal stability and flame retardation mechanism for each composition will be discussed in detail and the advantages as well as the disadvantages for lowering the fire hazards will also be compared.

Minimum Power Requirement for LNG Regasification

Harsha Nagesh Rao, Khai Hoe Wong, and I A Karimi

Department of Chemical and Biomolecular Engineering

National University of Singapore

Boil-off Gas (BOG)

- Natural gas is the cleanest fossil fuel
- Liquefied Natural Gas(LNG) is the most preferred option for long distance transport
- Heat leak during LNG storage and transport is inevitable
- Boil-off gas (BOG) generation is a major concern in LNG supply chain

LNG Regasification Terminal

- Imports and stores LNG, and sends regasified NG to consumers
- BOG management is important to reduce operating cost

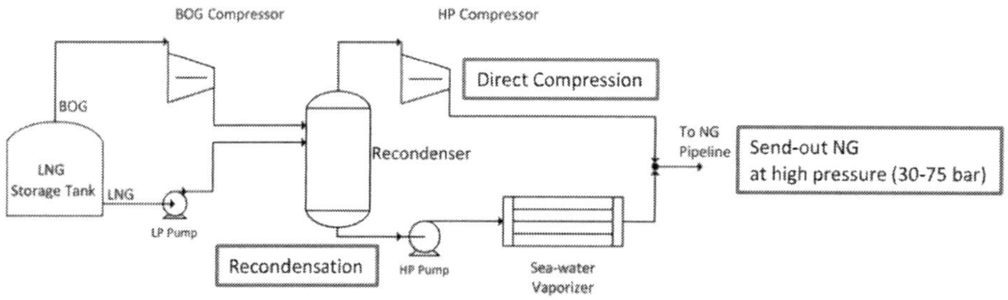

Literature

- Some works on BOG estimation and minimization from various sources
- Several studies on exergy/exergoeconomic analysis to link with power cycles
- <u>Retrofitting designs</u>: Most existing terminals may not invest in power cycles
- Liu et al. [1] studied multi-stage recondensation without HP-LNG cooling
- Park et al.[2] - Some degree of HP-LNG cooling but did not fully utilize

1. Chaowei Liu, J. Z., Qiang Xu,and John L. Gossage. (2010). Thermodynamic-Analysis-Based Design and Operation for Boil-Off Gas Flare Minimization at LNG Receiving Terminals. Industrial & Engineering Chemistry Research, 49, 7412–7420.
2. Park, C., Song, K., Lee, S., Lim, Y., & Han, C. (2012). Retrofit design of a boil-off gas handling process in liquefied natural gas receiving terminals. Energy, 44, 69-78.

Problem Statement

Given a terminal with

- Send-out rate $F = 1$ tonne/h (tph), temperature and pressure
- LNG composition and conditions (T, P)
- BOG composition, generation fraction β, and conditions (T, P)
- Ratio (p_{sw}) of the sea water pump power to regasifier duty (To quantify sea water pump power)

Obtain:

- Minimum thermodynamic power consumption of the terminal per F as a function of β
- Optimal process steps

Assumptions

- Terminal is at steady state
- Pumps and compressors are 100% efficient (Isentropic)
- Counter-current heat exchange with $zero$ minimum approach temperature
- No pressure drops in equipment and pipelines
- NPSH (Net Positive Suction Head) for all pumps is zero
- Normal range of BOG and LNG composition

Minimum Power Requirement for LNG Regasification

Highlights:

- Maximized the use of LNG cryogenic energy
- Fundamental thermodynamic analysis is presented (P-H Diagram)
- An evolving series of schemes is proposed
- Lower bound on power consumption is estimated for each scheme
- Schemes are compared with respect to minimum power consumption

Single Stage Compression Schemes

Scheme 1.1-Indirect Recondensation

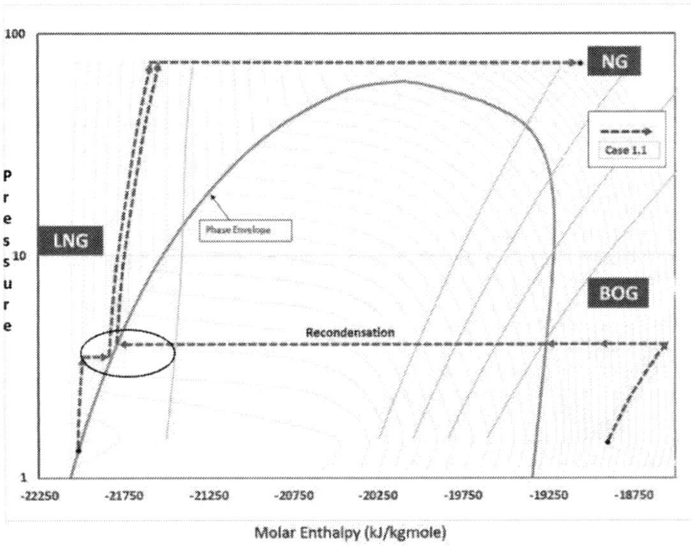

- Scheme features:
  - Counter current indirect heat exchange
  - HP-LNG cooling before recondensation
- BOG Compression pressure and LNG pump pressure can be different

Scheme 1.2-Direct Recondensation

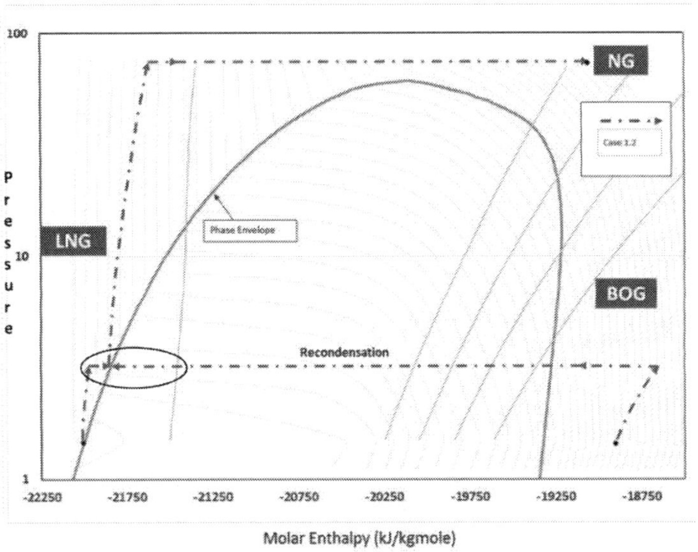

- Scheme features:
  - Direct mixing
  - HP-LNG cooling before recondensation
- Prevalent practice in industry
- BOG and LNG at same pressure level
- Reduces latent heat due to mixing
- Recondensation at a lower pressure

Scheme 1.3- HP Precooling

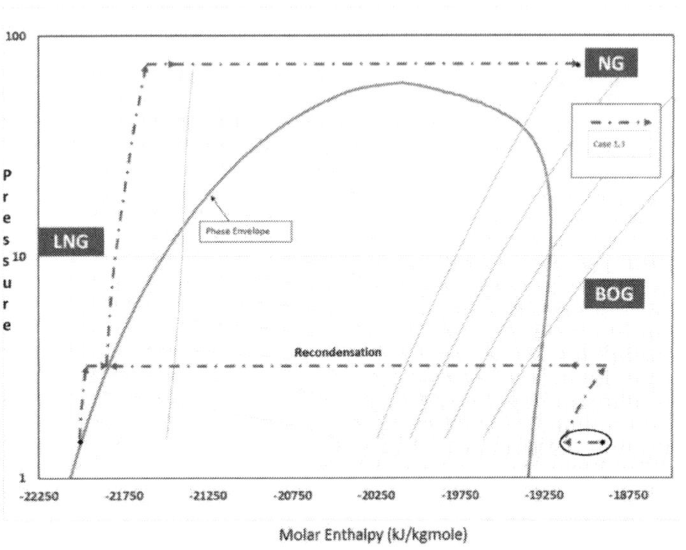

- Scheme features:
  - Direct recondensation
  - HP-LNG precooling before compression
- BOG prior to compression is **superheated**!
- Abundant HP-LNG can be used to precool BOG
- Reduces compression work significantly
- Ignored in current designs

Scheme 1.4- LP Precooling

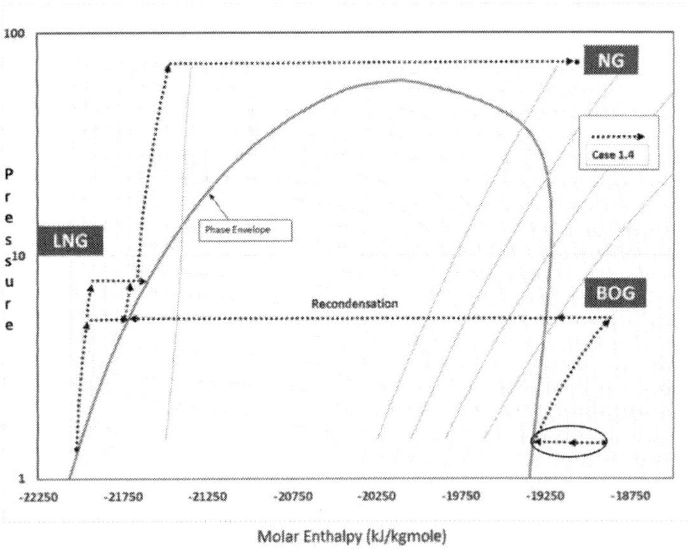

- Scheme features:
  - Direct recondensation
  - HP-LNG and LP-LNG precooling
- A fraction of LP-LNG can be used to precool BOG further
- More reduction in compression work
- Reduces LP-LNG available for recondensation
- Recondensation at a slightly higher pressure

Two-Stage Compression schemes

Scheme 2.1-HP Intercooling

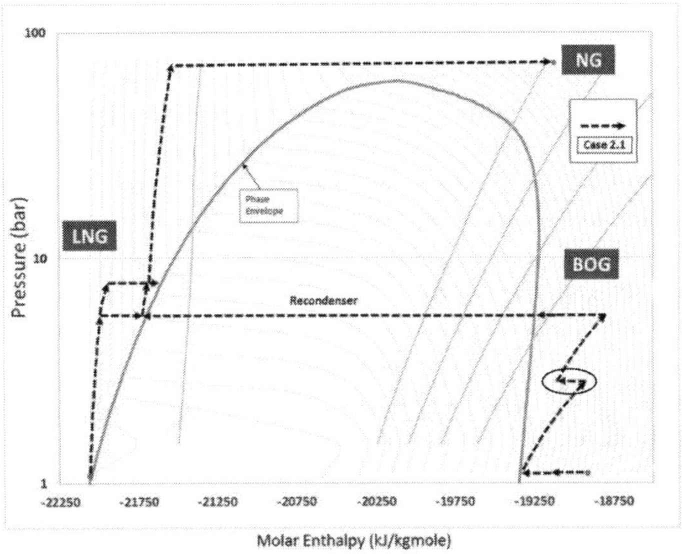

- Scheme features:
  - Two stage compression
  - HP-LNG and LP-LNG precooling
  - HP-LNG intercooling
- Multiple stages of compression needed for higher β
- HP-LNG can be used to intercool BOG between stages of compression
- Lower compression work

Scheme 2.2- LP Intercooling

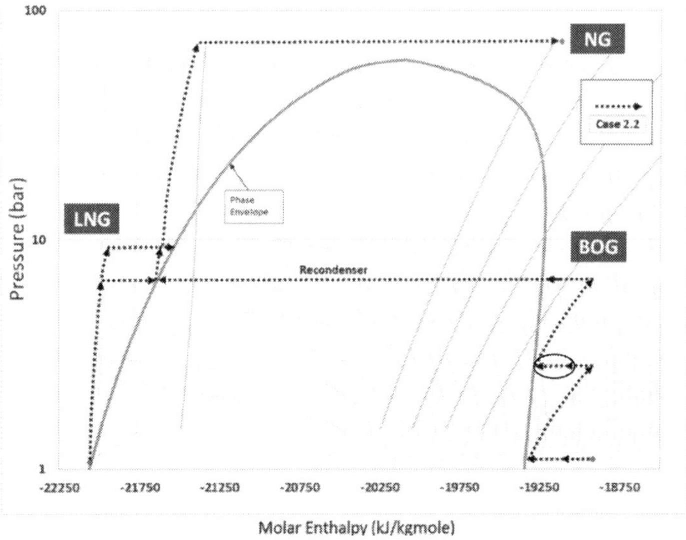

- Scheme features:
  - Two stage compression
  - HP-LNG and LP-LNG precooling
  - HP-LNG and LP-LNG intercooling
- Much cooler feed to second compression stage
- Further decrease in LP-LNG available for recondensation
- Recondensation at a higher pressure

Scheme 2.3- Two Stage Recondensation

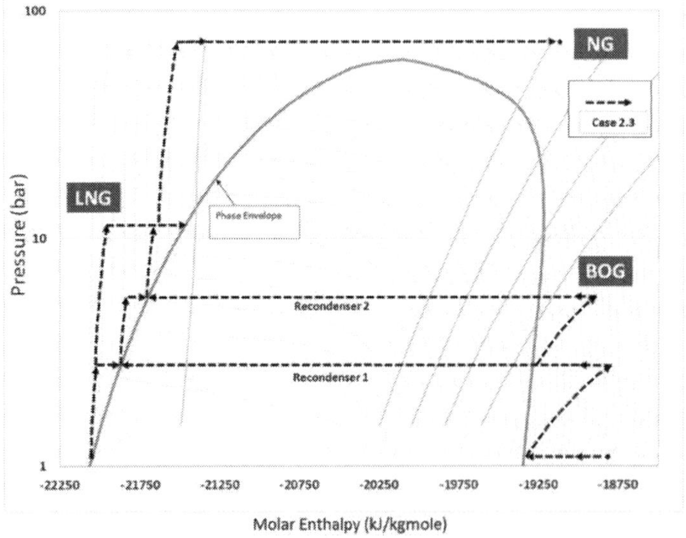

- Scheme features:
  - Two-stage direct recondensation
  - HP and LP-LNG precooling
- All of BOG in I stage is compressed to lower pressure
- Only uncondensed BOG is sent to II stage of compression
- Saturated feed to II stage of compression

Park et al. Scheme

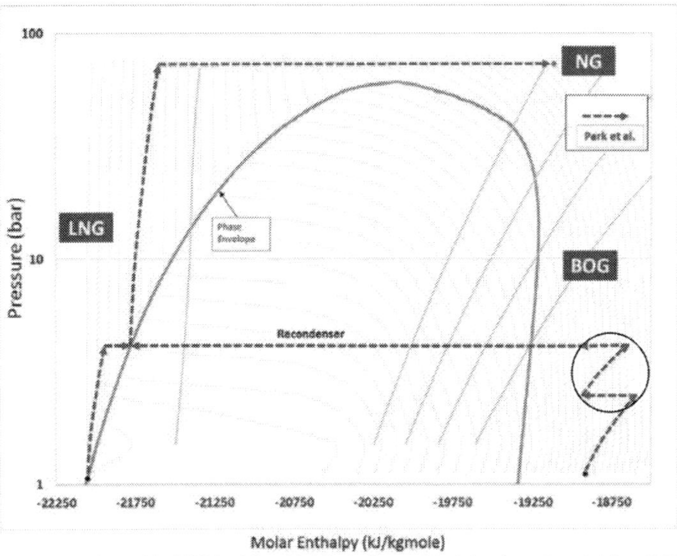

- Best reported scheme in literature
- <u>Scheme features:</u>
  - Two-stage compression
  - HP-LNG intercooling
  - HP cooler before recondensation

Park, C., Song, K., Lee, S., Lim, Y., & Han, C. (2012). Retrofit design of a boil-off gas handling process in liquefied natural gas receiving terminals. Energy, 44, 69-78.

Case Study Results

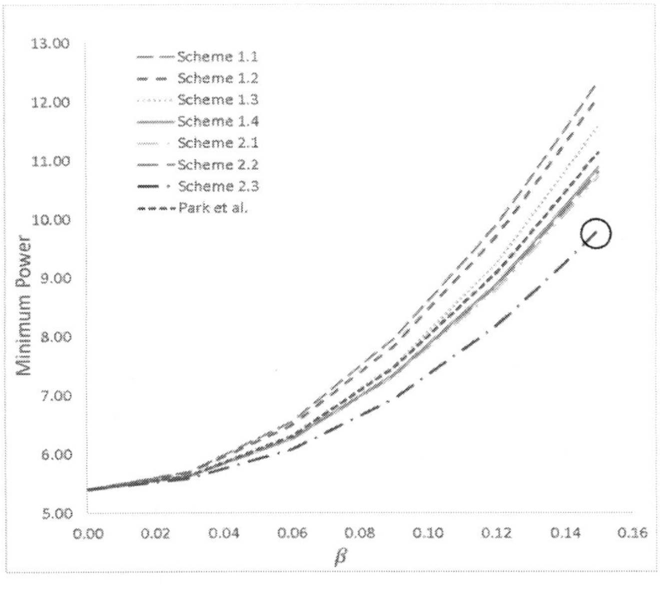

- Case Study on Pyeongtaek LNG receiving terminal in South Korea
- For β**=0.15**
  - **18.77%** lower than **Scheme 1.2**
  - **11.90%** lower than **Park et al. scheme**

Conclusions

- Use of direct recondensation is justified
- Precooling BOG before compression is very effective
- HP-cooling prior to recondensation is favourable
- Two-stage recondensation has a profound impact
- This study may help guide terminal design
- Cost optimization and non-ideal steps may yield different results

(88b) Heavy Hydrocarbons in Mixed Refrigerant Used in Natural Gas Liquefaction Process

Jia, R., Shanghai Jiao Tong University
Lin, W., Shanghai Jiao Tong University
Originally presented on: 11/9/2015 8:52:00 - 9:14:00

When lubricant is used in a compressor for mixed refrigerant (MR) used in natural gas liquefaction process, butane and pentane, the heavy hydrocarbons in MR, are more likely to dissolve in the lubricant, comparing to other hydrocarbon components. In such liquefied natural gas plants, MR containing less or even no heavy hydrocarbons may be adopted. In this paper, the single mixed refrigerant (SMR) liquefaction process with different butane or pentane content in MR was optimized. In order to compare the influence of MR composition, first set the suction and exhaust pressure of MR compressor, then change the percentage of MR composition sequentially, and automatically adjust the MR flow to guarantee the minimum temperature difference of LNG cold box no less than 3°C. Knowledge based algorithm (KBO) is applied to simulate the process. Based on the work, the refrigerant flow and specific power consumption of the processes are investigated. The result shows that both MR flow and specific power consumption first decrease, and then increase with the increase of pentane content. The optimal value of specific power consumption is 0.33kWh/kg(LNG). For butane content, similar results are obtained and the optimal value is 0.35kWh/kg(LNG). When the content of pentane or butane is changed, the variation of propane content is the largest, and that of nitrogen content is the least. This phenomenon is related to the boiling point and latent heat of each composition. Thermodynamic analysis is applied to explain the results.

432740 Techno-Economic Analysis of Carbothermal Reduction of Magnesia in a Reduced Pressure Hybrid Solar-Electric Reactor

Abstract:

A techno-economic analysis was performed on carbothermal reduction (CTR) of magnesia in a moving bed, reduced pressure, hybrid solar-electric reactor. Pretreatment unit operations were designed for a pelletized carbon-MgO feed into the primary CTR reactor. Post-reaction steps included high temperature condensation, vacuum distillation, and ingot casting. CTR reaction kinetics derived from laboratory-scale experiments were used to optimize temperature and pressure in the primary reduction reactor. An nth-plant with 20,00 tonne-Mg/year capacity was designed under two base cases where the primary reduction reactor was heated externally using 1) Electrical resistance heating, 2) Hybrid solar-electric heating. The solar tower height and heliostat field were optimized according to actual daily irradiance and supplemented with electrical resistance heating when the total net solar flux was insufficient for heating and reaction after accounting for heat losses. Economic profitability was driven primarily by the cost of magnesia ($0.30-0.80/kg) and the total fixed capital investment (-20%, +100%,). Operational energy requirements and emissions per kilogram of magnesium product were 10 kWh and 4.0 $kgCO_{2\text{-}eq}$, respectively, for Case 1. Case 2 showed operational energy requirements and emissions of 8.8 kWh and 2.9 $kgCO_{2\text{-}eq}$. These results are below DOE targets of 27 kWh and 10 $kgCO_{2\text{-}eq}$ and far below state-of-the-art Pidgeon specifications of 102 kWh and 37 $kgCO_{2\text{-}eq}$. The hybrid solar-electric case revealed a production cost of $1.96/kg-Mg using $150/$m^2$ for the heliostat field, or 20% cheaper than Pidgeon magnesium.

432357 Pressure-Driven Dynamic Simulation for Flare Minimization and Greenhouse Gas Reduction during an Ethylene Plant Startup

Abstract:

Flare emissions release volatile organic compounds (VOCs) and pollutants into the atmosphere, which react with nitrous oxide in sunlight to generate ground-level ozone and smog. A typical start-up of an ethylene plant flares approximately 5.0 million lbs of raw materials and generates at least 15.4 million lbs of CO_2, 40.0 Klbs CO, 7.4 Klbs NOx, 15.1 hydrocarbons, and 100.0 Klbs HRVOCs. Hence, managing flare emission during startup, shutdown and malfunction (SSM) of an ethylene plant are very important. Flare source reduction approach employs simulation and optimization, emphasizes the upstream section of flare system and the whole plant integration, in which more upstream modification and optimization contribute in limiting unwanted products.

In this study, rigorous flow-driven and pressure-driven dynamic models of a front-end De-Ethanizer ethylene plant with sulfur recovery unit are constructed to serve as an optimization platform during plant startup. Since off-specification streams are inevitable during plant turnaround operations, to significantly reduce flaring emission, they must be either recycled to the upstream process for online reuse, or stored somewhere temporarily for future reprocessing, when the plant manufacturing returns to stable operation. Thus, the off-spec products will be able to be reused instead of being flared. Besides, charged gas compressor (CGC) startup is reportedly the most critical step and has largest amount of vent gas. Several scenarios of CGC feed at different flow rates and compositions are examined to find the least flaring one. The pressure-driven model is utilized to investigate the inlet conditions at CGC suction, thus, guarantee the feasibility of the proposed startup strategy. These results serve as a basis for emission calculation and greenhouse gas study such as CO_2 control. The process can be considered from smaller scale (sectional model) to more comprehensive scale (whole plant model). The best results in these case studies are incorporated and validated in the integrated model of the whole plant in order to stabilize the system in the shortest time while minimizing material and energy loss.

Simultaneous Targeting and Scheduling for Water Minimisation in Batch Plants: A Polyvinyl Chloride (PVC) Manufacturing Case Study

Lee, J. Y., National Taipei University of Technology
Foo, D. C. Y., University of Nottingham Malaysia

Originally presented on: 11/9/2015 9:58:00 - 10:20:00

Batch processing has received considerable attention from both industry and academia for its flexibility and suitability for the production of low volume, high value-added products, also as a result of the recent trend towards product-centred manufacturing globally. With inherent complexity, the design and synthesis of batch processes necessitates concurrent development of effective scheduling techniques. Common objectives include profit maximisation and cost minimisation in the time horizon of interest. In addition to process scheduling, *process integration* may be considered to improve energy utilisation and/or reduce waste production. A particular case commonly found in the process industries is water integration for minimum fresh water consumption and wastewater generation. Various process integration techniques and tools have been developed for resource conservation in batch plants. Most early developed techniques assume the schedule to be fixed, while many of recent ones are combined with a scheduling model to exploit the flexibility of batch processes. The *automated targeting model* – ATM – is one of the established tools for *batch water network synthesis*. Its formulation is based on *pinch analysis* and thus enables flow targets to be identified prior to detailed network design. A great advantage of the ATM is its linearity, for which global optimality is guaranteed. Furthermore, it is more flexible than conventional pinch methods being able to incorporate cost considerations. However, the ATM was only developed for cases with fixed schedules. This drawback will be overcome in the present work by embedding the ATM in a scheduling formulation to allow flow/cost targeting and process scheduling to be performed simultaneously, and the true minimum flow/cost targets to be located. A case study on polyvinyl chloride (PVC) manufacturing is presented to demonstrate the application of the proposed simultaneous approach. In the batch-operated PVC plant, the upstream reactor is scheduled to match the downstream separator, while water minimisation through direct reuse/recycling is considered. The trade-off between fresh water (operating) and storage (capital) costs will also be analysed.

419301 Quick and Efficient Process Energy Integration in Microsoft Excel

Abstract:

Industrial sectors account for one third of global energy consumption. A common feature of industrial processes is reliance on fossil fuels as the primary source of energy, where a large part of the energy consumption is spent on production of utilities. Due to the huge negative impacts of fossil fuel combustion on the environment, the scientific world makes a significant effort to find alternative sources of energy. However, even by the most optimistic assessments, these alternatives are long-term solutions and many projections show that in the near future, fossil fuels will remain the primary sources of energy, in particular for the process industries.

Pinch analysis [1] is a well-known methodology to optimize process energy consumption. It allows the determination of the minimum required process utilities consumption. Simulis Pinch [2], software used directly in Excel, has been developed to perform energy integration calculations with the pinch method, while also systematically considering real-world engineering constraints. Based on a chosen pinch that is acceptable for the whole process, Simulis Pinch calculates the minimum utility needs and the maximum heat that can be recovered through energy integration. The software also plots the hot and cold composite curves, as well as the grand composite curve. The energy integration concept is used by the software to reach the optimal synthesis of a heat exchanger network that achieves the maximum energy recovery. In Pinch theory, the synthesis of the optimal heat exchanger network is subject to construction rules (for example, the division of the problem into two sub-problems, one below and the other above the pinch temperature). However, the ideal point may not be simple to reach (due to partial exchanges on streams, stream divisions…), and the number of heat exchangers needed may be significant. While optimisation tools can help to determine the optimal structure of the heat exchanger networks using MINLP-type algorithms, these tools often have calculation times that are very long, and their convergence on an industrial scale is not guaranteed. The approach of Simulis Pinch is more pragmatic: the aim is to quickly propose some good solutions with a limited number of heat exchangers. The method is not an "optimal" method in the mathematical sense of the term, as it systematically aims to exchange the maximum power between two streams. If a hot stream and a cold stream can exchange heat, the algorithm searches for the possible couplings that allow the exchange of the maximal thermal power under certain limitations or "boundary conditions". The automatic synthesis of a 'good' heat exchanger network takes into account the specific constraints on particular exchanges (fluids incompatibility, different plants, distance, difficulties…). Between two streams that can exchange heat, the combination of the possible exchanges is limited to a maximum of 3 possible couplings between these streams.

REFERENCES

[1] Bodo Linnhoff, PhD thesis "Thermodynamic Analysis in the Design of Process Networks", Leeds University, 1979

[2] www.prosim.net

420133 Novel Processes for the Production of Olefins and Aromatics from Methane Via Methanol: A Comprehensive Process Synthesis and Global Optimization Approach

Abstract:

Recently discovered sources of shale gas, along with the technological improvements within the shale gas industry, have greatly increased the supply of natural gas in the United States. However, several challenges have emerged due to these developments, a critical one being the proper utilization of this resource into more valuable and highly demanded products via routes that are both economical and profitable. Since natural gas is the cleanest fossil fuel, these challenges present tremendous opportunities for the proper utilization of methane and have significantly increased interest in C1 chemistry across industries and within academia and government agencies.[1] To this extent, we propose novel methods for the production of high-value aromatics and olefins from natural gas via methanol. In the mid-2000s, almost all of the approximately 4.5 million metric tons per year of methanol produced in the United States utilized the synthesis gas route.

Conversion of natural gas into methanol as an intermediate for chemicals production is exceptionally promising and remains the most probable route for the widespread commercialization of natural gas based refineries in the United States. Olefins and aromatics are valuable petrochemical products and in high demand. In 2007, the worldwide consumption of ethylene was 115 million tons and the global demand of propylene was 73.5 million tons.[2] Between 2005 and 2008, the global demand for benzene, para-xylene, ortho-xylene, and meta-xylene was approximately 40, 26, 6, and 0.4 million metric tons per year, respectively.[2] Thus, opportunities exist for widespread penetration of natural-gas based chemical refineries, albeit with challenges associated with which types of chemicals to produce and which technologies to invest in.

In this work, we propose a process synthesis superstructure that contains several direct and indirect commercialized natural gas conversion technologies, including autothermal reforming and steam reforming, among others.[3,4] To address the economic barriers often associated with the utilization of alternative feedstocks, we develop a novel branch-and-bound global optimization framework that is capable of determining the optimal technologies to produce aromatics and olefins from natural gas at the highest profit.[3-5] Several commercialized and promising olefins and aromatics production, upgrading, and separation alternatives are included within this process synthesis framework.[6-9] This approach provides an adequate baseline for comparing competing technologies and identifying bottlenecks for natural gas conversion based technologies. Several case studies are presented that explore the effect of refinery capacity and product output. The key topological decisions and analysis of uncertain parameters will be discussed.

1. Mokrani T., Scurrel M. Gas conversion to liquid fuels and chemicals: the methanol route-catalysis and processes development. Catalysis Reviews 2009, 51:1-145

2. de Klerk, A. Fischer-Tropsch Refining; Wiley-VCH Verlag & Co. KgaA: Weinheim 2011

3. Onel, O.; Niziolek, A. M.; Elia, J. A.; Baliban, R. C.; Floudas, C. A. Biomass and natural gas to liquid transportation fuels and olefins (BGTL+C2_C4): Process Synthesis and Global Optimization. Industrial & Engineering Chemistry Research 2014, 54, 359-385.

4. Niziolek, A. M.; Onel, O.; Elia, J. A.; Baliban R. C.; Floudas, C. A. Coproduction of Liquid Transportation Fuels and C6_C8 Aromatics from Biomass and Natural Gas. AIChE Journal 2015, 61, 831-856.

5. Baliban, R. C.; Elia, J. A.; Misener, R.; Floudas, C. A. Global Optimization of a MINLP Process Synthesis Model for Thermochemical Based Conversion of Hybrid Coal, Biomass, and Natural Gas to Liquid Fuels. Computers and Chemical Engineering 2012, 42, 64-86.

6. Niziolek, A. M.; Onel, O.; Hasan, F. M.; Floudas, C. A. Municipal solid waste to liquid transportation fuels – Part II: Process synthesis and global optimization strategies. Computers and Chemical Engineering 2014, 74, 184-203.

7. Niziolek, A. M.; Onel, O.; Elia, J. A.; Baliban, R. C.; Xiao, X.; Floudas, C. A. Coal and Biomass to Liquid Transportation Fuels: Process Synthesis and Global Optimization Strategies. Industrial Engineering & Chemistry Research 2014, 53, 17002-17025.

8. Baliban, R. C.; Elia, J. A.; Floudas, C. A. Novel natural gas to liquids (GTL) technologies: Process synthesis and global optimization strategies. AIChE Journal 2013, 59, 505–531.

9. Baliban, R. C.; Elia, J. A.; Floudas, C. A. Biomass and Natural Gas to Liquid Transportation Fuels: Process Synthesis, Global Optimization, and Topology Analysis. Industrial & Engineering Chem Research 2013, 52, 3381–3406.

422590 Process Design and Optimization of Chemicals' Production from Biomass Feedstocks

Abstract:

Process Design and Optimization of Chemicals' Production from Biomass Feedstocks

Zhaojia Lin[1], Abhay Athaley[1], Vladimiros Nikolakis[2], Marianthi Ierapetritou[1]

[1] Department of Chemical and Biochemical Engineering, Rutgers - The State University of New Jersey

[2.] Catalysis Center for Energy Innovation & Department of Chemical & Biomolecular Engineering, University of Delaware

This work focuses on the process design and optimization of processes for the production of chemicals from lignocellulosic biomass feedstocks. The acceptance of these products in the market place depends on their competitiveness compared with petroleum-based products not only in terms of economics but also in terms of process sustainability.[6] The development of efficient and economically sustainable routes for the production of bio-based platform chemicals such as 5-hydroxymethylfurfural (HMF), furfural, levulinic acid, etc. has attracted a lot of attention.[1,2] We investigate paths based on the non-enzymatic hydrolysis of the cellulosic and hemicellulosic fractions to produce sugar mixtures followed by "one-pot" reactive extraction to convert sugar mixtures to furan derivatives. The aforementioned chemicals can be produced from glucose or xylose solutions using reactive extraction. However, the requirement for the use of organic solvents as extractants and salts that promote the partition coefficients of HMF and furfural have a negative impact on the economics and sustainability. Moreover, xylose and glucose solutions from hydrolysis steps usually contain small amounts of glucose or xylose respectively, unless further purification process is implemented.[3] Therefore, it appears more economically and environmentally attractive to simultaneously manufacture HMF and furfural from C5 and C6 mixtures that are produced from concentrated-acid biomass hydrolysis.[4,5]

Previous studies evaluated the economics and environmental impacts of these bio-products independently;[7-9] however only few studies have explored the optimization of biobased production involving both reaction and separation steps.[10] It is essential to implement flowsheet optimization to obtain the optimal operating conditions that minimizes production costs and environmental impacts both at the production but also at the separation stage. Surrogate-based optimization is a useful tool for the cases that 1) original models are expensive to evaluate; 2) the original gradient information is inaccessible for those commercial simulation tools or the noise exists so that the gradient-based algorithms cannot be accurately used. [11,12]

This work uses techno-economic analysis and life cycle assessment to design and evaluate the production alternatives. The detailed kinetics models are included in the reaction step and the separation units are simulated using Aspen Plus. At the next stage a surrogate-based optimization approach is used to achieve the optimal alternative in terms of economic and environmental criteria. More specifically, the surrogate-based models are generated to represent one or groups of original reaction and separation units; next the surrogate-based models are connected and reformulated into multi-criteria nonlinear programing problem (NLP); and finally NLP problem is solved by a commercial solver to present the pareto optimal set of solutions.

References:

1. Bozell JJ, Petersen GR. Technology development for the production of biobased products from biorefinery carbohydrates-the US Department of Energy's "Top 10" revisited. *Green Chemistry.* 2010;12(4):539-554.

2. Dutta S, De S, Saha B, Alam MI. Advances in conversion of hemicellulosic biomass to furfural and upgrading to biofuels. *Catalysis Science & Technology.* 2012;2(10):2025-2036.

3. Harris JF, Laboratory FP. *Two-stage, dilute sulfuric acid hydrolysis of wood: an investigation of fundamentals*: U.S. Dept. of Agriculture, Forest Service, Forest Products Laboratory; 1985.

4. Chheda JN, Roman-Leshkov Y, Dumesic JA. Production of 5-hydroxymethylfurfural and furfural by dehydration of biomass-derived mono- and poly-saccharides. *Green Chemistry.* 2007;9(4):342-350.

5. Zhang J, Lin L, Liu S. Efficient Production of Furan Derivatives from a Sugar Mixture by Catalytic Process. *Energy & Fuels.* 2012/07/19 2012;26(7):4560-4567.

6. Hawkins ACG, David A.; Buelter, Thomas; Wade, James; Meinhold, Peter; Peters, Matthew W.; Gruber, Patrick R.; Evanko, William A.; Aristidou, Aristos A., Assignee, Inventors; Gevo Inc. USA, assignee. METHODS FOR THE ECONOMICAL PRODUCTION OF BIOFUEL PRECURSOR THAT IS ALSO A BIOFUEL FROM BIOMASS 2009.

7. Lin Z, Ierapetritou M, Nikolakis V. Aromatics from Lignocellulosic Biomass: Economic Analysis of the Production of p-Xylene from 5-Hydroxymethylfurfural. *AIChE Journal.* 2013;59(6):2079-2087.

8. Lin Z, Nikolakis V, Ierapetritou M. Alternative Approaches for p-Xylene Production from Starch: Techno-Economic Analysis. *Industrial & Engineering Chemistry Research.* 2014/07/02 2014;53(26):10688-10699.

9. Lin Z, Nikolakis V, Ierapetritou M. Life Cycle Assessment of Biobased p-Xylene Production. *Industrial & Engineering Chemistry Research.* 2015/03/04 2015;54(8):2366-2378.

10. Torres AI, Daoutidis P, Tsapatsis M. Continuous production of 5-hydroxymethylfurfural from fructose: a design case study. *Energy & Environmental Science.* 2010;3(10):1560-1572.

11. Caballero JA, Grossmann IE. An algorithm for the use of surrogate models in modular flowsheet optimization. *AIChE Journal.* 2008;54(10):2633-2650.

12. Henao CA, Maravelias CT. Surrogate-based superstructure optimization framework. *AIChE Journal.* 2011;57(5):1216-1232.

(91c) A Review on Life Cycle Assessments of Prioritized Renewable Chemicals and Their GHG Reduction Potential

Montazeri, M., Northeastern University

Zaimes, G. G., University of Pittsburgh

Khanna, V., University of Pittsburgh

Eckelman, M. J., Northeastern University

Originally presented on: 11/9/2015 9:10:00 - 9:30:00

Abstract:

The estimated biomass production in the world is 100 petagrams (100 Gt) of carbon per year, about half in the ocean and half on land.[1] Biomass has always been a major source of energy for humankind and is presently estimated to contribute on the order of 10?14% of the world's energy supply.[2] In today's fossil fuel-driven industrial economy, production of bio-based chemicals and fuels may have numerous benefits but should also aim to decrease life cycle environmental impacts (such as GHG emissions) to be considered as favorable alternatives compared to their conventional counterparts. The RFS2 (renewable fuel standard) program administered by the USEPA regulates minimum content of renewable fuels in regular blend and sets a threshold of greenhouse gas emissions reduction for renewable fuels of least 20% for corn ethanol and a 50-60% reduction for fuels from non-corn feedstocks, cellulosic and agricultural wastes, and biodiesel.[3] While fuels have been a focus of public policy, nearly 50 million tons of bio-based chemicals are also produced annually worldwide, but there is no commensurate threshold for GHG emissions reduction of bio-based chemicals in order to define what can be considered as a renewable chemical. This presentation will review available bio-based chemical pathways and associated non-renewable energy use and life cycle GHG emissions reductions.

The US Department of Energy has conducted two separate screening analyses [4] to identify priority bio-based chemicals based on available technologies and the market demands. These chemicals can be produced from both sugar and non-sugar components of biomass resources. In this study, we conducted a thorough literature review on life cycle assessments of 11 sugar-based and 12 lignin-based building blocks reported by USDOE. 11 other bio-based chemicals that are currently in active research and development but not listed by USDOE, are also studied. This additional group can either be produced from sugar or non-sugar components of biomass. Selected studies are comprised of cultivation, preparation, intermediate and final production processes for each chemical; environmental burdens from the use and end of life phases are excluded from this study and system boundary is set to be cradle to gate. Various sources of biomass (e.g, seeds, agricultural and woody waste, pulp and paper waste streams, and algae); technology pathways (e.g., fermentation, catalytic conversion, pyrolysis, and gasification); national or regional differences; and finally life cycle assessment allocation methods (e.g., economic and mass allocation and system expansion) were considered. Tables 1a-c show the chemicals under consideration.

Table 1. a) Sugar-based and b) lignin-based building blocks from DOE reports, and c) other bio-based building blocks of interest

Sugar-based Building Blocks	Lignin-based Building Blocks	Other Building Blocks
1,4-Succinic, Fumaric and Malic Acids	Methanol/Dimethyl ether (DME)	Adipic Acid
2,5-Furan Dicarboxylic Acid	Styrenes	Polylactic Acid (PLA)
3-Hydroxy Propionic Acid	Biphenyls	Polyhydroxyalkanoate (PHA)
Aspartic Acid	Phenol	Low/ High Density Polyethylene
Glucaric Acid	Catchols, Cresols, Resorcinols	Propylene Glycol
Glutamic Acid	Vanillin	Propanediol
Itaconic Acid	Vanilic Acid	Acrylic Acid
3-Hydroxybutyrolactone	Cyclohexane	Butanol
Glycerol	Aromatic Acids	Acetic Acid
Sorbitol	Aliphatic Acids	
Xylitol/Arabinitol	Wood Preservatives	
	Formaldehyde Free Adhesives and Binders	

Life cycle GHG emission results from each of these studies are collected and compared against common petrochemical counterparts. For the comparative LCAs, GHG changes were extracted from the articles but in cases where GHG emissions of the petrochemical counterpart was not reported, Eco-Invent unit processes were used for the estimation. Other results including non-renewable energy use was also included for most of the studied scenarios.

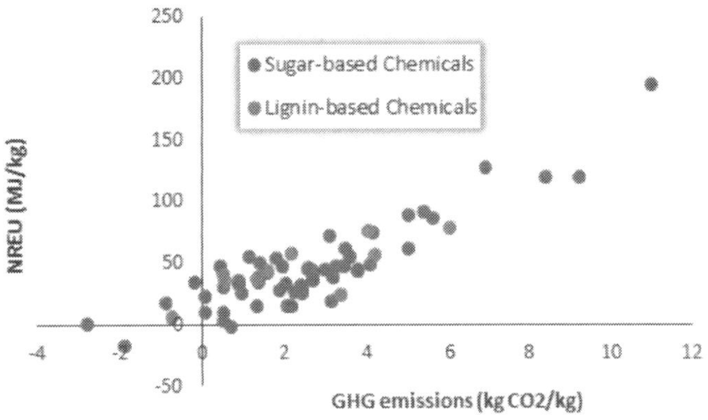

Figure 1. Non-renewable energy use versus GHG emission for sugar-based and lignin-based chemicals

Figure 1 shows the trend of non-renewable energy consumption versus greenhouse gas emissions in kg CO_2 equivalent for a functional unit of 1 kg of final product. As expected, there is a linear relationship between these parameters, as an increase in process energy consumption supplied by fossil sources will increase the emissions.

This study shows the gaps in LCA of bio-based chemicals that require more investigation, especially for lignin-based chemicals due to the complexity in primary structure of lignin and treatment methods; aspartic acid, glucaric acid, sorbitol and arabinitol are examples of sugar-based chemicals while styrene, biphenyl and cresols are examples of lignin-based chemicals included in the list but have not yet been studied on life cycle basis. Moreover, current results are dependent on processing pathways and display significant variability. The GHG results vary from >100% increase for acetic acid production from corn starch to >100% decrease for succinic acid production from corn compared to their counterparts. Even for a single chemical such as succinic acid, GHG emissions reductions range from 49% up to 100% depending on the choice of primary source, conversion and allocation methods. Bio-based chemicals that show reductions in GHG emissions also show lower fossil energy use in most cases, as shown in Figure 1.

Applying RFS2 requirements of at least 50% reduction in GHG emissions, succinic acid, PHB (polyhydroxybutyrate), xylitol, PLA (polylactic acid), polyethylene, propylene glycol and butanol show lower average values for sugar-based chemicals while methanol, PLA, PDO (propanediol), and butanol are preferable lignin-based products. Figure 2 shows these results. Average values are indicated with the full range of reported results also indicated. Blue dots show the average values while red triangles represent chemicals with no LCA studies found, indicating opportunities for further research.

Results of this study can be useful for policy makers in defining a threshold for environmental benefits should be met by renewable chemicals. There are limited data available right now to conduct a harmonization study, though. As shown in Figure 2-b, for most of the lignin-based chemicals one study was found, represented as a single point without error bars, and for the rest there are few life cycle assessment studies. On the other hand, above results are just GHG emission change, a full life cycle assessment accounting for other environmental impacts will assist finding overall benefits of renewable products. Some of the collected studies looked into other impact categories as well, but since most of the regulations are based on GHG emissions, this is the main impact category studied so far. Development of a consistent and comprehensive database for prioritized chemicals, including different pathways, sources and environmental burdens, is required for appropriate decision making.

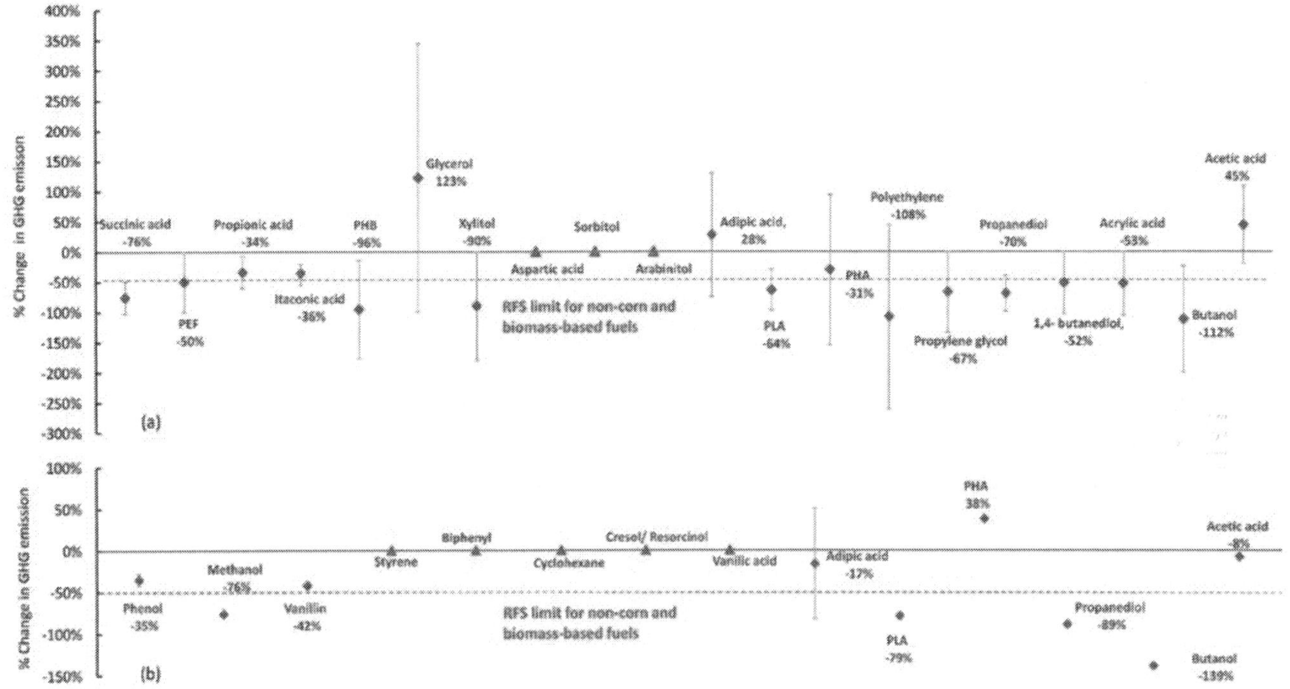

Figure 2. a) Percent change in life cycle GHG emissions of sugar-based chemicals and b) Percent change in life cycle GHG emissions of lignin-based chemicals

References:

1. Field, C. B.; Behrenfeld, M. J.; Randerson, J. T.; Falkowski, P., Primary production of the biosphere: integrating terrestrial and oceanic components. *Science* **1998,** *281* (5374), 237-240.

2. McKendry, P., Energy production from biomass (part 1): overview of biomass. *Bioresource Technology* **2002,** *83* (1), 37-46.

3. Schnepf, R.; Yacobucci, B. D. In *Renewable fuel standard (RFS): overview and issues*, CRS Report for Congress, 2010.

4. Werpy, T.; Petersen, G.; Aden, A.; Bozell, J.; Holladay, J.; White, J.; Manheim, A.; Eliot, D.; Lasure, L.; Jones, S. *Top value added chemicals from biomass. Volume 1-Results of screening for potential candidates from sugars and synthesis gas*; DTIC Document: 2004;

5. Holladay, J.; Bozell, J.; White, J.; Johnson, D., Top value-added chemicals from biomass. *Volume II?Results of Screening for Potential Candidates from Biorefinery Lignin, Report prepared by members of NREL, PNNL and University of Tennessee* **2007**.

428006 Process Simulation Models Parameterizing Multi-Sector Technoeconomic Framework to Test Viability of Emerging Bioprocesses

Abstract:

With recent advances in the development of bioprocesses, several new production schemes utilizing renewable feedstock have emerged for a wide range of chemicals. Many of these have demonstrated promising outcomes in a lab/small scale set up. Hence, lately many efforts were devoted to validate the feasibility of these processes using technoeconomic framework which intended to be used by researchers as one of the major tools to quantify the economic viability and test the potential market vulnerability.

While several exist, one of the major challenges in optimizing such a multifaceted framework is to obtain accurate process parameters, corresponding variable costs and emission estimates for the production of biochemicals. This contribution will demonstrate the manner by which these parameters are generated and subsequently linked to a Multi-Sector technoeconomic framework. Wholly, these models have the capability to capture various attributes of fundamental research and link those with the higher level economic sector, guiding the monetary investment towards the selection of robust candidate biochemicals. The resulting decision support tool can be used to study the existing market and test the viability of bioprocesses while accounting for inter and intra sector interactions in the existing value chain. The traits of this tool, bolstered with detailed process models, make it an apt venue to test various renewable process technologies, provide insights towards future production strategies and guide the research to produce bio-based products.

434302 Biolubricant: Lignin in Choline/Amino Acid Ionic Liquids

Abstract:

Lignocellulosic biomass is the most abundantly available raw resource on the earth for the production of valuable products. Utilization of cellulose in versatile promising applications has been intensely investigated in last a few decades. However, the utilization of lignin is greatly restricted due to the its resistant to degradation, being held together with strong chemical bonds and also appears to have a lot of internal hydrogen bonds. Lignin fractionation from biomass has been demonstrated in choline based ionic liquids(ILs), however, the high cost of the ILs necessitates its recycle in industrial process which is difficult due to the large energy consumption. In this work, we present a new application for lignin dissolved in Choline based amino acid ILs as a green bio-lubricant. The addition of lignin could improve choline based ILs' anti-wear properties and friction stability. Various characterization techniques including differential scanning calorimetry, thermogravimetric analysis, and rheology are used to study the effect of lignin on the thermal, physical and tribological properties of lignin/choline based ILs. The results show that the lignin could enhance the physiochemical properties simultaneously.

435428 Characterization of a Macrophyte Microcosm As a Surface Water Treatment System for Antibiotics

Abstract:

Natural water treatment systems provide for a means to manage nutrient (and carbon) cycles while improving water quality. We present chemical engineering design characteristics from an example flow through and model tertiary water treatment microcosm. This research demonstrates that aquatic macrophytes (*Pistia stratiotes* and *Eichhornia crasippes*) can degrade concentrations (0-10 mg/L) of oxytetracycline (OTC). First order OTC degradation rate coefficients (0.07 - 4.5/day) can be adjusted by altering; agitation, pH, temperature, and aeration mass flows in exudate treatment strategies when simulated wastewater was fed at 10 L/day. Further, we suggest that electronic paramagnetic resonance can be used to aid in hypothesis testing that external elicitation with phytohormones can be used to promote biological based oxidative bursts with methyl jasmonate during water treatment. Finally, we suggest there is a lack of research that considers macrophytes in more holistic natural and human use cycles.

421706 Effective Removal of Arsenic from Water with Zirconium Metal-Organic Framework

Abstract:

Zirconium metal-organic framework (Zr-MOF) with particle size in micrometre order was synthesized and applied as a novel adsorbent in this work to remove arsenic contamination. The adsorbent functioned excellently under a wide pH range of 1 to 10, with the best adsorption performance at pH 2. No inhibitory effect on the adsorption process was observed in the presence of such commonly existing anions as chloride, nitrate, carbonate and sulphate. The adsorption isotherm study indicated that the maximum arsenate adsorption capacity reached as high as 303 mg/g at optimal pH 2 and 148 mg/g at neutral pH condition 7, which greatly outperform most adsorbents reported in literature. The EDX, PXRD and FTIR studies indicated that the interaction between arsenic and Zr-MOF occurred at two possible sites within the adsorbent framework, i.e., hydroxyl group and benzenedicarboxylate ligand, by which Zr-O-As coordination bonds were formed. Compared to most conventional nanoparticle sorbents, the highly porous crystalline structure of Zr-MOF provides a larger contact area and more active sites in unit space, which facilitate the diffusion and interaction of arsenic species towards the adsorbents. The excellent adsorption performance in removal of arsenic from aqueous solutions using Zr-MOF would provide new insights to the application of MOFs for water treatment, and lead to a promising advanced adsorbent in arsenic decontamination industry.

430686 Residual Solvent Release from Froth Treatment Tailings-Part 2

Abstract:

In oil sand processing, bitumen froth treatment tailings originating from tailings solvent recovery unit (TSRU) contain trace amounts of solvent. Currently, the TSRU tailings are deposited into tailings ponds under water. However, in future operation, the tailings will also be deposited on a beach. When the TSRU tailings are deposited part of the residual solvent inevitably evaporates and is released to atmosphere. It is important to understand the behaviour of the solvent release from the TSRU tailings. We conducted a research program to study the solvent release from tailings for underwater deposition and aerial deposition (on beach). The study on solvent release from underwater deposition was presented in AIChE conference 2014. In this presentation, the study on solvent release from aerial deposition is reported.

In this work, we studied the solvent release from aerial deposited tailings in two approaches: 1) kinetic modeling, and 2) experimental testing. A kinetic model based on continuity equations was found to successfully model solvent release from unsaturated tailings under aerial conditions at ambient temperature. The calculated solvent mass transfer rate agreed well with the experimentally measured value. However, at temperatures above the solvent boiling point, boiling solvent generated bubbles and produced foam. The foam formation changed the characteristic of tailings and made the modeling very difficult. Therefore, at temperatures above boiling point of the solvent, the model becomes inapplicable due to foam formation and volume increase of vapour. It seems that the mechanism of solvent release from aerially deposited tailings at high temperature is based on formation of pentane vapour bubbles and subsequent plug flow of the bubbles to the top of the deposit.

405376 Biodegradation Characteristics of Wastewater Using Different E. coli Strains

Abstract:

Cheese whey is produced in huge amounts during cheese manufacturing causing serious pollution problems. The main objective of this study is to investigate the biodegradation of cheese whey lactose using different bacterial strains, namely: *E. coli* strain as parental cells, VHb-expressing *E. coli* that is transformed with *Vitreoscilla hemoglobin gene, vgb* (*E. coli*:pUC8:16), and *E. coli* harboring a pUC9 plasmid (*E. coli*:pUC9). Batch fermentation experiments were performed at controlled temperature with different conditions such as initial whey lactose concentration, carbon and nitrogen sources, and yeast extract concentration. The VHb strain was found to be more efficient in whey lactose biodegradation than the other *vgb*-lacking strains. The lactic acid was detected as a product of cheese whey lactose biodegradation. Maximum growth and lactic acid production of VHb-expressing *E. coli* cells was achieved by using a mixture of 20% whey and 80% minimal salt medium in the growth culture. Under optimum growth conditions, a *vgb*-bearing strain was slightly more efficient than *Lactobacillus acidophilus* based on lactic acid production measurements. Contrary to Monod's, Haldane's growth kinetics model gave a good fit to the growth kinetics data. Kinetic constants of the Haldane equation were $\mu_m = 0.5573\ h^{-1}$, $K_s = 4.8812$ g/L, $K_I = 53.897$ g/L. Biomass growth was well described by the logistic equation while Luedeking-Piret equation defined the product formation kinetics. Substrate consumption was explained by production rate and maintenance requirements. In simulation studies including the Haldane model, an evident agreement was observed between measured and calculated biomass, product, and substrate concentrations.

439912 Water Disinfection using Hydrodynamic Cavitation: Influence of orifice shape & inception

Monday, November 9, 2015: 2:10 PM
Salon J (Salt Lake Marriott Downtown at City Creek)

Vivek V. Ranade, Chemical Engineering and Process Development, National Chemical Laboratory, Pune, India and Vikrant Gaikwad, Maharashtra Institute of Technology, Pune, India

Water Disinfection using Hydrodynamic Cavitation: Influence of orifice shape & inception

Vivek Ranade* and Vikrant Gaikwad1

Chemical Engineering and Process Development Division

CSIR – National Chemical Laboratory

Pune 411008, INDIA; vv.ranade@ncl.res.in

1 Maharashtra Institute of Technology, Pune, India

Hydrodynamic cavitation offers distinct advantages for water disinfection and waste water treatment particularly at large scales of operations. For applying hydrodynamic cavitation for water disinfection/ treatment, water stream to be treated is passed through a fluidic device in which there exists a low pressure region. Cavities are generated in this region when the pressure falls below vapour pressure (in absence of any dissolved gases). These cavities are then collapsed when they travel to regions of higher pressure and lead to very high shear, temperatures and pressures in localized regions. These high pressure/ temperature as well as high shear facilitate disinfection as well as reduction in chemical oxygen demand, color and amonical nitrogen in the water (Ranade and Bhandari, 2014). In this work, we have investigated hydrodynamic cavitation generated by different orifices and have tested their effectiveness for water disinfection.

In this work, five different orifices have been studied which have different orifice shapes with same fractional open area (see Figure 1). Systematic experiments were carried out to identify inception of cavitation. Measured pressure drop results as a function of liquid velocity as well as acoustic signals were used to identify inception points. Different flow regimes such as no cavitation, inception of cavitation, developed cavitation and super cavitation were identified for varied inlet pressures for each orifice plate. The efficacy of generated cavitation on water disinfection was quantified based on measurements of non-pathogenic E Coli colony forming units (CFU/ml).

The presented results and analysis will be useful for designing hydrodynamic cavitation based systems for water disinfection.

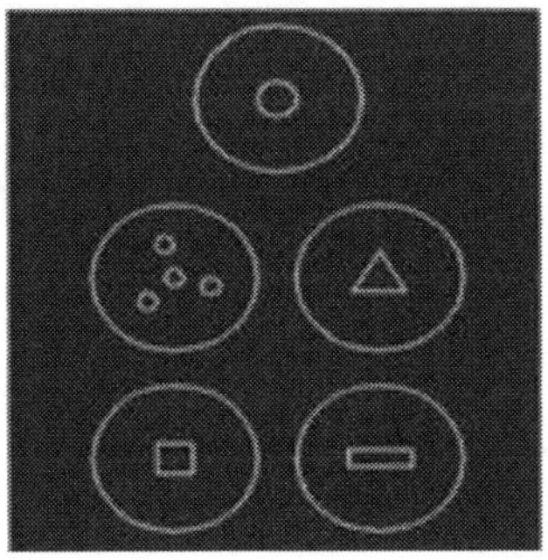

Five orifices with 10%

open area

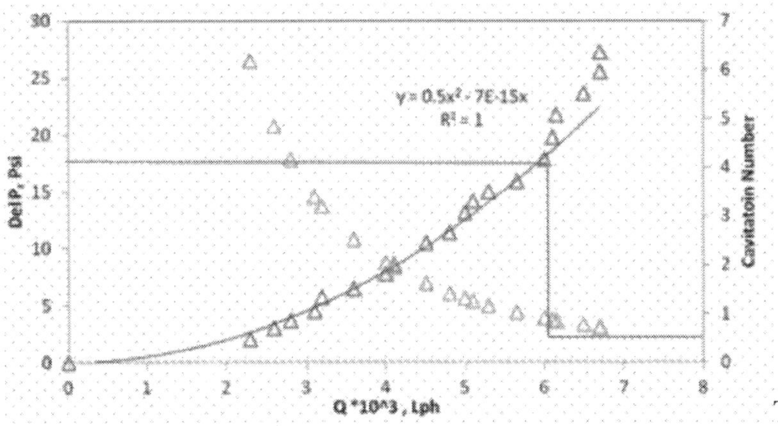

Typical cavitation inception

References:

Ranade, V.V. and V.M. Bhandari [Editors] (2014) Industrial Wastewater Treatment, Recycle & Reuse, Elsevier, Amsterdam.

435852 Potential of Microalgae Based Wastewater Treatment Process and Its Implications

Abstract:

Potential of microalgae based wastewater treatment process and its implications

Muhammad Ilyas*

*Center for Environment and Water Research Institute, King Fahd University of Petroleum and Minerals, Dhahran, Saudi Arabia

ABSTRACT

Microalgae based wastewater treatment coupled with CO_2 fixation is considered to be the future solution of wastewater treatment due to advantages of CO_2 fixation, biofuel production and environmental sustainability. The wastewater produced from different sources i.e. industrial, municipal, dairy and agricultural wastewater is rich in nutrients necessary for microalgae growth. Flue gas may act as large sources of CO_2 and sunshine a natural source of light for microalgae photosynthesis. The effectiveness of wastewater treatment depends upon on the uptake of different nutrients from wastewater by microalgae cells through phototrophic mechanism utilizing inorganic CO_2 sources and heterotrophic mechanism involving utilization of organic carbon present in wastewater. This method presents economical solution by saving fresh water and nutrients supply cost necessary for microalgae growth in biofuel production process. In this research detail mechanism of microalgae based wastewater treatment, working of photobioreactors for microalgae cells growth and factors affecting the performance and effectiveness of wastewater treatment are discussed.

Keywords: wastewater treatment, microalgae, CO_2 fixation

431829 The Role of Carbon Dioxide As a Solvent for Greener Biodiesel Production over a Heterogeneous Catalyst

Abstract:

Within the field of alternative energies, fuels produced from biological feedstocks have great potential to replace petroleum-based products. In order to sustainably produce fuels and materials, biorefineries must be designed for and operated with sustainable processes. Compressed carbon dioxide – subcritical, supercritical, and CO_2 expanded liquids – can potentially be used in a biorefinery as a green solvent. Key processes that may be improved via the use of CO_2 are extraction (including fractionation and refinement) and reaction. In this presentation, supercritical CO_2 ($scCO_2$) is shown to be an effective solvent for the selective conversion of lipids into biodiesel in combination with a heterogeneous catalyst. Successful transesterification of triglycerides into fatty acid methyl esters is demonstrated via a CO_2-expanded liquid phase. The role of CO_2 in reaction moderation is explored via kinetics with potential implications for selectivity and system optimization. While pitfalls do exist for the application of compressed CO_2 in commercial production, this versatile compound holds great promise for utilization as a green solvent in a biorefinery setting.

436989 New Process for Recovery of Renewable Energy and Chemicals from Black Liquor – A Review

Abstract:

In the pulp and paper industry, large quantities of forest biomass are being used. The by-products or residues which result include black liquor, bark, and forest logging residues. These can be used for energy purpose to produce electricity, heat, steam, and biofuels. The kraft pulping process accounts for almost 60% of all pulp production. Black liquor from kraft process represents a potential energy source of 250-500 MW per mill. This paper describes an exciting emerging technology that uses low temperature steam re-forming instead of traditional combustion to recover energy and chemicals from black liquor boosts high thermal efficiency, low emissions (including particulate matters and odorous TRS gases), and low power consumption. This versatile process, a very good example of pollution prevention at the source through process changes, is now available for commercial scale processing of black liquors generate at the pulp and paper industry. The black liquor steam reforming process utilizes indirect heating of a steam fluidized bed of sodium carbonate solids to recover cooking chemical and energy. Black liquor is sprayed directly into the bed where the liquor droplets uniformly coat the bed solids, resulting in high rates of heating, pyrolysis and steam reforming. The process steam reforms black liquor in the absence of oxygen, producing a hydrogen-rich, medium-Btu gas with a heating value of 300-350 Btu/SCF. Bed temperatures are maintained at 1060 to 1150 ^{0}F (565-620 ^{0}C), thereby avoiding liquid smelt formation and the associated smelt-water explosion hazards. In this dry (no smelt) recovery process, the sodium sulfate in the liquor is reduced by reaction with the steam-reforming products, principally carbon monoxide and hydrogen. The reduced sulfur form is unstable in the reformer environment and decomposes rapidly to gaseous hydrogen sulfite (H_2S) and solid sodium carbonate (Na_2CO_3). The product gas is scrubbed with a solution derived from the sodium carbonate bed solids to produce green liquor for re-use in the pulping cycle.

Heat required to reach operating temperature and for the endothermic steam-reforming reactions is supplied by heat exchangers immersed in the fluidized bed. The heat exchangers consist of bundles of pulsed heater resonance tubes. A portion of the product gas is burned in the pulsed heaters to supply the necessary heat, thus making the steam reformer self-sufficient on its own fuel. Pulsations in the resonance tubes produce a gas-side heat transfer coefficient which is up to five times greater than a conventional fire-tube heater. This efficient heat transfer reduces the size and cost of the heat exchangers and reformer vessel. The hot combustion gases leaving the pulsed heaters are sent to a waste heat boiler to generate steam and to preheat the pulsed heater combustion air. The product gas from the reformer is passed through a cyclone to remove particulate matter, then cooled in a waste heat steam generator, scrubbed and quenched.

By separating the sodium and sulfur elements, this process provides an imaginative and chemically attractive alternative to conventional kraft recovery. The process separates the individual unit operations of the conventional recovery boiler and allows independent control of each operation under its own optimized conditions, producing a well-controlled, more efficient modular process plant. It is applicable to both incremental and full-scale production facilities. The process also provides attractive energy alternatives, since the product gas is compatible with combined-cycle cogeneration operations.

434099 Study of Iron Oxide Based Mixed Oxygen Carriers for in-Situ Tar Cracking in Biomass Chemical Looping

Abstract:

A chemical looping process that uses biomass as the feedstock could prove to be a highly economical, environmental and sustainable means for electricity and/or chemical production with net negative carbon emissions. A chemical looping system converts carbonaceous fuels, such as biomass, using metal oxide based oxygen carriers circulating between three reactors - a reducer, an oxidizer and a combustor – to produce heat and high purity hydrogen with simultaneous carbon dioxide (CO_2) sequestration. In the reducer, the biomass fuel is converted to CO_2 and H_2O by oxygen transfer from the oxygen carriers. The reduced oxygen carriers are then partially oxidized by steam in the oxidizer producing high purity hydrogen and subsequently fully oxidized by air in the combustor. The energy generated due to the exothermic reactions in the combustor is sufficient to compensate for the heat required to perform the endothermic reactions in the reducer and to produce electricity via steam generation.

Tar evolved during biomass conversion in the reducer mainly consists of oxygen containing heavy hydrocarbons and complex polycyclic aromatics. It is a major concern as it can hinder the continuous operation and performance of a biomass chemical looping (BCL) system. Hence, it is important to design oxygen carriers that have the ability to eliminate the tars in the reducer. While Fe_2O_3 is an ideal candidate for oxygen carrier material for chemical looping processes, in this study, different tar cracking catalysts (Olivine, Dolomite, K_2CO_3 and NiO) have been used in conjunction with Fe_2O_3 to enhance the carriers' ability to crack tar. Thermogravimetric Analysis (TGA) has been used to study the ability of the oxygen carriers to undergo multiple redox cycles. A series of fixed bed experiments have been used to crack biomass-derived tar with the mixed oxygen carriers. The product gaseous species have been identified using Gas Chromatography–Mass Spectrometry (GC-MS) and fresh and reacted solids have been characterized using X-Ray Diffraction (XRD). Addition of NiO to the Fe_2O_3 based oxygen carrier was observed to greatly improved its tar cracking ability and maintained its redox reactivity over multiple cycles. The results from this study will help in the development of oxygen carriers towards designing an optimal BCL system.

427895 Synthesis of SBA-16 Supported Cesium Salt of Iron Substituted Lacunary Keggin Type Polyoxometalate Based Catalyst for Selective Thermal Catalytic Cracking of Low Density Polyethylene

Abstract:

Polyethylene is a non-biodegradable polymer which may cause serious environmental issues, but its cracking can produce fuel oil and may solve the exhausting fuel issue. Thermal degradation of waste plastics into fuel oil is one of the most promising methods for conversion of the plastic wastes into fuel. However, in catalytic cracking of polyethylene wastes more solid residue with wax is achieved. And the liquid product contains a range of hydrocarbons which affect the nature as well as yield of the fuel fractions. The development of a suitable catalyst with high selectivity is highly desirable. Polyoxometalates (POMs) are class of molecularly defined inorganic metal-oxide clusters formed from early transition metals (vanadium, tungsten etc.). Their use as catalysts can result in selective recovery of useful hydrocarbons from waste polyethylene. This research is based on synthesis of SBA-16 supported cesium salt of iron substituted lacunary keggin type polyoxometalate. The synthesized material was characterized using FTIR, TGA-DSC, XRD and SEM-EDX to investigate their structure and evaluate their properties. The material was thermally stable above 500 °C and therefore used as catalyst for thermal cracking of low density polyethylene. The cracking temperature was evaluated by TGA-DSC of physical mixture of POM and polyethylene and experimental conditions were designed accordingly. The catalytic cracking was carried out in a batch reactor and the product was analysed by GC-MS. The results showed that the synthesized catalysts exhibited excellent efficiency in conversion of maximum polyethylene into liquid oil. This research will contribute a lot to solve the natural resource depletion issues and environmental issues.

429411 Energy and Carbon Foot Print Analysis of the Production of a Liquid Organic Hydrogen Carrier

Abstract:

In a future energy system based on renewable energies, energy carriers and storage systems are required to store excess energy for times with low provision of energy from renewable resources. In this context, hydrogen is considered as a suitable energy carrier although the efficient, dense and secure storage of hydrogen is still a challenge.

A promising approach to overcome this limitation is the use of liquid organic hydrogen carriers (LOHCs). LOHCs are unsaturated organic compounds which can store hydrogen by reversible hydrogenation (e.g. dibenzyltoluene). The hydrogenated form of the LOHC exhibits high thermodynamic stability, enabling storage at ambient conditions. In contrast to the application of solid chemical carrier substances e.g. metal hydrides, the handling and transportation of the LOHC rather easy due to the liquid nature of the carrier substance.

A lot of research was performed concerning the LOHC storage concept in terms of applicability and efficiency but only little is known about the energy demand and the ecological impact of the production of the carrier as such. Especially the carbon foot print of an energy carrier substance is important to ensure the overall ecological benefit of the storage concept.

In this contribution, we present a process concept for the production of the LOHC dibenzyltoluene on technical scale. Based on the process concept, the heat and electricity demand is determined and potential energy savings are identified by a pinch analysis. Moreover, a carbon foot print analysis is performed to reveal the ecological impact related to the production of the LOHC.

First results show that the main energy demand of the overall process is linked to the product purification steps. Especially for low conversions in the synthesis reactor this energy demand increases remarkably. By rigorous heat coupling of the overall process, the overall heat demand can be lowered by about 35 %.

428468 Process Synthesis of Flexible and Sustainable Processes Consisting of Membrane Separations and Reactive Distillation for the Production of Biodiesel

Abstract:

Due to global competition and an increasing demand for reduced emissions the development of sustainable and efficient processes based on renewable resources becomes more and more important. Furthermore, the fluctuating qualities of renewable feedstock necessitate highly flexible processes. Fatty esters are promising new building blocks for the chemical industry and can be produced by the transesterification of virgin vegetable oils. However, nowadays fatty esters are mainly used as biodiesel, an alternative fuel in future based on renewable resources [1,2]. Due to the high raw material costs for the virgin vegetal oils [3], no economic production is possible. A promising approach to lower the production costs is the application of cheap low-quality oils, such as waste cooking oil or crude non-edible oil [4,5]. These low quality oils contain a high amount of free fatty acids, which lead to undesired side reactions in the alkaline catalyzed transesterification reaction [6,7]. Consequently, new process concepts are necessary to enable the application of low-quality oils as a feedstock.

In order to achieve an efficient processing of the low-quality oils this study investigates intensified process concepts, such as reactive distillation and membrane-assisted reactive distillation. One possibility to reduce the free fatty acid content in the low-quality oils is the acid catalyzed esterification of the free fatty acids with an alcohol. Since this reaction is equilibrium limited the application of reactive distillation is favorable to enable higher conversion. Furthermore, it is indicated that the transesterification of the oils could take place simultaneously. To further enhance the conversion in the reactive distillation the combination with vapor permeation is promising. The selective removal of the by-product water from the condensate stream facilitates a shift of the reaction equilibrium towards the product side. Another possibility to reduce the free fatty acid content is the separation of the free fatty acids from the low-quality oils by means of organic solvent nanofiltration. The application of organic solvent nanofiltration enables an energy efficient separation due to the mild operation conditions, which also ensure consistent quality of the fatty acids and oils. The coupling of organic solvent nanofiltration with reactive distillation provides the potential to an even more energy efficient process. Besides the processing of the low-quality oils, the profitability of the production process might be increased because of the generation of new products, as e.g. the free fatty acids, which can be used as building blocks in the chemical industry.

The current study investigates the process synthesis of different flow sheet configurations of reactive distillation, vapor permeation and organic solvent nanofiltration. Therefore, models for each unit operation are developed and validated with own experimental results and literature data. Hence, the potential of the different intensified process configurations is evaluated and it is determined if they are capable to handle fluctuating feedstock compositions. Finally, the most promising configurations are identified with regard to product specifications, energy demand and production costs.

References:

[1] Demirbas, A., "Biodiesel: A realistic fuel alternative for diesel engines," Springer, London (2008).

[2] Santacesaria, E., Vicente, G.M., Di Serio, M., and Tesser, R., "Main technologies in biodiesel production: State of the art and future challenges," *Catalysis Today*, 195, pp. 2–13 (2012).

[3] Lim, S., and Teong, L.K., "Recent trends, opportunities and challenges of biodiesel in Malaysia: An overview," *Renewable and Sustainable Energy Reviews*, 14, pp. 938–954 (2010).

[4] Phan, A.N., and Phan, T.M., "Biodiesel production from waste cooking oils," *Fuel*, 87, pp. 3490–3496 (2008).

[5] Yaakob, Z., Mohammad, M., Alherbawi, M., Alam, Z., and Sopian, K., "Overview of the production of biodiesel from Waste cooking oil," *Renewable and Sustainable Energy Reviews*, 18, pp. 184–193 (2013).

[6] Freedman, B., Pryde, E.H., and Mounts, T.L., "Variables affecting the yields of fatty esters from transesterified vegetable oils," *Journal of the American Oil Chemists Society*, 61, pp. 1638–1643 (1984).

[7] Kumar Tiwari, A., Kumar, A., and Raheman, H., "Biodiesel production from jatropha oil (Jatropha curcas) with high free fatty acids: An optimized process," *Biomass and Bioenergy*, 31, pp. 569–575 (2007).

437483 Designing Software for Chemical Engineers – a Product Developer's Perspective

Abstract:

Process simulators are tools widely used for the design of chemical processes and chemical products. These products are ubiquitous in process engineering work and a normal part of the engineer's work flow. These tools and associated tools like the software plug-ins designed to provide highly specialized services such as mass transfer separation modules, flare systems and advanced chemical reactors. Although the tools are widely used little is said about the process used to develop these tools, the challenges the software product developer faces and how these challenges are met. In this presentation we take an introspective look at the development of software products commonly used in process design and process simulation.

425315 Designing Chimeric Peptides Using Swarm Intelligence Optimization for the Protein Folding Problem

Abstract:

Chimeric peptides offer biological engineering methodologies adapted in materials science for wide range of applications including assembly and synthesis. Tailored chimeric functionality depends not only the sequence but also the molecular structure of the resulting molecules. Designing new chimeric peptides therefore relies on developing a better understanding of the protein folding. Here, a new *ab initio* method for finding peptide backbone angles was developed with a swarm intelligence algorithm to find peptide folds faster than using a molecular dynamics approach. Currently, many approaches to the protein folding problem have been investigated. The first approach is to experimentally determine the structures. This is the most accurate way of determining the structure, but the process is slow and requires commitments of large resources. Determining a structure experimentally without knowing that the peptide design is already successful is prohibitive. The second approach is to use molecular dynamics simulations. The main limitation of these simulations for peptide design is the low number of peptides that can be analyzed. Having larger peptide design candidate pools should help to improve chimeric peptide design. Using existing data avoids the cost of determining new structures and statistical analysis avoids the computational cost of simulating the protein folding process in the second approach. However in swarm intelligence algorithm, statistical analysis of determined folding structures may avoid the challenges of the first two approaches. In previous work[1], an antimicrobial peptide that self-assembles onto titanium implants has been developed. Here, we applied two optimization strategies to these peptide sets for finding chimeric peptide folds more effectively. We next compared their folding patterns to engineer sequence-structure function relations. The first optimization strategy uses a swarm intelligence algorithm inspired by bee hive coomunication to coordinate the selection of highly likely folds. The second approach is to match the peptide fragments with historical data of protein folds[2]. PyRosetta software[3] was used to score the likelihood of structures and to find the bond angles for the amino acid side chains. The likelihood of the structures for each approach was compared, and a time analysis was performed. Our approach offers to expand the pool of new chimeric peptide design. *The NIH-NIAMS (7R21AR062249-03) is acknowledged.*

[1] Yucesoy D, Hnilova M, Boone K, Arnold P, Snead M, Tamerler C. Chimeric Peptides as Implant Functionalization Agents for Titanium Alloy Implants with Antimicrobial Properties. JOM 2015;67:754-66.

[2] Kim DE, Chivian D, Baker D. Protein structure prediction and analysis using the Robetta server. Nucleic Acids Research 2004;32:W526-W31.

[3] Chaudhury S, Lyskov S, Gray JJ. PyRosetta: a script-based interface for implementing molecular modeling algorithms using Rosetta. Bioinformatics 2010;26:689-91.

15AIChE
Annual Meeting, Salt Lake City, UT

DTU

A New Optimization Model for Computer-Aided Molecular Design Problems

Lei Zhang, **Stefano Cignitti**, Rafiqul Gani*

PSE for SPEED Project, Department of Chemical and Biochemical Engineering, Technical University of Denmark

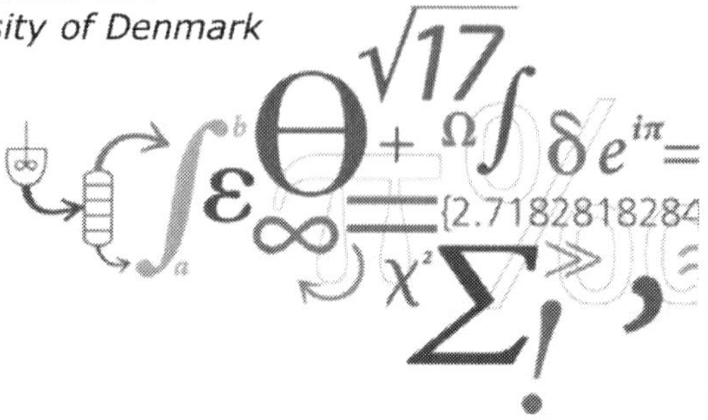

DTU Chemical Engineering
Department of Chemical and Biochemical Engineering

15AIChE
Annual Meeting, Salt Lake City, UT

DTU

Outline

- Introduction
- Computer-aided Framework for Product Design Problems
- New Optimization Model for CAMD
- Case Studies
  - Surfactant Design
  - Refrigerant Design
  - Die Attach Adhesive Design
- Conclusion

15AIChE
Annual Meeting, Salt Lake City, UT

DTU

Chemical Product Design

- The modern society needs many products for its survival.
- Which are the important products, which resources should be used to produce them and what is the role of the chemicals based manufacturing process?
- With the help of computer-aided chemical product design, the optimal, more sustainable product-process will be designed.

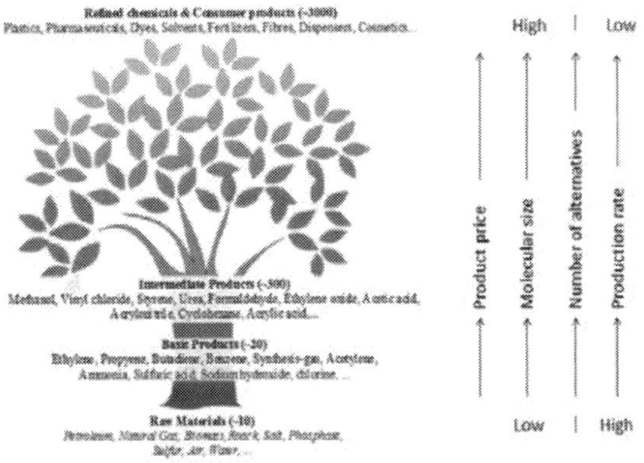

3 DTU Chemical Engineering, Technical University of Denmark

17 December 2015

15AIChE
Annual Meeting, Salt Lake City, UT

DTU

Computer-Aided Chemical Product Design

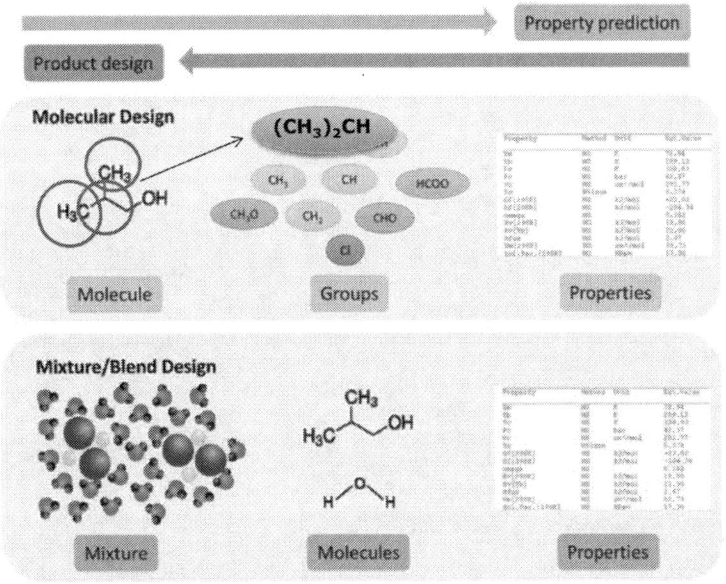

4 DTU Chemical Engineering, Technical University of Denmark

17 December 2015

Computer-Aided Framework for Product Design Problems

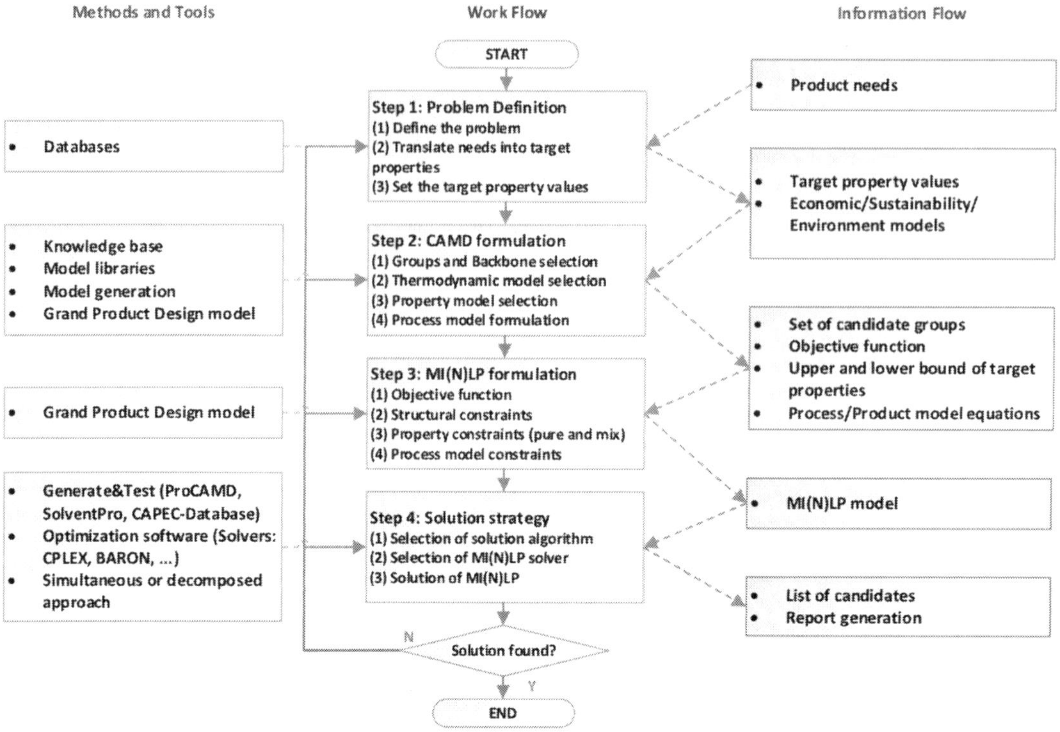

15AIChE
Annual Meeting, Salt Lake City, UT

DTU

New Optimization Model for CAMD: MINLP Formulation

$$\min / \max f_{obj}(\mathbf{X}, \mathbf{N})$$

$$\text{Structural constraints:} \quad g_1(\mathbf{N}, \mathbf{Y}) \leq 0$$

$$\text{Pure component property:} \quad g_2(\mathbf{N}, \mathbf{Y}) \leq 0$$

$$\text{Mixture property:} \quad g_3(\mathbf{X}, \mathbf{N}) \leq 0$$

$$\text{Process model and others:} \quad g_4(\mathbf{X}, \mathbf{N}) = 0$$

$$\mathbf{X} \in \mathbb{R}^n; \mathbf{N} \in \mathbb{Z}_+^m; \mathbf{Y} \in \{0, 1\}^q$$

$\mathbf{N}$: group number;

$\mathbf{Y}$: adjacency matrix;

$\mathbf{X}$: continuous variables related to the mixture and process.

- Adjacency matrix is introduced to predict the detailed structure of the molecule.
- New constraints are involved to find the number of 2nd order groups from the adjacency matrix using binary variables.

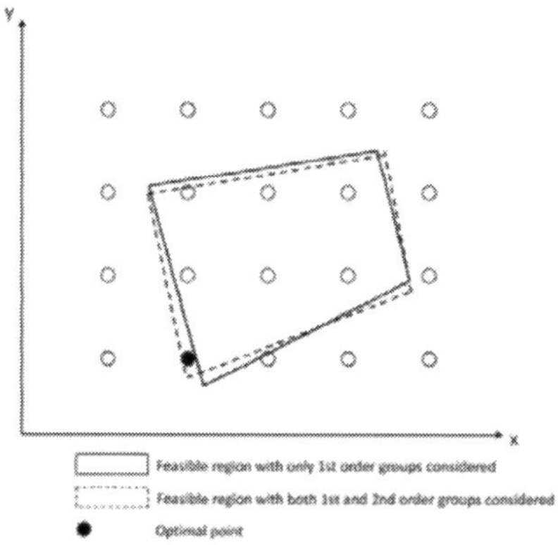

Figure: Higher order groups need to be considered simultaneously to obtain "real" optimal product

Generic mathematical programming formulation

1. Structural constraints:

1.1 Octet rule

$$\sum_{i \in G_1} (2-\nu_i)\, n_i^{(1)} = 2q$$

$$\sum_{i_1 \neq i_2; i_1 \in G_1} n_{i_1}^{(1)} \geq n_{i_2}^{(1)} (\nu_{i_2} - 2) + 2 \quad \forall i_2 \in G_1$$

1.2 Constraints ensure only one molecule is formed (Churi and Achenie (1996)).

1.3 Group number constraints:

$$n_i^L \leq n_i^{(1)} \leq n_i^U \quad \forall i \in G_1$$

$$n^{min} \leq \sum_{i \in G_1} n_i^{(1)} \leq n^{max}$$

1.4 Adjacency matrix constraints:

$$y_{i_1,id_1,i_1,id_1} = 0 \quad \forall i_1 \in G_1, id_1 \in ID$$

$$y_{i_1,id_1,i_2,id_2} = y_{i_2,id_2,i_1,id_1} \quad \forall i \in G_1; id \in ID$$

$$\sum_{id_1 \in ID} z_{i_1,id_1} = n_{i_1}^{(1)} \quad \forall i_1 \in G_1$$

$$\sum_{i_2 \in G_1} \sum_{id_2 \in ID} y_{i_1,id_1,i_2,id_2} = \nu_{i_1} z_{i_1,id_1}$$

$$\forall i_1 \in G_1, id_1 \in ID$$

1.5 Second-order group constraints:

$$N_B^J - M\left(1 - b_{id_1,id_2,\cdots,id_{N_B}^J}\right) \leq \sum_{\text{If these two grps connect in 2nd grp. } J} y_{i_1,j_1,i_2,j_2} \leq N_B^J - 1 + M\left(b_{id_1,id_2,\cdots,id_{N_B}^J}\right) \quad \forall J \in G_2$$

$$n_j^{(2)} = \frac{1}{T} \sum_{id_1,id_2,\cdots,id_{N_B}} b_{id_1,id_2,\cdots,id_{N_B}^J} \quad \forall J \in G_2$$

2. Property constraints:

$$p_k^L \leq p_k \leq p_k^U \quad \forall k \in P$$

$$p_k = \sum_{i \in G_1} n_i^{(1)} p_k^{(1)} + \sum_{j \in G_2} n_j^{(2)} p_k^{(2)} \quad \forall k \in P$$

3. Process model, economic model and other constraints.

[1] Lei Zhang, Stefano Cignitti, Rafiqul Gani. *Computers & Chemical Engineering*, **78**, 79-84

Churi N, Achenie LE., Ind Eng Chem Res 1996;35(10):3788–94.

Surfactant Design - Product Design (1/3)

- Step 1: Design of a UV sunscreen, in the emulsified form, with a high sun protection factor. The phenol ring is fixed as backbone structure in this case study.
- Step 2:
  - **Objective:** Toxicity should be minimized.
  - **Group selection:** CH_3 , CH_2 , CH , C , aCH , aC-OH , CH_2COO , CH_3O , CH_2O , aC-O , OCH_2CH_2OH from 220 groups (Marrero and Gani, (2001)).
  - **Backbone selection:** -C_6H_4OH

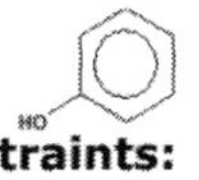

  - **Structural and property constraints:**

Need	Lower bound	Upper bound
N_G	10	15
N_F	0	8
$lc50$	3.16log(mol/L)	-
Sp	-	25MPa$^{0.5}$
clp	343.15K	-
Fp	343.15K	-
Vm	0.1cc/mol	0.3cc/mol

LC_{50}: Fathead Minnow 96-hr LC_{50}; *Sp*: Hildebrand solubility parameter; *clp*: cloud point; *Fp*: Flash point; *Vm*: Liquid molar volume at 298K.

15AIChE
Annual Meeting, Salt Lake City, UT

DTU

Surfactant Design (2/3)

- Step 3:

$$\max lc50 = lco + \sum_{i_1} n_{i_1}^{(1)} lc50_{i_1}^{(1)} + \sum_{i_2} n_{i_2}^{(2)} lc50_{i_2}^{(2)}$$

1. **Structural constraints:**

   1.1 Octet rule and group number constraints:

$$\sum_{i} (2 - \nu_{i_1}) n_{i_1}^{(1)} = 0$$

$$\sum_{i_1, i_1 \neq i_1'} n_{i_1}^{(1)} \geq n_{i_1'}^{(1)} \left(\nu_{i_1'} - 2\right) + 2 \quad \forall i_1'$$

$$1 \leq n_{i_1}^{(1)} \leq 5 \quad \forall i$$

$$10 \leq \sum_{i_1} n_{i_1}^{(1)} \leq 15$$

   1.2 Adjacency matrix constraints. (Slide 6)

   1.3 Second-order group constraints. (Slide 6)

2. **Pure component property constraints:**

$$lco + \sum_{i_1} n_{i_1}^{(1)} lc50_{i_1}^{(1)} + \sum_{i_2} n_{i_2}^{(2)} lc50_{i_2}^{(2)} \geq 3.16$$

$$Spo + \sum_{i_1} n_{i_1}^{(1)} Sp_{i_1}^{(1)} + \sum_{i_2} n_{i_2}^{(2)} Sp_{i_2}^{(2)} \leq 25$$

$$\sum_{i_1} n_{i_1}^{(1)} clp_{i_1}^{(1)} \geq 343.15^2$$

$$Fpo + \sum_{i_1} ni_1{}^{(1)} Fp_{i_1}^{(1)} + \sum_{i_2} n_{i_2}^{(2)} Fp_{i_2}^{(2)} \geq 343.15$$

$$0.1 \leq Vmo + \sum_{i_1} ni_1{}^{(1)} Vm_{i_1}^{(1)} + \sum_{i_2} n_{i_2}^{(2)} Vm_{i_2}^{(2)} \leq 0.3$$

- #Equations: 1,324,764; #Variables: 1,286,278 (1,286,266 Integer var.).

- Step 4: This model can be solved directly using GAMS CPLEX solver.

15AIChE
Annual Meeting, Salt Lake City, UT

DTU

Surfactant Design (3/3)

Table: Optimization results of surfactant design problem

n_1	1 CH_3, 5 CH_2, 1 CH, 4 aCH, 1 aC–OH, 1 CH_2COO, 1 aC–O, 1 OCH_2CH_2OH
n_2	1 AROMRINGs1s4
$lc50$	4.124log(mol/L)
Sp	24.732MPa$^{0.5}$
clp	300.016K
Fp	552.383K
Vm	0.281cc/mol

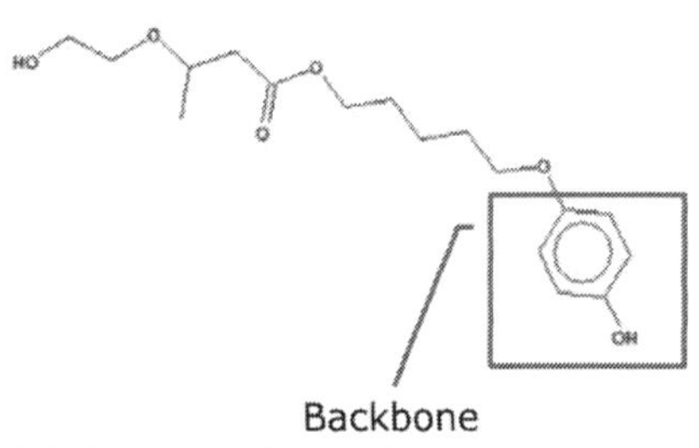

Refrigerant Design - Product-Process Design (1/3)

- Step 1: Design a new refrigerant for the replacement of R-134a, the new fluid should have (1) similar/better properties and thermodynamic behaviour; (2) reduced environmental impact.
- Step 2:
  - **Objective:** Coefficient of performance (COP) should be maximized.
  - **Group selection:** Fluids can contain C, H, F and Cl. 30 molecular groups are selected from 220 groups (Marrero and Gani, (2001)).
  - **Target properties for product and process:**

Target Property	Lower bound	Upper bound
Molar mass	-	110g/mol
Normal boiling point	-	250K
Critical temperature	350K	400K
Critical pressure	30bar	50bar
Thermal conductivity (liquid, Tb)	0.08W/m K	-
Enthalpy of Vaporization at Tb	200KJ/Kg	-
ODP	-	0
GWP	-	1400
Atmospheric Lifetime *	-	14years
Process needs	Lower bound	Upper bound
Evaporation temperature	313.15K	333.15K
Condensation temperature	265.65K	280.15K

Refrigerant Design (2/3)

$$\max COP = \frac{H_1(Z_{V1}, T_1, P1) - H_3(Z_{L3}, T_3, P_3)}{H_2(Z_{V2}, T_2, P_2) - H_1(Z_{V1}, T_1, P_1)}$$

s.t.

1. Structural constraints.
2. Property constraints.
3. Process constraints:

$$\ln \psi_{V1} = \ln \psi_{L1}$$
$$\ln \psi_{V2} = \ln \psi_{L2}$$
$$H_{L3}(T_3, P_3) = H_{L_4}(T_4, P_4)$$
$$S_{V2}(T_2, P_2) = S_{V1}(T_1, P_1)$$
$$P_1 = P4 \quad P_2 = P_3 \quad T_1 = T_4$$
$$P_2 \leq 0.8P_c \quad T_2 \leq 0.8T_c$$
$$265.65K \leq T_1 \leq 280.15K$$
$$313.15K \leq T_3 \leq 333.15K$$
$$Z_{V1}^3 - Z_{V1}^2 + Z_{V1}(-B^2 + A - B) - AB = 0$$
$$Z_{V2}^3 - Z_{V2}^2 + Z_{V2}(-B^2 + A - B) - AB = 0$$
$$Z_{V3}^3 - Z_{V3}^2 + Z_{V3}(-B^2 + A - B) - AB = 0$$

[1]Stefano Cignitti, et al., CAST Division Plenary

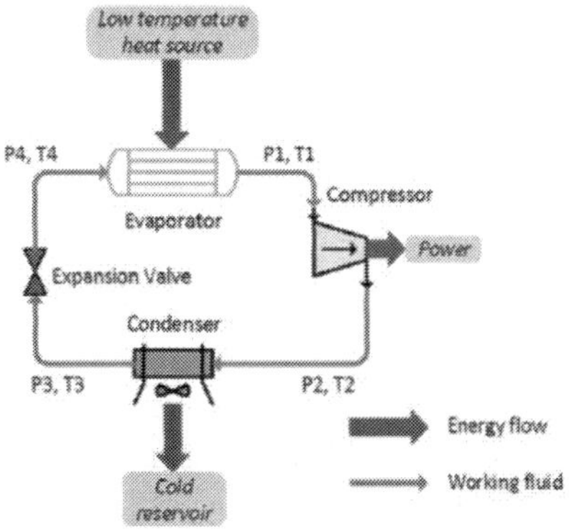

Constraints to ensure correct cubic EOS root selection (Kamath et al., 2010):

$$f(Z_{V1}) = 0 \quad f'(Z_{V1}) \geq 0 \quad f''(Z_{V1}) \geq 0$$
$$f(Z_{V2}) = 0 \quad f'(Z_{V2}) \geq 0 \quad f''(Z_{V2}) \geq 0$$
$$f(Z_{V3}) = 0 \quad f'(Z_{V3}) \geq 0 \quad f''(Z_{V3}) \geq 0$$

Refrigerant Design (3/3)

Table: Optimization results for refrigerant design problem

	r134a	r152a	3,3,3-diFluoropropyne	3,3-diFluoropropene

Mw (g/mol)	66	102	94	92
Tb (K)	230	236	242	246
Tc (K)	366	374	369	392
ω	0.27	0.33	0.28	0.14
k (W/m K)	0.097	0.082	0.086	0.093
ΔH_v at Tb (J/mol)	28562	26523	18635	21312
ODP	0	0	0	0
GWP	**1370**	**133**	**134.6**	**2.4**
Atmospheric lifetime (years)	**13.4**	**1.5**	-	-
COP	**10.1**	**12.8**	**9.5**	**11.0**

Die Attach Adhesive (DAA) Design - Product Design with Economic Analysis (1/3)

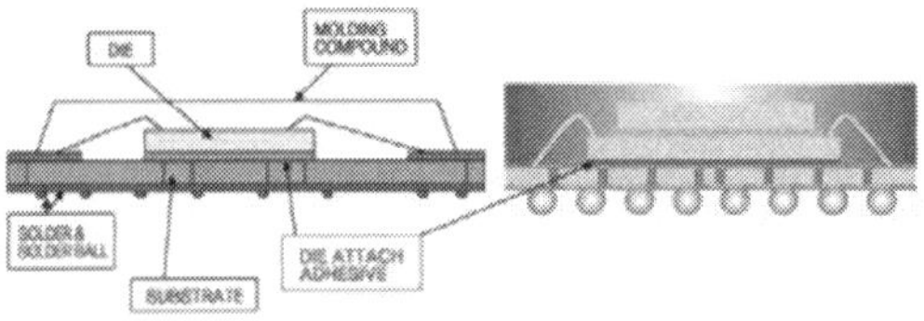

- Step 1: Die attach adhesive design used in LED lamp. It is crucial to remove the heat from the chip as fast as possible (high thermal conductivity)
- Step 2:
  - **Objective:** Net present value (NPV) should be maximized.
  - **Ingredients:** Polymer (designed), microparticle (Ag or Cu) and nanoparticle (Ag or Cu).
  - **Group selection for polymer:** -C6H4-, -O-, CH2, CH, C=O, -S-, -SO2-, from 100 groups (D.W. van Krevelen, 2009)
  - **Property constraints:** The thermal conductivity of the product should equal 30 W/m K

*Collaboration with HKUST (Kelvin Zhang and Prof. Ka M. Ng)

15AIChE
Annual Meeting, Salt Lake City, UT

DTU

Die Attach Adhesive Design (2/3)

$$\max npv = \sum_{j=-n}^{m} \frac{pcf_j}{(1+R)^j}$$

s.t.

1. Structural constraints.
2. Property constraints (Group contribution method).
3. Economic model:

$$pcf_j = \begin{cases} -TDC_j - TCI_j - \Delta NWC_j & j \leq 0 \\ (p_j d_j - c_{m,j} - Adv_j - FC_j)(1-t) + tD_j + \Delta NWC_j & j > 0 \end{cases}$$

$$c_{m,j} = (w_{mp}c_{mp} + w_{np}c_{np} + c_{m,pc})d_j \quad j \geq 1$$

$$p_j d_j = \left(\frac{\alpha_j}{\beta}\right)^{\rho} P_j^C \left(\frac{Y - p_j d_j}{P_j^C}\right)^{1-\rho} d_j^{\rho} \quad j \geq 1$$

$$Y = p_j d_j + P_j^C d_j^C \quad j \geq 1$$

15AIChE
Annual Meeting, Salt Lake City, UT

DTU

Die Attach Adhesive Design (3/3)

Optimization results of die attach adhesive design problem

	Polymer	Microparticles	Nanoparticles
Chemical	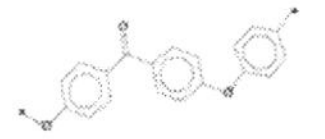	Cu	Cu
Weight fraction	0.1187	0.7974	0.0839
NPV	Net present value (MM$)		80.1024

j	Years	-1	0	1	2	3	4	5
PCF_j	Project cash flow (MM$)	-1	-2.5	27.8867	27.9587	27.8819	27.8358	28.8358
p_j	Selling price ($/Kg)	-	-	1351	1351	1351	1351	1351
d_j	Demand (t)	-	-	55.4647	55.4647	55.4647	55.4647	55.4647
c_j	Product cost (MM$)	-	-	27.654	27.654	27.654	27.654	27.654
Adv_j	Advertisement cost (MM$)	-	-	0.5	0.5	0.5	0.5	0.5
$d_{c,j}$	Competitor's demand (t)	-	-	8.3561	8.3561	8.3561	8.3561	8.3561

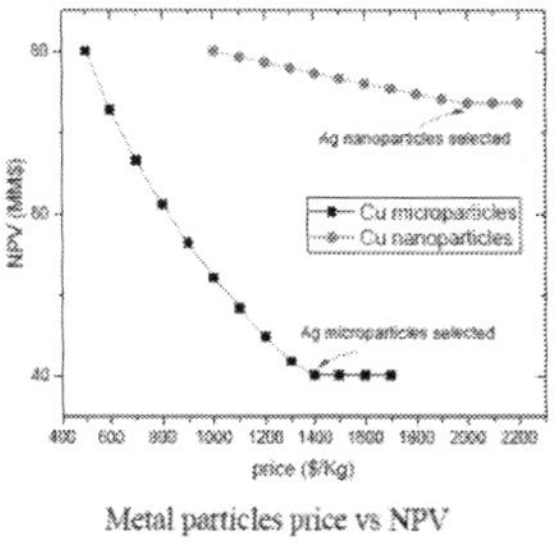

Metal particles price vs NPV

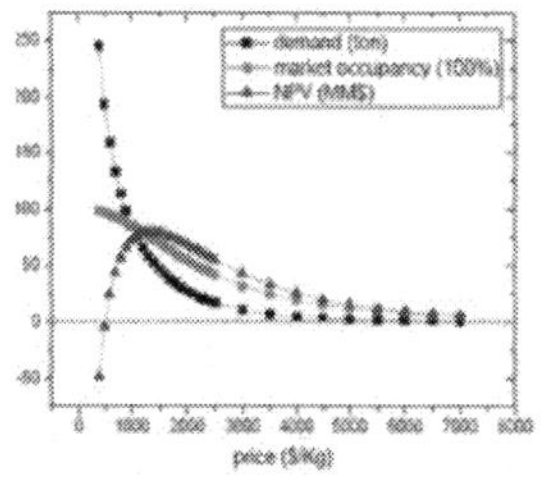

Product price vs demand, market occupancy and NPV

15AIChE
Annual Meeting, Salt Lake City, UT

DTU

Conclusion

- A computer-aided framework was developed in this work.
- This framework has been shown to systematically define and solve a large MINLP product design, product-process design, and product-process design with economic analysis problems.
- Generate and test methods can only partially solve these problems.
- Higher order description of groups and molecular structure prediction is integrated in this new MINLP mathematical programming model.
- This model is tested using a large range of product design problems (surfactants, refrigerants and polymers) and for consideration of sustainability inclusion through economic and environmental models.

17 **DTU Chemical Engineering, Technical University of Denmark** 17 December 2015

427253 Designing Optimal Mixtures Using Generalized Disjunctive Programming: Convex Hull Reformulation

Abstract:

The design of mixtures is a topic of great importance across many industrial sectors, as part of process development and of product design. Computer-Aided Mixture/blend Design (CAM^{b}D) [1] has emerged as an attractive methodology which has been successfully applied to designing various types of mixtures. Within the CAM^{b}D framework, these design problems have often been posed as Mixed-Integer Nonlinear Programming (MINLP) problems, where the optimal molecular structure is determined with respect to a set of property constraints. In most existing methodologies a reduced version of the general CAM^{b}D problem has been addressed: the number of mixture ingredients is fixed (usually a binary mixture is considered) and the identity of a compound (or of all compounds) that can participate in the mixture is chosen from a given set of candidate molecules. Solving the resulting optimization problems can be quite challenging, as these are combinatorial in nature due to the presence of binary variables and to the highly nonlinear expressions that relate composition, structure and physical properties.

In working towards a generalised CAM^{b}D problem, a novel methodology where the number, identity and compositions of mixture constituents are optimised simultaneously, was recently proposed [2,3]. In order to address the difficulties arising from the complexity of expressing the problem within a mathematical framework, a logic-based methodology, Generalized Disjunctive Programming (GDP) [4], was used to formulate the discrete choices of a mixture design problem. The design methodology was first developed for a standard formulation with a fixed number of components (restricted problem) and then generalised to allow the number of mixture ingredients to vary up to a maximum number (general problem). In this approach, the designed components in the mixture are selected from a list of candidate pure compounds. The GDP problems are formulated as MINLP problems in order to exploit the available algorithms for solving optimisation problems. The formulations are converted into MINLP problems by replacing the Boolean variables with binary ones and converting the logic propositions into linear inequalities. In this initial work [2,3], conditional constraints inside the disjunctions were formulated using the Big-M (BM) approach [4]. The BM approach is the simplest representation of a GDP problem in a mixed-integer form. However, it introduces a large parameter (Big-M) which in most cases is specified intuitively and can yield to poor relaxations. Here, we extend this work by converting the GDP models into MINLP problems using the Convex Hull (CH) reformulation [5]. The CH reformulation introduces a new set of disaggregated variables and new constraints, increasing the size of the problem. It can however be shown [6] that when the discrete domain of the CH formulation is relaxed, it gives at least as tight or tighter relaxations than BM.

The design methodology is applied to two case studies: the first one aimed to design an optimal solvent mixture to dissolve ibuprofen and the second one to design the most effective solvent mixture to extract acetic acid from water. The results of the problems studied show that the proposed formulations offer a promising approach to the problem of mixture design, as the simultaneous design of the optimal number, identity and compositions of the components that participate in a mixture, is achieved. The CH reformulation is compared to the Big-M approach and both methods appear suitable for formulating mixture problems effectively. Finally, better performance is obtained with a solvent mixture in both examples, showing the benefit of using mixtures instead of pure solvents to attain enhanced properties.

References

1. L.E.K. Achenie, R. Gani, and V. Venkatasubramanian. *Computer aided molecular design: theory and practice*. Elsevier, Amsterdam, Netherlands, 2003.
2. P.T. Akula, P.M. Kleniati, and C.S. Adjiman. On the design of optimal solvent mixtures using generalised disjunctive programming. In *22nd European Symposium on Computer Aided Process Engineering*, pages 1-5, 2012.
3. S. Jonuzaj, P.T. Akula, P.M. Kleniati, and C.S. Adjiman. The formulation of optimal mixtures using generalized disjunctive programming: choosing components from a pre-defined list. *Journal paper in preparation*, 2015.
4. R. Raman and I.E. Grossmann. Modelling and computational techniques for logic based integer programming. *Computers & Chemical Engineering*, 18(7):563-578, 1994.
5. S. Lee and I.E. Grossmann. New algorithms for nonlinear generalized disjunctive programming. *Computers & Chemical Engineering*, 24(9-10):2125-2141, 2000.
6. S. Lee and I.E. Grossmann. Generalized convex disjunctive programming: convex hull relaxation. *Computational Optimization and Applications*, 26:83-100, 2003.

432235 A COSMO-Based Approach to Computer-Aided Mixture Design

Abstract:

A relatively well-known field in the literature, Computer-Aided Molecular Design (CAMD) defines a class of problems aimed at designing a single, optimal molecular structure to best fit some purpose. While CAMD[1] has seen some success in the literature, its mixture analogue, Computer-Aided Mixture Design (CAMxD), still has some significant limitations. CAMxD is, unsurprisingly, much more complicated. This class of problems now involves the design of a blend of compounds – both every compound's molecular structure and mole fraction in the composition – and this greatly increases the number of variables in the problem and adds many more nonlinear constraints in the form phase equilibria equations, miscibility checks, and safeguards against including two molecular species which might react.

The traditional approach to CAMxD in the literature has two main shortcomings: (1) many approaches rely on enumerating every possible structure for every possible component and then checking every combination for mixture feasibility; and (2) the search space is inherently curtailed because these approaches often solve the mixture part of the problem with UNIFAC[2] method, which can only work when a regressed interaction parameter is known for every pair of molecular substructures in the blend. These issues have resulted in approaches to the CAMxD problem that can only investigate a very limited area of the molecular search space.

Our approach to the CAMxD problem circumvents these two issues by: (1) using a derivative-free optimization scheme to design for structures and test their mixture properties on-the-fly; and (2) deriving a group-contribution method to estimate COSMO-RS[3] and COSMO-SAC[4] sigma profiles, which can be used to calculate mixture properties without the need for binary interaction parameters. These two steps enable a much more efficient optimization of a much larger chemical search space. In this presentation, we will outline our approach in more detail and discuss how the COSMO group contribution methods were made. We will also demonstrate our algorithm as applied to a solvent design problem and a reaction rates optimization problem.

[1] Group-contribution based estimation of pure component properties. Jorge Marrero and Rafiqul Gani.*Fluid Phase Equilibria.* 2001. 183, 183-208.

[2] Group-contribution estimation of activity coefficients in nonideal liquid mixtures. Aage Fredenslund, Russel L. Jones, and John M. Prausnitz. *AIChE Journal.* 1975. 21(6), 1086-99.

[3] Fast solvent screening via quantum chemistry: COSMO RS approach. Frank Eckert and Andreas Klamt. " *AIChE Journal.* 2002. 48(2), 369-385.

[4] A priori phase equilibrium prediction from a segment contribution solvation model. Shiang-Tai Lin and Stanley I. Sandler. " *Industrial & engineering chemistry research.* 2002. 41(5), 899-913.

(173f) Coupling Experiments and Computational Tools to Estimate Thermal Properties for Product Design in the Fat-Based Food Industry

Teles dos Santos, M., Universidade de São Paulo - USP
Viana, I. S., Universidade de São Paulo
Ract, J. N. R., Universidade de São Paulo - USP
Le Roux, G. A. C., Polytechnic School of the University of São Paulo
Originally presented on: 11/9/2015 14:15:00 - 14:36:00

Product design has been evaluated as one of the challenges that chemical engineering has to face to help industry meet market demands for value-added products with higher performance [1]. In the food and pharmaceutical industry, the design of fat-based products is of great importance, since many foods have their desired properties related to a complex phase behavior involving lipids and many cosmetic and pharmacological products use vegetable oils as a lipidic medium in which active principles and fine chemicals are dispersed. Attempting to use Process Systems Engineering tools (namely thermodynamic modeling and computational/numerical approach), for product design, the present work coupled experimental studies and computational simulations to access the thermal properties of different scales of triacylglycerol mixtures, ranging from simple binary mixtures to ternary blends of vegetables oils.

In order to test the developed tools in raw materials used by industry, different blends formed by palm stearin (PS), fully hydrogenated soybean oil (FHSO) and canola oil were studied. The solid fat content at different temperatures, melting points and changes in heat capacity due to solid-liquid transitions are covered by the present work, both experimentally and computationally. Computational simulations and experimental results are compared and discussed. The following experiments were conducted: fatty acids composition identification, stereo distribution of fatty acids in the glycerol structure, softening point and thermal analysis (differential scanning calorimetry). The computational simulations were performed by the following steps: prediction of triacylglycerol composition of the blends by using statistical methods (random and non-random distribution of fatty acids in the glycerol structure); predictions of the distribution of molecules among solid and liquid phases by direct minimization of Gibbs free energy function, using a Generalized Reduced Gradient Method (CONOPT 3 GAMS solver). The optimization algorithm was then coupled with the main program written in FORTRAN 90, which handles the calculation of interaction parameters, melting temperature and melting enthalpy and the generation of triacylglycerols from fatty acids data [2]. The Solid-liquid equilibrium problem was then solved over a temperature range to be compared with experimental data in different temperatures and at different blend ratios (single vegetable oil, binary blends and ternary blends).

As the demand for more sophisticated fat products increases and the simple physical mixture (blending) is not sufficient for matching the desired properties, new technologies have emerged. The chemical interesterification reaction is one important technique that changes the triacylglycerol profile of the blends and, thus, modifies their physical properties. This reaction uses catalysts (e.g. sodium methoxide) and heat to promote a random distribution of fatty acids among the three positions of glycerol [3]. In order to predict how the reaction affects the final solid fat content of the blends, this work simulates the effect of the reaction by using random distribution of fatty acid in the glycerol chain and computing the melting profile of the reaction product.

REFERENCES

[1] Hill, M. (2009). "Chemical product engineering — the third paradigm." *Comput. Chem. Eng.*33, 947–953.

[2] Teles dos Santos, M.; Gerbaud, V.; le Roux, G.A.C. (2013). "Modeling and simulation of melting curves and chemical interesterification of binary blends of vegetable oils." Chem. Eng. Sci. 87, 14–22.

[3] Rodrigues-Ract, J.N.; Cotting, L.N.; Poltronieri, T.P.; da Silva, R.C.; Gioielli, L.A. (2010). "Crystallization behavior of structured lipids by chemical interesterification of milkfat and sunflower oil." Food Sci. Techn. 30, 258-267.

433479 Multi Objective Molecular Design of Reactants and Products

Abstract:

In reactive systems, computer-aided molecular design (CAMD) is performed to design solvents that enhance reaction kinetics, improve catalyst activity/lifetime as well as reactants and products. CAMD of reactants and products helps in their selection such that target properties of interest are optimized with satisfaction of property constraints. Owing to the scarcity of CAMD methodologies that design reactants and products, there has been an increased interest to address this issue. The few recently available methodologies to design reactants and products however have their shortcomings. The earliest available methodologies are restricted to cases where only structures of a single reactant and product are variable. Also, employment of linear property models and lack of a methodology to treat property constraints have been some of the other restrictions. Dev et al. [1, 2] recently developed two algorithms to design reactants and products which are unrestricted by the number of reactants and products whose structures are uknown. They also made provisions for non-linear property models and property constraints. In their first algorithm [1], only the dominant properties of the products are optimized. Also, only the products are subjected to their individual set of property constraints. In their second algorithm [2], properties that are dependent on the structures of both reactants and products are optimized. Also, all reactants and products are subjected to their respective set of property constraints. In this work, we have developed an algorithm that generates structures of reactants and products irrespective of their numbers such that each reactant and product's target property(s) is to be optimized. Also, each reactant and product is subjected to a set of property constraints. All properties have been expressed in terms of molecular structure using different types of property models. Since there exists a need to treat different types of property models on a single platform, signature descriptors have been utilized. A variety of property models can be expressed in terms of signature descriptors which are molecular building blocks. They enable the simultaneous utilization of Quantitative Structure Activity/Property Relationships (QSARs/QSPRs) and Group Contribution Models (GCMs) [3]. In the formulated problem, all property models are expressed in terms of the occurrence number of signature descriptors. Relationships have been developed that relate the occurrence number of signatures of reactants and products. This is achieved by tracking the atoms in the reactants to the products by taking into consideration the reaction mechanism. Thus the reactants and products have been structurally related. Due to this structural relation, property constraints acting on reactants and products influence the selection of both optimal reactants and products. The objective functions of each of the reactants and products display a similar influence. Since the selection of each reactant and product is dependent on the property constraints and property objective functions of the other reactants and products in addition to its own, the problem is formulated as a multi-objective optimization problem when objectives are conflicting. This problem is solved using the Augmented Epsilon-Constraint Method (AUGMECON) developed by Mavrotas [4]. AUGMECON generates pareto optimal solutions while completely avoiding weakly pareto optimal solutions. The generated solutions also include unsupported pareto optimal solutions. Specifically for multi-objective integer programming (MOIP) problems, an improved version of AUGMECON, AUGMECON2 [5] can also be utilized. AUGMECON2 is very efficient in generating the entire pareto optimal set for MOIP problems compared to AUGMECON and some earlier available methods. In this contribution, the utility and novelty of the developed methodology has been highlighted through a case study.

[1] V.A. Dev, N.G. Chemmangattuvalappil, M.R. Eden, 2014, Computer Aided Chemical Engineering, 34, 291-296

[2] V.A. Dev, N.G. Chemmangattuvalappil, M.R. Eden, 2015, Computer Aided Chemical Engineering, 37, 1445-1450

[3] N.G. Chemmangattuvalappil, C.C. Solvason, S. Bommareddy, M.R. Eden, 2010, Computers & Chemical Engineering, 34, 12, 2062-2071

[4] G. Mavrotas, 2009, Applied Mathematics and Computation, 213, 455-465

[5] G. Mavrotas, K. Florios, 2013, Applied Mathematics and Computation, 219, 9652-9669

414742 Efficient Calibration for in-Situ Measuring L-Glutamic Acid Solution Concentration Via ATR-FTIR

Abstract:

In this presentation, an efficient calibration method is proposed for in-situ measuring the concentration of L-glutamic acid (LGA) solution via ATR-FTIR. We clarify that using the unsaturated zone spectra as commonly practiced for establishing a concentration calibration model may lead to evidently biased measurement of the solution concentration. By comparison on using the spectra of unsaturated and metastable zones for calibrating the concentration of LGA aqueous solution, it is demonstrated that the use of metastable zone spectra to establish a calibration model can guarantee prediction accuracy for the solution concentration via real-time ATR-FTIR monitoring. Experiments under different concentrations and temperatures are performed to verify the efficiency of the proposed calibration method. In addition, it is shown that the proposed calibration model can also be used for nucleation detection, therefore facilitating real-time monitoring of batch cooling crystallization in practice.

415213 Optimization of a Reactive Crystallization Using a Pharmaceutical Engineering Approach

Abstract:

Reactive crystallizations are complex and difficult to control due to the number of competing parameters influencing the chemical transformation and particle growth. Here we will discuss a salt-to-salt transformation in the presence of multiple reactive chemical species. The crystallization is further burdened by the potential to create a poorly soluble free form that poses a risk to both product quality and yield. The presentation will highlight the use of FBRM in combination with other analytical techniques to identify and elucidate the side reactions responsible for observed variability in product potency. Detailed understanding of the upstream composition and the seed point proved vital to delivering a robust process. To offer reasonably broad parameter ranges to manufacturing and further mitigate risk, the refined process incorporates a crude crystallization focused upon yield and a re-crystallization that assures high quality. A fundamental understanding of the interactions of the chemical species and the root causes of variability were needed to deliver a suitable process for commercial manufacturing.

417528 Monitoring Co-Crystal Purity with on-Line Raman Spectrascopy

Abstract:

Pharmaceutical co-crystals are defined as stoichiometric molecular complexes that contain an active pharmaceutical ingredient (API) and a coformer in crystal lattice via noncovalent interactions, predominantly hydrogen bonds. Compared with APIs themselves, pharmaceutical co-crystals have advantages in solid state properties, such as solubility, melting point, chemical interaction, stability and bioavailability. Solution crystallization is an important method to produce co-crystals. However, crystals of constituent components, either API or co-former, may crystallize out concomitantly due to operational fluctuations in starting composition or temperature, compromising the purity of co-crystals. Therefore, it is desirable to detect crystals of single components when they appear during co-crystallization, whereby corrective operation can be taken to ensure co-crystal purity. In this study, a methodology is developed to monitor co-crystal purity using Raman spectroscopy with caffeine-glutaric acid-acetonitrile as the model system. Caffeine and glutaric acid can form co-crystal from solution in acetonitrile at the appropriate conditions. Raman spectra were collected in three classes of suspensions with a solid mixture of caffeine crystals and co-crystals, pure co-crystals, and a mixture of glutaric acid crystals and co-crystals, respectively at different temperatures. Principal component analysis (PCA) was performed for these spectra and the resulting first few principal components were put to discriminant analysis. It was found that Raman spectroscopy is able to distinguish the three classes of suspensions clearly.

Key Words, co-crystal, PAT, Raman spectroscopy, crystallization, principal component analysis

425352 Automated Multi-Loop Direct Nucleation Control Using Focused Beam Reflectance Measurement in Multistage Continuous Mixed Suspension Mixed Product Removal Crystallizer

Abstract:

Batch crystallization is an economical and widely used separation and purification unit operation especially in pharmaceutical industry. It has been extensively studied in literature and very well understood due to the development of various process analytical technology (PAT) tools like focused beam reflectance measurement (FRBM), particle video microscope (PVM), UV/Vis spectroscopy, IR spectroscopy and Raman spectroscopy. In addition, some of these tools have been successfully applied in feedback control frameworks in order to control the batch crystallization process in real time. Direct nucleation control (DNC) is a methodology that uses the chord count information from in situ FBRM to actuate adaptive heating or cooling rate to have controlled dissolution in the batch cooling crystallization process. It has been proved being able to produce much larger and more uniform crystal as compared with simple temperature control (T-control) approach in batch process.[1-2] Supersaturation control (SSC) and active polymorphic feedback control (APFC) apply UV/Vis or/and Raman spectroscopy in similar feedback control loops to maintain controlled supersaturation level or polymorphic transformation during batch cooling process.[3-4] These model-free feedback control approaches are proved very robust to automatically reject disturbances or to adjust the process when the required target product quality is changed. They can provide automated fast recipe design and only require simple mathematical tools without any modeling effort or much system knowledge.[5]

Continuous Crystallization is of great interest in recent decades due to consistent product quality, high process and equipment efficiency and reduced operation cost compared with batch crystallization. Mixed suspension mixed product removal (MSMPR) crystallizer is one of the most important continuous crystallization systems due to its simple operation and capability for long residence time. There are some published papers that investigate the recipe design for continuous MSMPR crystallization systems via modelling or experiment.[6-9] However, according to the authors' best knowledge, currently there are no published works that study the application of previously mentioned feedback control approaches (i.e. DNC, SSC, APFC) in any type of continuous crystallization systems.

In this work, a multi-loop DNC approach is experimentally investigated in a multistage MSMPR cooling crystallization process in order to automatically control the nucleation rate at each MSMPR stage. The feed solution is free of crystals. Each MSMPR stage is equipped with a heating circulator, a temperature sensor and a FBRM probe. A Labview based software Crystallization Monitoring and Control (CryMOCO) is installed on the supervisory computer to collect the chord count information from each individual FBRM probe. The information is then compared with the desired set-point of the corresponding MSMPR stage in CryMOCO, which remotely controls each temperature circulator and actuates a predefined adaptive heating or cooling rate in the corresponding MSMPR stage. In each DNC loop, the total chord count is targeted as the controlled variable and the set-point since it indicates information about nucleation, generation of fine particles as well as the average particle size. For example, in a certain stage the adaptive heating rate is actuated automatically once the measured FBRM total count is larger than the set-point of the corresponding stage, and vice versa. Paracetamol and ethanol are used as the model compound and solvent, respectively. The proposed multi-loop DNC approach is found being able to reach desired steady state and produce crystals of consistent desired size distribution after a few heating/cooling cycles with or without disturbances. The proposed approach has the following advantages: (1) It can automatically help avoid nucleation in the second stage and after when fine particles are not desired in the final product. This is achieved by designing a proper set-point for each MSMPR stage so that limited amount of nucleation or breakage is allowed. (2) It can automatically achieve optimal operating recipes to produce desired crystals of various sizes (i.e. either small or big). (3) It can automatically reject disturbances such as accidental seeding or nucleation. It is faster than simply washing out without any control. (4) It can reduce startup duration time and startup waste as compared with simple T-control that uses constant temperatures. Simulation work is also carried out in the group to validate and explain some of the experimental observations, and therefore to optimize the control performance.

References

[1] Saleemi, A.; Rielly, C.; Nagy, Z. K. Automated direct nucleation control for in situ dynamic fines removal in batch cooling crystallization. *CrystEngComm* **2012**, 14(6), 2196-2203.

[2] Abu Bakar, M. R.; Nagy, Z. K.; Saleemi, A. N.; Rielly, C. D. The impact of direct nucleation control on crystal size distribution in pharmaceutical crystallization processes. *Cryst. Growth Des.* **2009**, 9(3), 1378-1384.

[3] Saleemi, A. N.; Rielly, C. D.; Nagy, Z. K. Comparative investigation of supersaturation and automated direct nucleation control of crystal size distribution using ATR-UV/vis spectroscopy and FBRM. *Cryst. Grwoth Des.* **2012**, 12(4), 1792-1807.

[4] Simone, E.; Saleemi, A. N.; Tonnon, N.; Nagy, Z. K. Active polymorphic feedback control of crystallization processes using a combined Raman and ATR-UV/Vis spectroscopy approach. *Cryst. Growth Des.* **2014**, 14(4), 1839-1850.

[5] Nagy, Z. K.; Fujiwara, M.; Braatz, R. D. Modelling and control of combined cooling and antisolvent crystallization processes. *J Process Control* **2008**, 18, 856-864.

[6] Zhang, H.; Quon, J.; Alvarez, A. J.; Evans, J.; Myerson, A. S.; Trout, B. Development of continuous anti-solvent/cooling crystallization process using cascaded mixed suspension, mixed product removal crystallizers. *Org. Process Res. Dev.* **2012**, 16, 915-924.

[7] Vetter, T.; Burcham, C. L.; Doherty, M. F. Regions of attainable particle size in continuous and batch crystallization processes. *Chem. Eng. Sci.* **2014**, 106, 167-180.

[8] Yang, Y.; Nagy, Z. K. Advanced control approaches for combined cooling/antisolvent crystallization in continuous mixed suspension mixed product removal cascade crystallizers. *Chem. Eng. Sci.* **2015**, 127, 362-373.

[9] Yang, Y.; Nagy, Z. K. Combined cooling and antisolvent crystallization in continuous mixed suspension mixed product removal cascade: steady state and startup optimization. *Ind. Eng. Chem. Res.* **2015**, DOI: 10.1021/ie5034254.

423519 Digital Phase Diagram of K2SO4-KOH-H2O and Crystal Growth of K2SO4 in the Strong Alkali Solution

Abstract:

Potassium sulfate has always been an important and superior potash fertilizer. In this paper, we investigate the crystallization process of potassium sulfate from the strong alkali solution that is involved in the clean production process of raw alunite mineral. The research focuses on the digital phase diagram for a ternary system of K_2SO_4-KOH-H_2O and the crystal growth of K_2SO_4 in the strong alkali solution.

Firstly, the phase equilibrium data for the ternary system of K_2SO_4-KOH-H_2O are measured from 40°C to 80°C, where the contents of potassium and sulfur in the aqueous solutions are detected by ICP"COES and the composition of solid phases are tested by X-ray diffraction, respectively. Based on the measured phase equilibrium data, origin is used to present the 3D colormap and ternary contour drawings, as shown in Figure 1. With the interpolation and fitting results obtained from the software, the saturation temperature of a given K_2SO_4-KOH-H_2O content can be predicted from the smooth surface of the ternary phase diagram. It is found that the K_2SO_4 solubility decreases with the increase of the KOH concentration in the aqueous solution, and the K_2SO_4 solubility becomes smaller in the strong alkali solution which is beneficial to separate K_2SO_4 from the KOH solution.

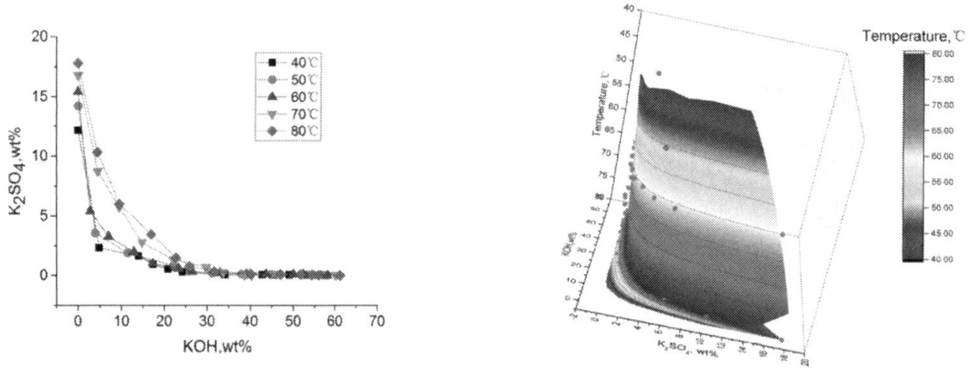

Figure 1. Solubility of K_2SO_4 in KOH solution and digital phase diagram of K_2SO_4-KOH-H_2O at different temperatures

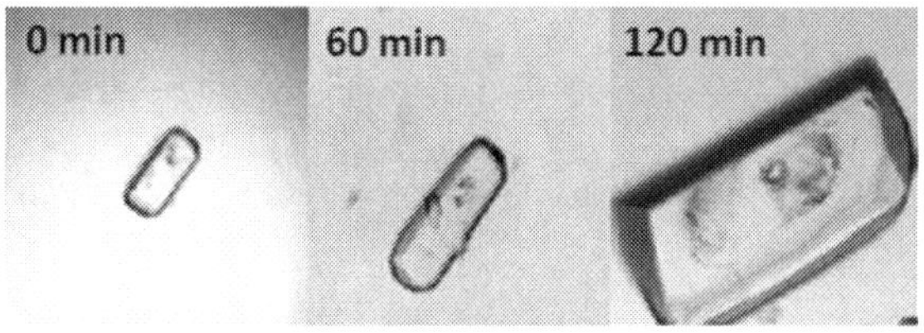

Figure 2. Crystal growth of K_2SO_4 in the strong KOH solution measured using a reflection microscopic technique at 313K

The supersolubility and metastable zone width of K_2SO_4 in the concentrated KOH solution are investigated by focused beam reflectance measurement (FBRM) and ultrasonic sensor online monitoring for different operating conditions, which will provide the important information for the design and optimization of K_2SO_4 crystallization process. For the system of K_2SO_4-KOH-H_2O, the FBRM and the ultrasonic sensor can provide a high sensitivity for nucleation detection, which detects changes related to the amount of grown nuclei and changes in solution concentration, respectively. Furthermore, the crystal growth rates of K_2SO_4 in the strong KOH solution are measured using a reflection microscopic technique, where the saturated solution is poured into a thermo-controlled microscopic cell with a volume of 5 mL and a diameter of 3.5cm. Experiments are conducted by decreasing the solution to different temperatures to create different supersaturation levels, and then the crystal growth rates of K_2SO_4 are calculated by measuring the changes in the dimensions of the crystals with the analysis Image Processing Program (Olympus). Figure 2 shows the crystal growth process of potassium sulfate in the strong KOH solution measured under the microscope at 313K. The crystal shape of K_2SO_4 is a typical rectangular shape, which is convenient for easy identification and measurements of specific crystallographic faces. The influence of impurities from the alkaline solution on the crystal shape will be investigated using the reflection microscopic technique. Moreover, the two-dimensional population balance equation is proposed for the calculation of the crystal growth rate of K_2SO_4 with an acceptable accuracy. From this study of crystal growth, a given content of solution can be designed with a desired growth mechanism in order to effectively separate K_2SO_4 from the strong KOH solution by crystallization.

435128 Controlling Crystal Size, Shape and Purity during Crystallization By Using Additives

Abstract:

Crystallization which is a purification and separation unit operation plays an important role in the industrial production of pharmaceuticals. The properties of the produced crystals such as purity or size and shape distribution have strong impact on the downstream processes as well as on the end usage. It is known, that chemicals are generally contaminated by different amounts of impurities, which can come from different sources, such as from upstream processes or chemical side reactions [1]. In most of the industrial processes, the effect of impurities may result e.g. in needle shape crystals which impeding manufacturability of the solid dosage form. While crystallization is often used as a purification process, trace amount of impurities generally exist in the product. While in some cases impurities of various amounts are acceptable for certain highly toxic impurities the acceptable limits are very low and difficult to achieve using a single crystallization step. For example genotoxic impurities carry significant health risk and strict regulations are applied on the concentration of these materials [6,8,9]. Impurities, even if it is in trace amounts, can influence the crystal properties like crystal shape. In this case, impurities are called as crystal growth modifiers (CGMs) [2-6]. In this work, a method is presented, which reduces the possible undesired effect of the CGMs during crystallization by using a control additive (CA). The control method is based on a chemical reaction between the control additive and the growth modifier, which results in a non-adsorptive reaction product. Thus, the concentration of CGM in the solution and thus in the solid phase is the function of the additive concentration in the solution. It is also known that the concentration of the CGMs in the solution affects the rate of the change of crystal growth.

This work deals with the investigation and control of batch cooling crystallization in the presence of crystal growth modifiers. The objectives of the control method are to maintain the purity of the crystal product and also to control the crystal size and shape distribution. A mathematical model including nucleation and growth of two characteristic crystal sizes as well as impurity adsorption and chemical reaction model is developed in order to investigate and control the purity and the crystal shape. The growth modifiers influence directly the growth kinetics. This inhibitory effect on the crystal surfaces leads to changed crystal size and shape distribution [2-6]. On the other hand, growth modifiers are built in the crystal structure according to surface adsorption mechanism and so CGMs also change the purity of the crystal product [7]. The proposed purity control method is based on chemical reaction between the undesired impurity and the control additives which provides an alternative transformation route for the compound against the surface adsorption. Hence, the chemical reaction inhibits the adsorption rate by lowering the concentration in the solution.

Utilizing the benefits of the competitive adsorption and reaction mechanisms, model based control for crystal shape and purity is developed. The simulator is suitable to calculate the optimal concentration of control additive in order to achieve desired crystal shape and/or reducing the impurity concentration below the desired limits. The results demonstrate the efficiency of the proposed approach for shape and purity control.

Acknowledgement: Financial support provided by the European Research Council grant no.[280106-CrySys] is acknowledged.

References:

[1] N., Variankaval, A.S., Cote, M.F., Docherty, 2008. From form to function: Crystallization of active pharmaceutical ingredients. AIChE Journal, 54(7), 1682-1688.

[2] A., Majumder, Z.K., Nagy, 2013. Prediction and control of crystal shape distribution in the presence of crystal growth modifiers. Chemical Engineering Science, 101, 598-602.

[3] R.J., Davey, 1976. The effect of impurity adsorption ion the kinetics of crystal growth from solution. Journal of Crystal Growth, 34, 109-119.

[4] N., Kubota, J.W., Mullin, 1995. A kinetic model for crystal growth from aqueous solution in the presence of impurity. Journal of Crystal Growth, 152(3), 203-208.

[5] N., Kubota, S., Sasaki, N., Doki, N., Minamikawa, M., Yokota, 2004. Absorption of an Al(III) impurity onto the (100) face of a growing KDP crystal in supersaturated solution. Crystal Growth and Design, 4(3), 533-537.

[6] A., Borsos, A., Majumder, Z. K., Nagy, 2014. Model development and experimental validation for crystal shape control by using tailored mixtures of growth modifiers. Computer Aided Chemical Engineering, 33, 781-786.

[7] K., Maeda, R., Tabuchi, Y., Asakuma, K., Fukui, 2006. Distribution of metallic ions in single KDP crystal grown from aqueous solution. Cryst. Res. Technol., 41(10), 955-960.

[8] A., Garcia-Arieta, 214. Interactions between active pharmaceutical ingredients and excipients affecting bioavailability: impact on bioequivalence. European Journal of Pharmaceutical Sciences, 65, 89-97.

[9] S., Gorog, 2006. The importance and the challenges of impurity profiling in modern pharmaceutical analysis. Trends in Analytical Chemistry, 25 (8), 755-757.

433667 Reverse Design of Ionic Liquids for CO2 Absorption

Abstract:

Chemical engineering design problems have evolved in recent years from large scale bulk processes to targeted small batch approaches. As a result, product design is reaching beyond the traditional solutions to classic chemical engineering problems such as extraction, purification, removal of sulfur from fuel, CO_2 capture and sequestration, energy storage and biomass pre-treatment. Ionic liquids can provide environmentally benign solutions that can be tailored to specific process requirements. Due to the unique structure of ionic liquids, the cation, anion and length of the alkyl chain can be varied to create a molecule with specific physical properties. It has been estimated that over 10^{14} unique molecules exist at room temperature, so experimental trial and error methods will not be effective. More advanced methods must be created determine the ideal combination for a specific set of process requirements.

The purpose of this work is to determine the ideal ionic liquid to capture CO_2, a question which has been the topic of many experimental laboratory studies. However, only a small percentage of the potential ionic liquids have been synthesized and tested for the solubility of CO_2. This innovative approach decouples the solution from time-consuming laboratory study and streamlines the search for the most promising candidates that should be synthesized.

A characterization based group contribution method is combined with density functional theory to determine the ionic liquid that can most effectively absorb CO_2. Infrared spectra data contains descriptor data that can be used to estimate properties of ionic liquids, but does not exist for all ionic liquids. Density functional theory is used to create IR data based on a training set of experimental data. Principal component analysis and partial least square techniques will be employed to reveal important features and patterns in the molecular architecture. A characterization based group contribution method is used to estimate the properties of the new structures. The reverse design of potential ionic liquid molecules will be completed by an exhaustive search of combinations with various cation, anions and lengths of alkyl chains until a candidate molecule is found that provides the highest solubility of CO_2.

428564 Economic and Environmental Potentials for Natural Gas to Enhance Biomass to Liquids Technologies

Abstract:

With the increased availability of low cost natural gas, co-conversion of natural gas with biomass (GBTL) is one of the promising routes being investigated by the U.S. Department of Energy (DOE) [1], in order to increase liquid fuel yields while lowering greenhouse gas emission relative to petroleum. For example, Praxair has developed a natural gas enhanced biomass to liquids process that increases liquid fuel yield by 2.5 times versus the base case of straight biomass to liquids [2]. It is necessary to fully understand how the use of natural gas and biomass may be optimized and integrated into a hydrocarbon liquid fuel conversion process.

The conceptual process design and cost comparison of liquid biofuel using both biomass derived biogas intermediate and natural gas have been studied in this work. This GBTL techno-economic analysis (TEA) provides baseline cost results comparable produce hydrocarbon liquid fuels as well as can be viewed as a near term deployment opportunity for renewable jet fuel production. This paper offers insights on the current understanding of techno-economic performance analysis associated with the GBTL conversion process. The GBTL process consists of the key process steps of gasification, gas conditioning, reforming, Fischer-Tropsch, separation processes and energy integration. The associated Life Cycle Assessment is included to provide the insights of on the sustainability metrics for the liquid fuels production from GBTL technology. The quantification and assessment of the sustainable metrics associated with the GBTL conversion process will be discussed including life-cycle greenhouse gas (GHG) emissions and fossil energy consumption. Sensitivity analysis and uncertainty analysis will be analyzed and discussed to understand the uncertainties of process variations, as well as impact from the variations of biomass to natural gas ratios, design assumptions and natural gas prices on the economics. Technologies based on the GBTL concept offer opportunities in optimizing the usage of natural gas along with biomass. Therefore, different blending ratios of natural gas/biomass could have a large effect on the economic and environmental performance, which would be investigated though scenario analysis. In addition, different operating parameters could also have a significant impact on equipment design, raw materials, utilities consumption, and overall process economic performance for GBTL system. Monte Carlo analysis is performed to assess the probability distribution of the minimum fuel selling price (MFSP) in order to investigate potential interactions among parameters studied. The goal of the sensitivity analysis is to provide an understanding of key cost drivers with significant environmental impacts.

Reference:

[1] Bioenergy Technolgies Office, "The Potential for Natural Gas to Enhance Biomass Technologies," 2014. Available at:

http://energy.gov/sites/prod/files/2014/04/f14/february2014_gbtl_webinar.pdf

[2] S. Chakravarti, D. P. Bonaquist, R. F. Drnevich, and M. M. Shah, "Natural gas enhanced biomass to liquids: Project development and modeling," *Computers & Chemical Engineering,* vol. 47, pp. 67-75, 12/20/ 2012.

435800 Amyloid Fibrils As Reinforcement Filler of Polymeric Nanocomposite Materials

Abstract:

Fibers have fascinated mankind since the very beginning. Most of textile products are made of fibers, which for centuries have been obtained from natural sources. Many biological materials use nano fibers as building block. One of the best examples is spider webs. The incredible properties of spider silk fibers, with their exceptional toughness, have prompted scientist to investigate their structure in the hope of creating light weight but highly resistant materials. Given the difficulty in working with silk proteins, we used instead amyloid fibrils, which are less known but have similar structure to silk and are much easier to prepare, to create an entire new class of reinforced rubber materials.

In this project, a novel type of composite polymeric materials reinforced by protein fibrils has been prepared. Protein fibrils are self-assembled structures resulting from partial denaturation of certain proteins, and the resulting aggregation mediated by beta-sheets interactions. These structures are characterized by an extremely large aspect ratio, with a diameter that changes depending on the age of the fibrils, and on the degree of association of the different strands. For the fibrils used in our experiments, which are prepared by partial denaturation of beta-lacto globulin at pH 2.0 and 80°C for 24 hours, and at a concentration of 10mg/ml, typical diameters are about 25nm with a length of several micrometers. For the first time high concentrations of fibrils have been incorporated into a polymer. The process used to incorporate fibrils is unique. We started from a suspension of polymer particles in water, which are mixed with the fibrils in the aqueous phase. It has been observed that not only the elastic modulus of composite materials with fibrils is increased, but also the ductility of the materials is enormously improved.

409240 Fluorine-Free Superhydrophobic Film Based on Polybenzoxazine

Abstract:

Bioinspired superhydrophobic surfaces which possess the merit of a water contact angle (CA) larger than 150° have numerous applications such as in self-cleaning, oil/water separation and liquid transportation. As is well known, chemical modification of a rough surface is rewarding to superhydrophobicity. Conventionally, the modification agent is a fluorine-containing low surface free energy material such as perfluorocarbon or fluoroalkylsilane. And roughening of a surface involves sophisticated processes such as etching and templating. Therefore, it has been desired to prepare superhydrophobic films by a cost-efficient and an easier process. As a new type of low surface free energy material, polybenzoxazine (PBZ) is promising for a superhydrophobic film due to its cross-linking structure, low water absorption and low cost. In our previous work, a fluorine-containing and fluorine-free silane-functionalized PBZ film with surface free energy of 15.5 mJ/m^2 and 14.91 mJ/m^2 has been synthesized respectively. (J. Liu, Z. Xin et al, Langmuir, 2013; L. Qu and Z. Xin, Langmuir, 2011)

In this paper, two important properties are combined: the low surface free energy property of polybenzoxazine (PBZ) and the photo-induced superhydrophilicity property of TiO_2. As a result, a multifunctional superhydrophobic PBZ/TiO_2 surface with a water contact angle of ~160 ° and sliding angle less than 2 ° is prepared by using spin coating method without any fluorine-containing surface modification agents. The as-prepared superhydrophobic film has good adhesion to substrate, and its superhydrophobicity is stable even after heat treatment at 300 °C and environmentally durable for more than half a year. When the content of TiO_2 is increased, the consequent nanocomposite film exhibits hydrophobicity-superhydrophilicity transitions with a variation of around 120° in water CA upon ultraviolet (UV) exposure-heat treatment cycles.

(235f) Synthesis and Properties of p-(Lauryl)-Benzyl Polyoxyethylenate Ethers

Sun, M.,
Fang, Y., Jiangnan University

Originally presented on: 11/9/2015 17:00:00 - 17:21:00

The benzene ring in alkylphenol ethoxylates (APEO) provides smooth polar transition between hydrophobic alkyl chain and hydrophilic ethylene oxide (EO) chain, imparting emulsification, solubilization, wetting, penetration, dispersion, washing and other outstanding properties to these surfactants. APEO was originally widely used in the field of domestic, civil and industrial cleaning, and in the fine chemicals and polymer industry. It had a large annual consumption worldwide. In recent years, the poor biodegradability of APEO itself and the toxicity of its degradation products have caused widespread concern. European Union countries have banned the use of APEO, and China has also been restricting its application in many areas.

With the growth of environmental security awareness, an all-around ban of APEO will become an inevitable trend. Therefore, the development of a new surfactant molecule as the substitute of APEO is desired. A surfactant molecule that not only retains the benzene ring structure with excellent performance, but also avoids the phenol structure with poor biodegradability and reproductive toxicity, has great theoretical significance and broad market prospects.

In this paper, six kinds of *para*-lauryl-benzyl polyoxyethylenate ether (LBEO*n*) with different number, *n*, of ethylene oxide (EO) units (n = 4, 5, 7, 9, 11, or 13) were synthesized starting from dodecylbenzene by a three-step process including chloromethylation, hydroxylation and ethoxylation. Chemical structure of the LBEO*n* was characterized by FTIR and ^{1}H NMR spectroscopy. Average EO mole adduct numbers in the LBEO*n* were characterized by ESI-MS. Critical micelle concentration (cmc) of the product was measured by surface tension method at 25 °C and other performance of applications of the products were also determined.

The experimental results showed that with an increasing number of EO in the surfactant, cloud points and wetting ability increased; cmc, emulsifying ability, and solubilization ability decreased when the EO mole adduct numbers equal to or greater than 9. Compared to the most commonly used alkylphenol ethoxylate, nonylphenol ethoxylate (trade name TX-10), the LBEO*n* with 9 EO has similar cloud point, cmc, and emulsifying and solubilizing power as TX-10. This LBEO may be an effective alternative to TX-10, with no concerns about environmental and reproductive toxicity.

(235g) Application Performance of Extended Surfactants Useful for Personal and Domestic Cleanser

Liu, H., Jiangnan University
Chen, J., Jiangnan University
Fang, Y., Jiangnan University

Originally presented on: 11/9/2015 17:21:00 - 17:42:00

Abstract: Extended surfactants are anionic surfactants containing groups with intermediate polarity (e.g., polypropylene oxide) between the hydrophilic anionic head group and the hydrophobic alkyl chain in the surfactant molecule. Past research has reported detailed information on good solubilization of polar oils and ultra-low interfacial tension caused by extended surfactants, but less attention has been paid to their potential application in household chemicals depending on their unique molecular structure. In this work, new extended surfactants were synthesized from several isomeric alcohols by propoxylation, followed by sulfation and neutralization. Their physicochemical and application properties such as γ_{cmc} , C_{cmc} , detergency, foaming power, wetting power, salt-tolerance, hard water-tolerance□Krafft point, and lime soap dispersing power were measured. The new extended surfactants have molecular structures similar to a commonly-used surfactant, sodium alcohol polyethyleneoxide ether sulfate (AES), but do not contain the carcinogenic 1,4-dioxane (an ethoxylation byproduct). They have excellent solubility at low temperature and good salt-tolerance and calcium-tolerance, and some of them possess good detergency and foaming power, comparable to AES. Thus, these extended surfactants show the potential to be used in personal and domestic cleaners to replace surfactants with environmental and toxicity concerns.

410668 In-Vitro Dewetting Dynamics on Silicone Hydrogel Contact Lenses

Abstract:

The tear film on our eyes provides a smooth optical interface as well as protects the eyes against environmental challenges. Of particular interest in this presentation is the meibomian lipid layer, secreted by glands lining our eyelids, that has been shown to possess remarkable interfacial viscoelastic properties. By using a newly developed instrument called the Interfacial Dewetting and Drainage Optical Platform(i-DDrOP) developed in the Fuller laboratory at Stanford University, we investigate the *in-vitro* dewetting dynamics of thin aqueous films laden with meibomian lipid layer on silicone hydrogel contact lenses at physiological conditions. Our *in-vitro* measurements show dewetting times that are comparable to the *in-vivo* tear break up times. This highlights the utility of the i-DDrOP as a tool to study biomimetic tear film solutions and their interactions with contact lens surfaces.

438452 Prediction of Vapor/Liquid Characteristics By Quantum Mechanics Data

Abstract:

A large number of chemical processes involve the purification of mixtures in industry, such as liquid extraction, distillation, and recrystallization. Infinite dilution coefficients can be used to characterize the vapor-liquid, solid liquid, and liquid-liquid phase behavior that underlies these unit operations. The largest deviation from ideality occurs in the dilute solution system. The ability to predict infinite dilution coefficients based on simple guidelines of hydrogen bonding could provide valuable design formulations and useful thermodynamic insights. The MOSCED model is sufficiently simple to fill this need. Unfortunately, a lack of characterized parameters for many compounds limits the capabilities of the MOSCED model. The goal of this paper is to develop a simple method to estimate the missing parameters based on readily available quantum mechanical results.

Sigma profiles have been calculated using quantum mechanics and tabulated at a website hosted by Virginia Tech. The charge density of the molecule over the molecular surface is described by the values where the projected surface charge is opposite of that of the molecule. Hydrogen bonding can occur when the charge density exceeds a threshold value. These observations can be used to calculate acidity and basicity parameters for the MOSCED model. Given that the bonding starts at a specific point it was assumed the rest of the contributions could be interpreted as another parameter, polarizability factor.

The acidity and basicity parameters were characterized by fitting the data from the area beneath the sigma profile at a certain distance and correlating the values to pre-determined values from Lazzaroni et al. The fit for the polarizability factor was determined by using a weighted function to give greater emphasis on values further from the center of the sigma profile.

Correlations of acidity and basicity were characterized based on experimental data for roughly 4500 binary solutions. A comparison of the results for the fit vs. the MOSCED with Lazzaroni parameter values and UNIFAC model for the solutions was used to determine the accuracy of the model. Several specific solutions were poorly characterized in comparison to the MOSCED model, but the correlation provides qualitative estimates of activity coefficients with in roughly 0.280 log units of the experimental data, compared to 0.106 log units for the original MOSCED method and 0.183 log units for UNIFAC.

Effect of Electrical Field on the Effective Parameters of Non-Newtonian Fluid

11/9/15
STEPHEN DUECK
STEFFANO OYANADER
DR. MARIO OYANADER

Outline

- Physics of the Problem & Applications
- Description of System
- Analytical Model
- Computational Model
- Graphical Illustrations
- Conclusions
- Future Work
- Acknowledgements

cbu | Gordon and Jill Bourns COLLEGE OF ENGINEERING
AIChE The Global Home of Chemical Engineers

Physics of the Problem & Applications

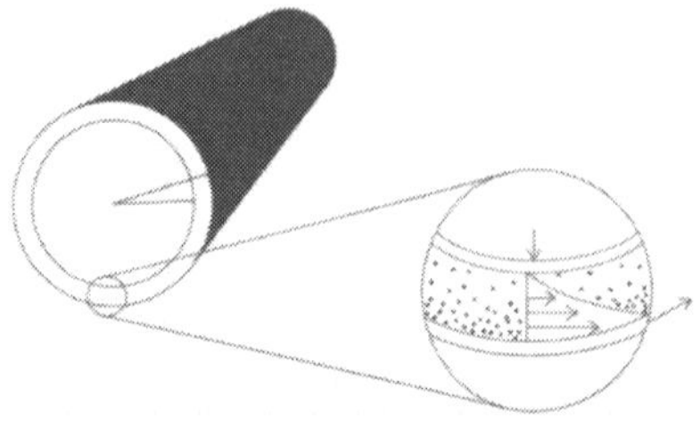

Applications:

- Environmental and toxicological investigations of nanomaterials
- Determination of size and shape of environmental and biological particles
- Separations
  - high and ultra high molar mass polymers
  - Macromolecules
  - Particulates
  - Nanoparticles
  - Gels
- Replacement of centrifugation
- Blood Cleansing (dialysis)
- Separation of pharmaceuticals and metabolites for treatment and diagnosis from biosamples

cbu | Gordon and Jill Bourns COLLEGE OF ENGINEERING — AIChE The Global Home of Chemical Engineers

Rectangular Simplification Does not Capture Non-Newtonian Velocity Field "n"

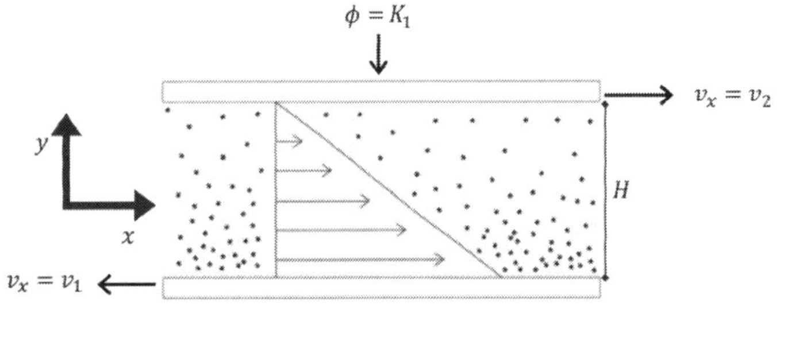

Governing Equation:

$$0 = \frac{\partial}{\partial y} \mathrm{T}_{y_x}$$

Boundary Conditions:

$$v_x = v_1 \ @ \ y = 0$$
$$v_x = v_2 \ @ \ y = H$$

Shear Stress Equation:

$$C_1 = \mathrm{T}_{y_x} = \mu \left| \frac{\partial v_x}{\partial y} \right|^n$$

Fluid Velocity Equation:

$$v_x(y) = \left(\frac{C_1}{\mu_0}\right)^{\frac{1}{n}} y + C_2 \rightarrow v_x(y) = \frac{v_2 - v_1}{H} y + v_1$$

cbu | Gordon and Jill Bourns COLLEGE OF ENGINEERING — AIChE The Global Home of Chemical Engineers

Cylindrical Captures "n" Parameter

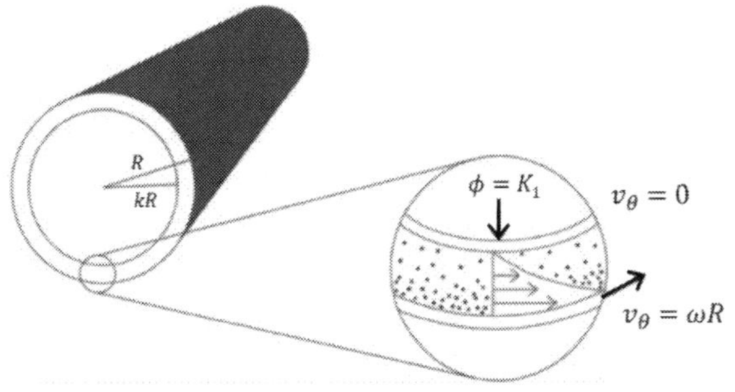

Governing Equation:

$$\frac{-1}{r^2}\frac{\partial}{\partial y}(r^2 T_{r\theta}) = 0$$

Boundary Conditions:

$$v_\theta = 0 \;@\; r = kR \;(inner)$$
$$v_\theta = \omega R \;@\; r = R \;(outer)$$

Shear Stress Equation:

$$T_{r\theta} = -m\left(r\frac{\partial}{\partial y}\left(\frac{v_\theta}{r}\right)\right)^n$$

Fluid Velocity Equation:

$$v_\theta = -\frac{n}{2}r^{1-\frac{2}{n}}C_1^{\frac{1}{n}} + rC_2 \rightarrow \frac{v_\theta}{\omega r} = \frac{1-\left(\frac{kR}{r}\right)^{\frac{2}{n}}}{\left(1-k^{\frac{2}{n}}\right)} = v_\theta^*$$

Cylindrical Captures "n" Parameter

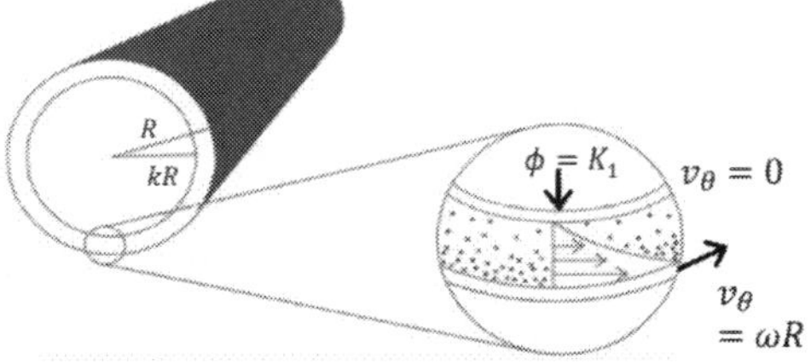

Governing Equation:

$$\frac{-1}{r^2}\frac{\partial}{\partial y}(r^2 T_{r\theta}) = 0$$

Boundary Conditions:

$$v_\theta = 0 \;@\; r = kR \;(inner)$$
$$v_\theta = \omega R \;@\; r = R \;(outer)$$

Shear Stress Equation:

$$T_{r\theta} = -m\left(r\frac{\partial}{\partial y}\left(\frac{v_\theta}{r}\right)\right)^n$$

Fluid Velocity Equation:

$$\frac{v_\theta}{\omega r} = \frac{1-\left(\frac{kR}{r}\right)^{\frac{2}{n}}}{\left(1-k^{\frac{2}{n}}\right)} = v_\theta^*$$

$$\langle v_\theta^* \rangle = \frac{\int_{kR}^{R} v_\theta^*(r)\,r\,dr}{\int_{kR}^{R} r\,dr}$$

$$\langle v_\theta^* \rangle = \omega\left\{\frac{1-k^2-nk^{\frac{2}{n}}+2nk^2-n}{\left(1-k^{\frac{2}{n}}\right)(n-1)(k^2-1)}\right\}$$

Electrical Field Governing Equation

$$\frac{1}{r}\frac{\partial}{\partial r}\left(r\frac{\partial \phi}{\partial r}\right) = 0$$

Boundary Conditions:

$\phi = 0 \quad @\, r = kR$

$\phi = K_1 \quad @\, r = R$

$$\phi = K_1\left(1 - \frac{\ln\left(\frac{r}{R}\right)}{\ln(K)}\right)$$

$$\langle \phi \rangle = \frac{\int_{kR}^{R}\phi(r) r\,dr}{\int_{kR}^{R} r\,dr} = \mathrm{K}_1\left\{\frac{1}{1-k^2} + \frac{1}{lnk^2}\right\}$$

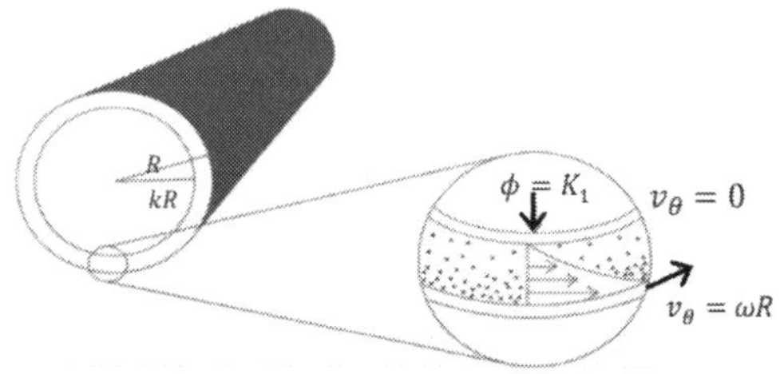

cbu | Gordon and Jill Bourns COLLEGE OF ENGINEERING

AIChE The Global Home of Chemical Engineers

Molar Species Continuity Governing Equation

$$\frac{\partial C}{\partial t} + \frac{v_\theta}{r}\frac{\partial}{\partial \theta} = D\left[\frac{1}{r}\frac{\partial}{\partial r}\left(r\frac{\partial C}{\partial r}\right) + \frac{1}{r^2}\frac{\partial^2 C}{\partial \theta^2}\right] + \mu^* \frac{1}{r}\frac{\partial}{\partial r}\left(rC\frac{\partial \phi}{\partial r}\right)$$

Boundary Conditions:

$$D\frac{\partial C}{\partial r} + \mu^*(C * \frac{\partial \phi}{\partial r}) = 0 \;@\; r = kR$$

$$D\frac{\partial C}{\partial r} + \mu^*(C * \frac{\partial \phi}{\partial r}) = 0 \;@\; r = R$$

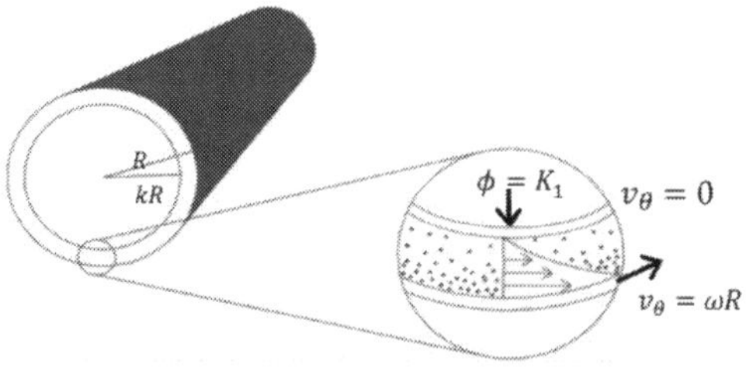

cbu | Gordon and Jill Bourns COLLEGE OF ENGINEERING

AIChE The Global Home of Chemical Engineers

Modeling Approach

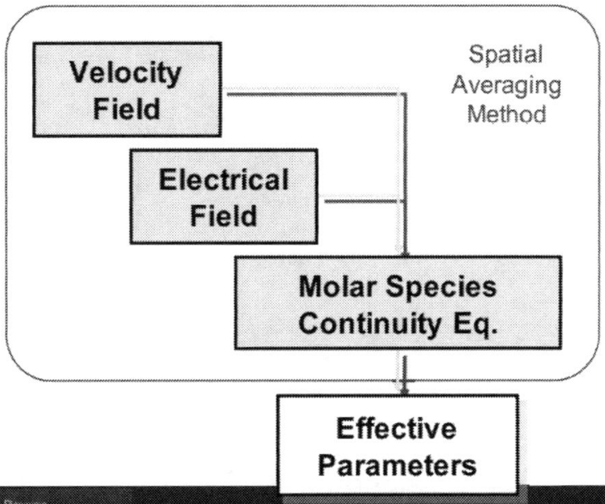

cbu | Gordon and Jill Bourns COLLEGE OF ENGINEERING   AIChE The Global Home of Chemical Engineers

$$\frac{\partial C}{\partial t} + \frac{v_\theta}{r}\frac{\partial}{\partial \theta} = D\left[\frac{1}{r}\frac{\partial}{\partial r}\left(r\frac{\partial C}{\partial r}\right) + \frac{1}{r^2}\frac{\partial^2 C}{\partial \theta^2}\right] + \mu^* \frac{1}{r}\frac{\partial}{\partial r}\left(rC\frac{\partial \phi}{\partial r}\right)$$

Effective Parameters Using the Area Averaging Method

$$\hat{v}_{eff} = \frac{v_{eff}}{\langle v_\theta^* \rangle} = 1 + \langle \tilde{v}_\theta * f(r) \rangle$$

$$\Theta = \frac{\langle \tilde{v}_\theta^* * g(r) \rangle}{\langle v_\theta^* \rangle R^2} = (1 - \hat{D}_{eff} / \mathrm{D})/Pe^2$$

f(r) and g(r) from Cwirko & Carbonell (1989)

cbu | Gordon and Jill Bourns COLLEGE OF ENGINEERING   AIChE The Global Home of Chemical Engineers

Effective Parameters Using the Area Averaging Method

Effective Velocity

$$\hat{v}_{eff} = 1 + A_0 + A_1 * f$$

$$\Omega \equiv \frac{\mu * K_1}{D}$$

Effective Dispersion Coefficient

$$\Theta = \frac{A_0}{2(2-\Omega)}\left\{\frac{(k^4 - k^4)}{(k^2 - k^4)}\frac{(A-2)}{(A-4)}A_0 - \frac{1}{2}(k^2+1)A_1 f\right\} - \left\{\frac{A_0}{2\Omega} + \frac{A_1}{(2-\Omega)\Omega}\right\}\{A_0 + A_1 f\}$$
$$- \frac{A_1}{(2-A)(2-A-\Omega)}\left\{\frac{(k^{2-A}+1)A_0}{2} - \frac{2(k^{4-A}-1)}{(k^2-1)(A-4)}A_1 f\right\}$$

cbu | Gordon and Jill Bourns COLLEGE OF ENGINEERING — AIChE The Global Home of Chemical Engineers

Effective Parameters Using the Area Averaging Method

$$A_0 = \frac{2k^2 - 2k^{\frac{2}{n}}}{\frac{2}{n} - 2 - \frac{2}{n}k^2 + 2k^{\frac{2}{n}}}$$

$$A_1 = \frac{\left(\frac{2}{n} - 2\right)(k^2 - 1)k^{\frac{2}{n}}}{\frac{2}{n} - 2 - \frac{2}{n}k^2 + 2k^{\frac{2}{n}}}$$

$$f = \frac{\left(k^{2+\frac{\Omega}{\ln(k)}} - k^{\frac{2}{n}}\right)\left(2 + \frac{\Omega}{\ln(k)}\right)}{k^{\frac{2}{n}}\left(k^{2+\frac{\Omega}{\ln(k)}} - 1\right)\left(2 + \frac{\Omega}{\ln(k)} - \frac{2}{n}\right)}$$

$$\Omega \equiv \frac{\mu * K_1}{D}$$

cbu | Gordon and Jill Bourns COLLEGE OF ENGINEERING — AIChE The Global Home of Chemical Engineers

V_{eff} Analysis

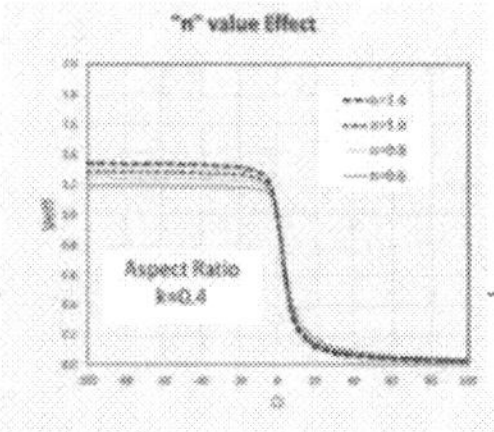

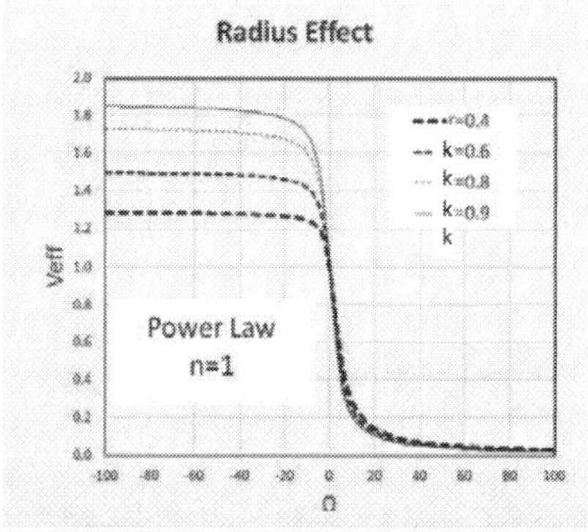

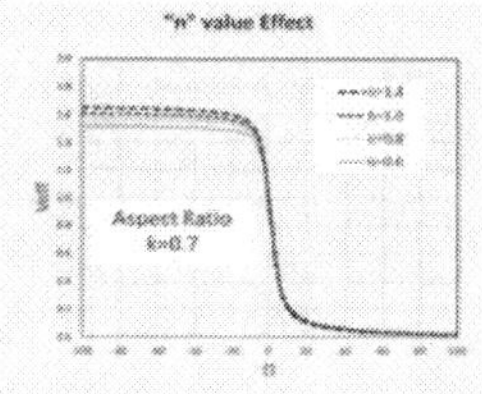

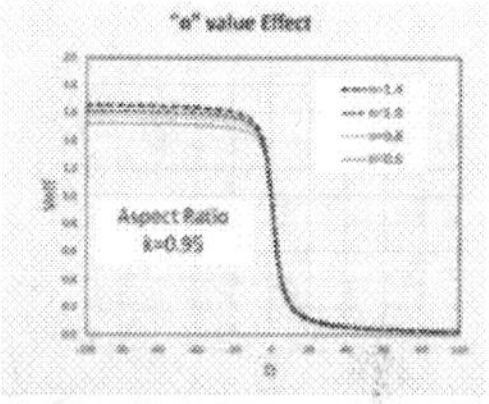

cbu | Gordon and Jill Bourns COLLEGE OF ENGINEERING

AIChE The Global Home of Chemical Engineers

D_{eff} Analysis

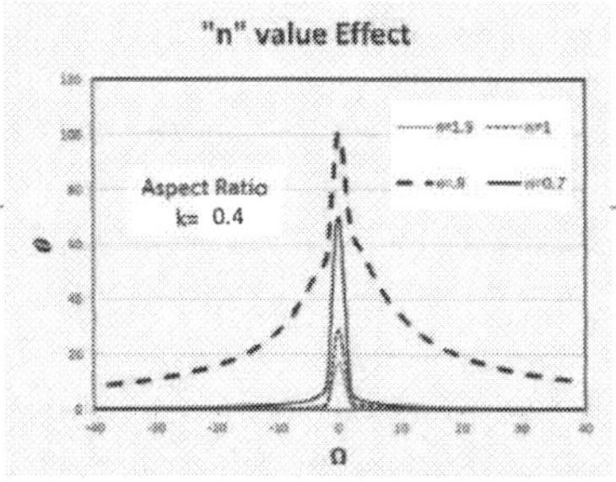

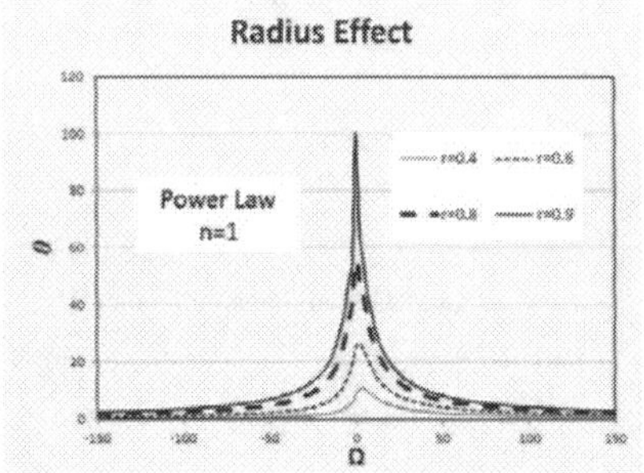

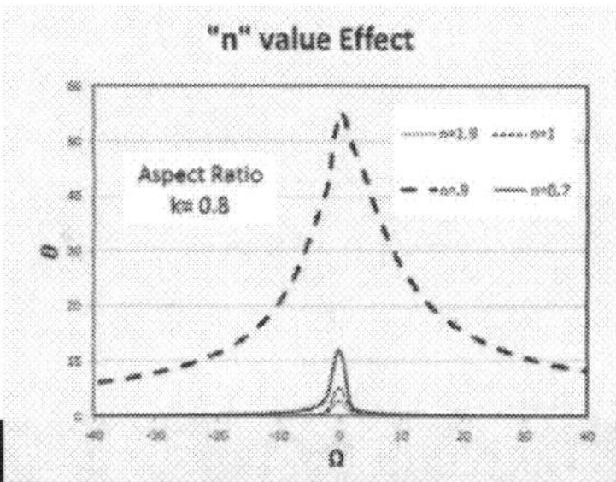

cbu | Gordon and Jill Bourns COLLEGE OF ENGINEERING

AIChE The Global Home of Chemical Engineers

General Conclusions

- Effective Dispersion Coefficient values decrease as the Aspect Ratio parameter (k) increases.
- Effective Velocity values increase as the Aspect Ratio parameter (k) increases.
- Values of the Power Law model parameter "n" produces lower Effective velocity values as it decreases.
- Inversion of the electrical Field Direction could bring Effective Velocity from maximum values to a minimum of zero.

cbu | Gordon and Jill Bourns COLLEGE OF ENGINEERING — AIChE The Global Home of Chemical Engineers

Future Work

- Computation of Optimal Time of Separation
- New Non-Newtonian Fluid Model
- Collaborative Paper Writing

Optimal Separation Time

$$\tau_{op} = \left[\frac{\sqrt{\dfrac{D_{effA}}{D_A}} + \sqrt{\dfrac{D_{effB}}{D_B}}}{\hat{v}_{effA} - \hat{v}_{effB}} \right]^2$$

cbu | Gordon and Jill Bourns COLLEGE OF ENGINEERING — AIChE The Global Home of Chemical Engineers

Acknowledgements

- California Baptist University, College of Engineering
- Department of Chemical Engineering

cbu | Gordon and Jill Bourns COLLEGE OF ENGINEERING

AIChE The Global Home of Chemical Engineers

438449 Degradation of a Thermoplastic Polyether Ester Elastomer in Downhole Conditions

Abstract:

Thermoplastic elastomers (TPEs) combine the strength of thermoplastics with the flexibility of elastomers, a characteristic potentially useful in downhole applications. We studied the degradation process of a particular TPE of interest from both a mechanical and chemical perspective, and have determined the Fickian diffusion constant, the chemical degradation rate, the change in crystallinity, and the storage modulus with respect to varying temperature and time. Techniques used to determine such rates include: dynamic mechanical Aanalysis (DMA), differential scanning calorimetry (DSC), and end group analysis. Applying Arrhenius' empirical relationship to the determined rates gives rise to a temperature-dependent model that predicts the degradation behaviour of the TPE outside of the experimental temperature range. Our results indicate that the chemical degradation via hydrolysis follows pseudo first order kinetics, and the degradation results in an increase of crystallinity and thus storage modulus and a decrease of tensile strength and strain. The polymer eventually deteriorates due to brittleness.

438453 Alternative Sampling Densities in Non-Uniform Sampling

Abstract:

Non-uniform sampling (NUS) is more widely practiced, but many implementations either do not weight the distribution of samples, or follow exponential sampling densities. Weighting the sampling density in proportion to the signal envelope provides intrinsic sensitivity enhancements (MRC, V49, 483-491, 2011). However sinusoidal weighting is a compelling alternative to exponential weighting matched to the signal decay (JBNMR, V58: 303-314, 2014). We now describe families of alternative densities as candidates to replace common exponential NUS weightings. Yielding equal intrinsic sensitivity to their exponential counterparts, these alternatives possess improved constraints on line shapes by distributing more samples beyond 1xT2. An algorithm to generate properly gapped sampling densities (JBNMR, V58: 303-314, 2014) is extended, and is also contrasted with deterministic NUS schedules.

438447 Dynamic Light Scattering Study of Sulfonated Polystyrene Ionomers in Benzene

Abstract:

From golf ball coatings (Surlyn®) to fuel cell membranes (Nafion®), ionic polymers (ionomers) are widely used. Ionomers are molecular compounds that contain long hydrocarbon chains with ions randomly distributed along the chain. These polymers form ionic aggregates when ionic groups attract each other in nonpolar environment, and these interactions dictate the properties exhibited. We studied the aggregation of sulfonated polystyrene (SPS) ionomers of varying molecular weights and ion contents in benzene. Dynamic light scattering (DLS) experiments determine the effective diameter and the polydispersity of the aggregates and molecules in the solution. To do this, we created solutions of the SPS and its acidic equivalent in benzene at different concentrations and analyzed the scattering. We also studied the effects of an ionic liquid, BMIM-PF_6 (1-butyl-3-methylimidazolium hexafluorophosphate) on SPS ionomers of higher ion contents. It was found that the ionic liquid weakens the ionic bonds and partly destroys ionic aggregates. We will study the properties of ionomers with high ion contents with other ionic liquids and with toluene as a solvent, for comparisons.

438443 A DFT Study of Methanol Reforming on C- and N-Doped TiO2 (110) Rutile Surfaces

Abstract:

: Biomass reforming is a sustainable method for hydrogen gas production. Hydrogen gas is an increasingly desirable clean burning fuel source that is commercially produced by methane reforming, which is a high energy process. Methanol reforming, a model for biomass reforming, over TiO_2 offers a lower energy alternative with a greater hydrogen gas yield; however, the process is not well understood, and thus it is not economically viable with current technologies. Nonmetal dopants on TiO_2 have been shown to reduce band gaps of the TiO_2 for visible light photocatalysis, but the mechanism on these altered catalysts are not well understood. We apply periodic DFT+U to gain a better mechanistic understanding of methanol reforming over C- and N-doped TiO_2 in order to inform future catalyst design. We compare binding energies and reaction barriers for both lattice and interstitial doping on rutile TiO_2 (110).

438313 Thermochemical Analysis of Intermolecular Vs. Intramolecular Reactions in Iridium Complexes

Abstract:

Transition metals are commonly used in catalytic reactions in academic and industrial settings. Transition metals have a variety of oxidation states and can bind ligands in a variety of geometries, including octahedral, tetrahedral, or square-planar coordination to name a few. The Cp^*IrPPh_3 complex (**1**) was studied for its ability to activate C-H bonds under mild conditions. As observed in experimental work, complex **1** in benzene has the ability to undergo both intermolecular and intramolecular reactions in a nearly 1:1 ratio. Literature suggests that the intramolecular reaction should be favored. Computational methods were used to investigate the details of the intermolecular and intramolecular reaction mechanisms. With the detailed knowledge of the reaction mechanism, we can determine if the 1:1 ratio is a kinetic or thermodynamic effect.

437858 Quantitative and Contextual Characterization of Plasmid Copy Number

Abstract:

Intracellular plasmid copy numbers directly affect the dosage of genetic constructs expressed from extrachromosomal replicons. This is particularly relevant for complex, multigene systems that require precise gene expression levels to be functional (e.g., genetic circuits, metabolic pathways). Furthermore, mathematical models of plasmid-based gene expression require more than a qualitative or semi-quantitative description of plasmid copy number to be accurate or meaningful. We have developed a straightforward method for determining plasmid copy number using a simple mass balance-based model of gene expression with two measurable inputs, steady-state fluorescence and growth rate. Using our approach, we have investigated various genetic, cellular and environmental context dependencies of plasmid copy number in commonly used laboratory strains of *Escherichia coli*.

(237j) A Novel Approach to Data Processing in High Throughput Material Science Experimentation

Albing, M.,

Originally presented on: 11/9/2015 17:30:00 - 17:45:00

Advancements in materials modeling techniques lead to rapid development of new materials. To test the properties of these modeled materials, a more rapid approach to material science research is required. High throughput experiments (HTE) in material science research accomplishes this challenge by providing a methodology for streamlining and optimizing the efficiency of experiments. In current HTE experiments large amounts of data are collected from a range of material compositions. While in the past HTE was limited by data collection, advances in computing have helped overcome these issues. To address the burgeoning big data in materials science problem we have been working to develop novel data processing algorithms. A typical experiment requires that a sample be scanned with a synchrotron light source, such as the one at the Stanford Linear Accelerator Center (SLAC) to gather information about the underlying crystal structure and composition. The resulting large volume of data would take a long time to process manually. An algorithm can process the data in seconds. A typical algorithm begins with a background subtraction. This is done quickly with a cubic spline fit. After the background is subtracted, peaks are fit with a pseudo-Voigt profile. Using the full width half maximum, peak location and peak height for each sample, the features across a sample can be analyzed quickly[J1] [s2] . This type of signal processing is widely used and greatly increases the data interpretation step of an experiment. We have been developing assisted machine learning algorithms to more [s3] quickly analyze crystal phase with respect to sample composition and other properties. The algorithm first separates data into many groups, called clusters. This initial clustering procedure does not require human input and forms clusters based on common traits in the XRD spectra. The next part of the algorithm requires that a human identify features on characteristic data picked by the algorithm from each cluster. The algorithm uses human input to create standard criteria for crystal phase identification in the rest of the data. This greatly reduces the time from data collection to sample characterization, and ultimately new material discovery. By developing new data processing techniques, experiments are optimized for efficiency and precision.Advancements in materials modeling techniques lead to rapid development of new materials. To test the properties of these modeled materials, a more rapid approach to material science research is required. High throughput experiments (HTE) in material science research accomplishes this challenge by providing a methodology for streamlining and optimizing the efficiency of experiments. In current HTE experiments large amounts of data are collected from a range of material compositions. While in the past HTE was limited by data collection, advances in computing have helped overcome these issues. To address the burgeoning big data in materials science problem we have been working to develop novel data processing algorithms. A typical experiment requires that a sample be scanned with a synchrotron light source, such as the one at the Stanford Linear Accelerator Center (SLAC) to gather information about the underlying crystal structure and composition. The resulting large volume of data would take a long time to process manually. An algorithm can process the data in seconds. A typical algorithm begins with a background subtraction. This is done quickly with a cubic spline fit. After the background is subtracted, peaks are fit with a pseudo-Voigt profile. Using the full width half maximum, peak location and peak height for each sample, the features across a sample can be analyzed quickly[J1] [s2] . This type of signal processing is widely used and greatly increases the data interpretation step of an experiment. We have been developing assisted machine learning algorithms to more [s3] quickly analyze crystal phase with respect to sample composition and other properties. The algorithm first separates data into many groups, called clusters. This initial clustering procedure does not require human input and forms clusters based on common traits in the XRD spectra. The next part of the algorithm requires that a human identify features on characteristic data picked by the algorithm from each cluster. The algorithm uses human input to create standard criteria for crystal phase identification in the rest of the data. This greatly reduces the time from data collection to sample characterization, and ultimately new material discovery. By developing new data processing techniques, experiments are optimized for efficiency and precision.

403712 A Multi Criteria Design Approach Regarding the Economic Impact of Heat Recovery on Chemical Processes

Abstract:

Energy efficiency is a mayor global concern of latest research as well as industrial application. Limited fossil fuels lead to constantly rising energy costs. Tightening environmental regulations necessitate the reduction of CO_2-emissions and waste water generation. All these issues underline the importance of the efficient use of energy and resources likewise. In this context process integration and optimization are powerful approaches that allow process industries to increase their effectivity and profitability.

In order to advance the state-of-the-art technology and accelerate a move to low carbon manufacturing industries several research projects were funded by the European Commission via the seventh framework program. One of those projects, namely EFENIS, aims for a fundamentally improved process integration by applying novel methods of total site targeting, heat recovery, intensified heat transfer, CO_2 and waste water management. The EFENIS project is composed of 17 partners including 10 academic institutions and 7 industrial companies.

Within EFENIS an efficient and sustainable workflow for assuring a long term validity of total site analysis results will be developed and implemented into a decision supporting software tool. This Energy Integration Manager has to be based on sophisticated interfaces between state-of-the-art process simulation, process optimization and data management. The operability of this software tool will be tested with datasets and process models of actual operating industrial plants.

Utilizing the above mentioned software solutions for process simulation enable the parallelization of process integration and process optimization as a simultaneous design step. This concept is in contrast to the traditional workflow of sequentially doing the heat integration after finishing the process design. In terms of heat integration, the design of a comprehensive heat exchanger network will be the focus. In terms of process optimization, the optimal parameterization of relevant process variables is the core objective.

In the presentation a concept of a genetic algorithm for process parameter optimization using a multi-criteria solution approach is shown. The impact of process parameters on heat integration measures will be presented via suitable pareto illustration. As a case study a process for the synthesis of dimethyl ether from methanol is used.

414817 Simultaneous Optimization of the Design and Operation of Batch Reactive Distillation Processes

Abstract:

Batch reactive distillation (BREAD) is an integrated process which combines the advantages of reactive distillation and the flexibility of batch processes. In most studies of BREAD processes, a fixed column design is assumed and the operation of the process is optimized by determining parameters such as the arrangement of product cut and off-cut collection, reflux ratio profile and the switching time for collections. However, if new process equipment is going to be designed and purchased, the column design must be considered simultaneously in the optimization to determine the most profitable process.

Therefore simultaneous optimization of equipment design and process operation is studied for BREAD processes in this work. The problem of determining the design and operation with the minimal total annual cost is illustrated for two processes with realistic kinetics and vapor-liquid equilibrium: hydrolysis of methyl lactate and esterification of formic acid. The optimal design is obtained by exhaustive gridding of the column design space coupled with an inner loop that determines the optimal values of the length of the product period and the reflux ratio during the offcut period (Al-tuwaim and Luyben, 1991). Although the optimization is somewhat time consuming, the systematic analysis results give more valuable information than a final set of optimal design and operating variables by gradient-based or stochastic methods (Mujtaba and Macchietto, 1996; Low and Sorensen, 2004).

To determine the minimum total annual cost, the annual product yield is specified and the optimal operation is determined by maximizing the batch capacity. The effect of process design variables including total number of stages, vapor boilup rate, catalyst loading, and process scale on TAC are investigated. The optimization results suggest that the process scale should be specified first. The column should be designed with adequate number of stages so that the CAP improvement by further increasing the number of stages is insignificant. Vapor boilup rate should be specified to make the best use of the available operating time, and the optimal vapor boilup rate is primarily affected by the catalyst loading. This insight into the effect of process design variables on design performance is used to develop an efficient algorithm for determining simultaneously the optimal column design and operating policy.

References

Al-tuwaim, M. S.; Luyben, W. L., Multicomponent batch distillation .3. Shortcut design of batch distillation-columns. Industrial & Engineering Chemistry Research 1991, 30, (3), 507-516.

Mujtaba, I. M.; Macchietto, S., Simultaneous optimization of design and operation of multicomponent batch distillation column - Single and multiple separation duties. Journal of Process Control 1996, 6, (1), 27-36.

Low, K. H.; Sorensen, E., Simultaneous optimal design and operation of multipurpose batch distillation columns. Chemical Engineering and Processing 2004, 43, (3), 273-289.

Dynamic Heat Exchanger for Flow Transients

Edris Ebrahimzadeh, Paul Wilding, David Frankman, Farhad Fazlollahi, and Larry Baxter
Chemical Engineering Department, Brigham Young University, Provo, UT, USA
Sustainable Energy Solutions LLC, Orem, UT, USA

Background

- Two pending patented configurations of heat exchangers that decrease or even eliminate heat exchanger losses in performance and efficiency associated with transient flow rates of the incoming streams.
- Heat exchanger constraints are some of the most restrictive transient response constraints in a process.
- These constraints include temperature changes, expansion, or thermal stresses in the heat exchangers or neighboring process equipment.
- Despite any changes in inlet conditions, the proposed configurations are capable of leveling the varying parameters such that the exit temperatures remain fixed.

Heat Exchanger Efficiency

$$\eta = 1 - \frac{\sum_{outlet} m_i \hat{b}_i - \sum_{inlet} m_i \hat{b}_i}{\sum_{eq} m_i \hat{b}_i - \sum_{inlet} m_i \hat{b}_i} = 1 - \frac{\Delta_{obs} m_i \hat{s}_i}{\Delta_{eq} m_i \hat{s}_i} = \frac{\sum_{eq} m_i \hat{s}_i - \sum_{outlet} m_i \hat{s}_i}{\sum_{eq} m_i \hat{s}_i - \sum_{inlet} m_i \hat{s}_i}$$

- $\hat{b}_i = \hat{h}_i - T_0 \hat{s}_i$
- $\Delta_{eq} m_i \hat{s}_i$ is the difference between the entropy in all the streams at equilibrium temperature and entropy of the streams at the inlet temperature.

Heat Transfer Rate Ratio

$$\zeta(x) = \frac{\sum_{hot} A_i v_i \rho_i c_{p,i}}{\sum_{cold} A_i v_i \rho_i c_{p,i}}$$

- Heat exchanger is balanced if $\zeta(x) = -1$.
- Heat exchanger is imbalanced if $\zeta(x) \neq -1$.
- The highest efficiency in heat exchangers are achieved when the flows are balanced.

Experimental

Retrofit Configuration

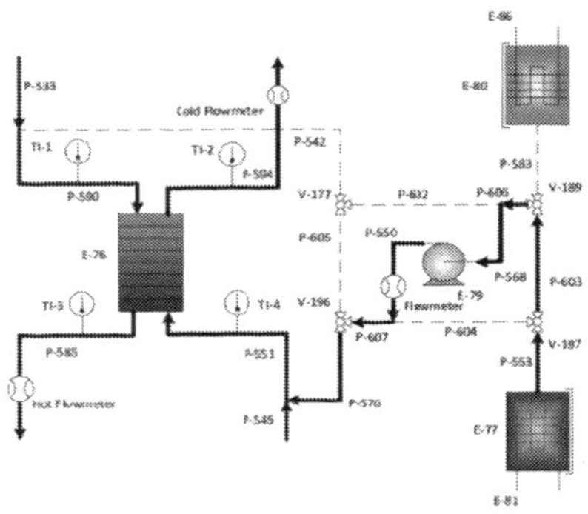

Non-retrofit Configuration

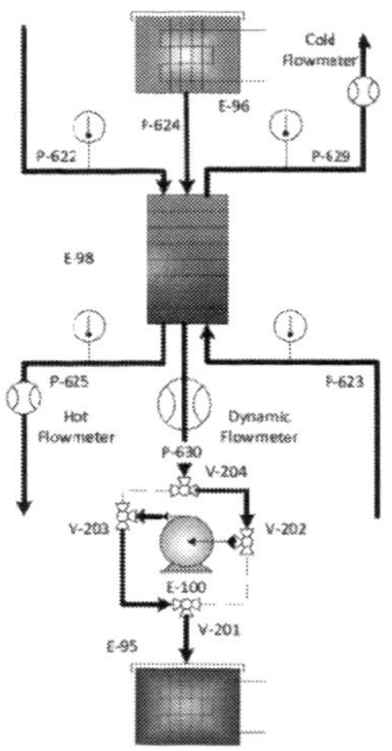

Results

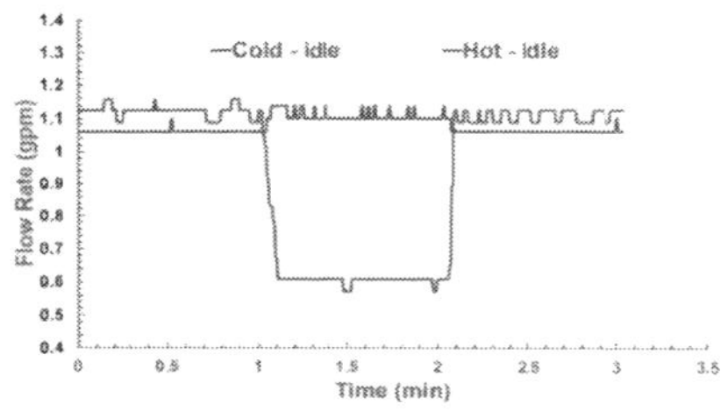

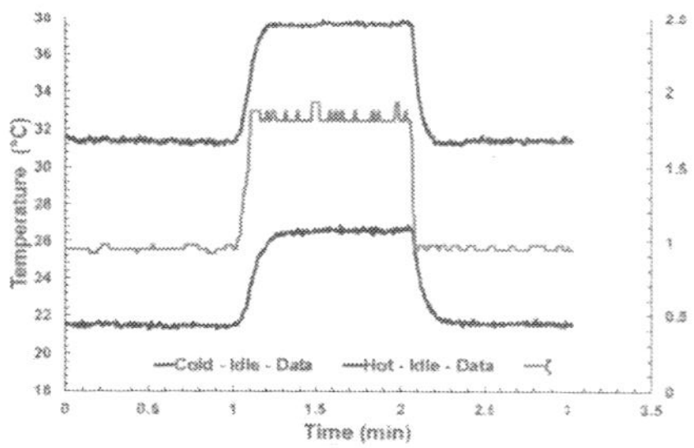

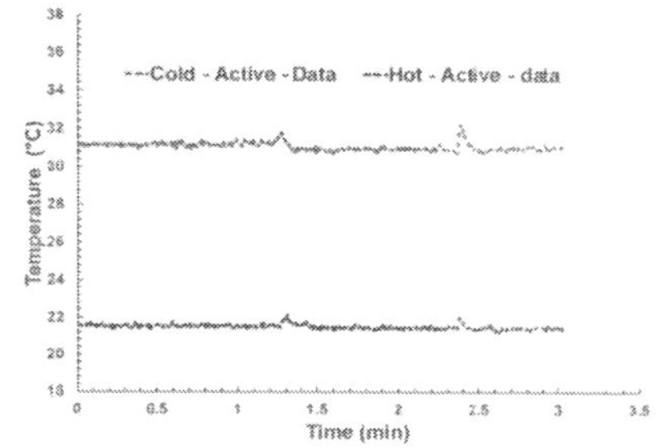

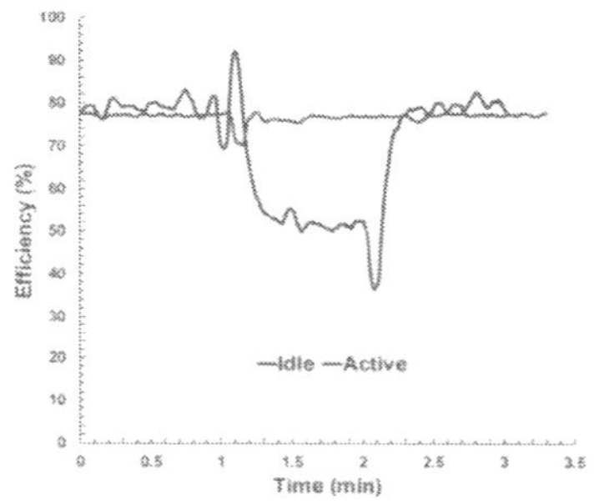

Conclusions

- Greatly reduces or eliminates compromises in heat exchanger exit temperatures and performance associated with transient process perturbations.
- responds to changes in process conditions with a near-zero time constant and very short-lived transients in the exiting flows.
- The dynamic heat exchanger technology represents a potentially large contribution to process flexibility and load change ability.

Acknowledgements & References

The financial and technical support of Sustainable Energy Solutions (SES) is gratefully acknowledged. This project was partially funded by SES projects sponsored by Advanced Conversion Technologies Task Force in Laramie, Wyoming through the School of Energy Resources and by the Climate Change and Emissions Management Corporation (CCEMC) of Alberta, Canada CCEMC.

- Ebrahimzadeh, E., Wilding, P., Frankman, D., Fazlollahi, F., and Baxter, L., "Theoretical and Experimental Analysis of Dynamic Plate Heat Exchanger: Non-retrofit Configuration", *Applied Thermal Engineering*, 2015.

Optimal Process Design Based on Energy and Exergy Analyses of Co-Gasification Power Plant with CO_2 Utilization

Speaker: Po-Chih Kuo

Department of Chemical Engineering

National Cheng Kung University, Tainan, Taiwan, ROC

Green Energy and Systems Engineering Laboratory (GESE)

E-mail: martin2292373@hotmail.com

1

Process simulation using Aspen Plus

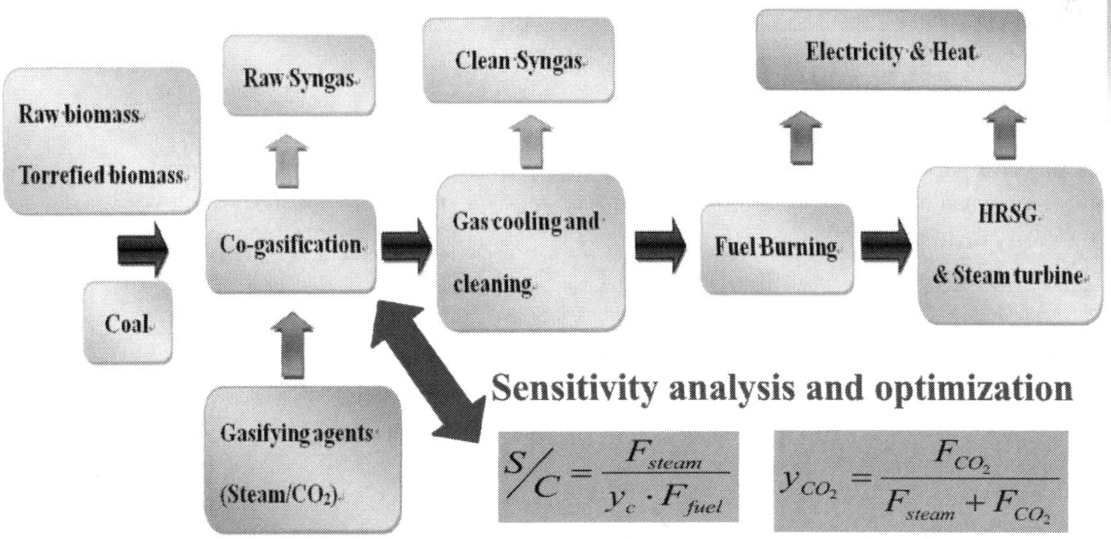

Sensitivity analysis and optimization

$$S/C = \frac{F_{steam}}{y_c \cdot F_{fuel}}$$

$$y_{CO_2} = \frac{F_{CO_2}}{F_{steam} + F_{CO_2}}$$

$$Blending\ ratio\ (wt\%) = \frac{F_{biomass}}{F_{biomass} + F_{coal}} \times 100\%$$

Optimization and CO_2 utilization

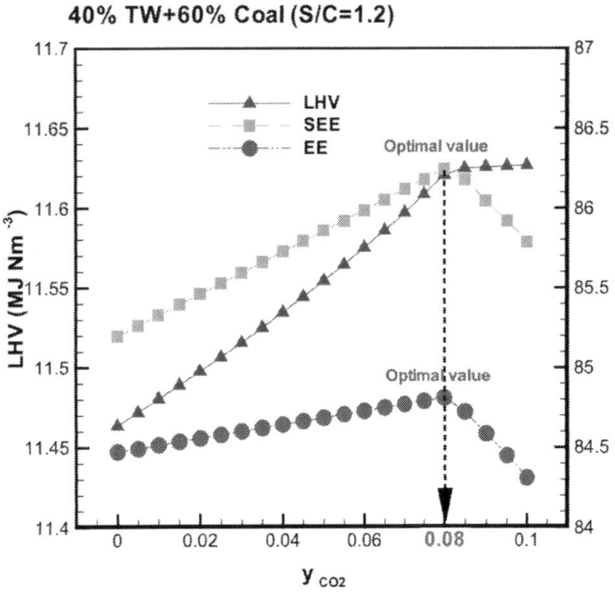

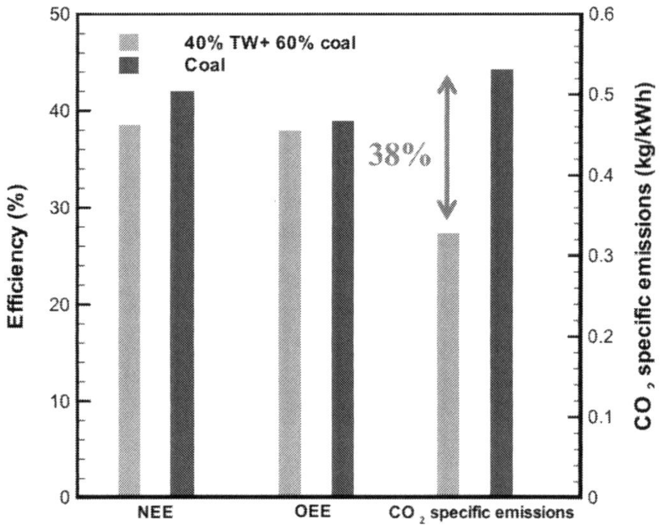

LHV: lower heating value of the product gas
SEE: system energy efficiency
EE: system exergy efficiency

NEE: Net energy efficiency
OEE: Overall exergy efficiency

424991 Smart Process Operations in Fuels Industries: Applications and Opportunities

Abstract:

The smart industrial revolution, the 4th after the mechanical, electrical, and digital ones, is a today's process industry frontline research in terms of technology, but also in terms of the use of technology that implies in human behavior and resource issues to deploy new process-of-work. Smart operation makes use of new information and communication technologies (ICT) and advanced algorithms such as optimization[1], therefore there are requirements for high qualified and trained teams to handle such technologies[2].

Smart Process Manufacturing (SPM) also known as Industry 4.0 is an emerging field of research and refers to a design and operational paradigm involving the integration of measurement and actuation, safety and environmental protection, regulatory control, high fidelity modeling, real-time optimization and monitoring, and planning and scheduling[3]. It is the enterprise-wide application of advanced technologies, tools, and systems, coupled with knowledge-enabled personnel, to plan, design, build, operate, maintain, and manage process manufacturing facilities, where is expected reduced costs in inventories, manufacturing, logistics, maintenance, etc[4].

We present several applications and opportunities of smart operations to be explored in fuels industries. These operations leverage the decision-making by (i) clustering crude oil arriving in the refineries as much as different they can be in terms of quality[5]; and (ii) integrating cutpoint temperature optimization of the distillates (initial and final boiling points) with the final blending producing specified fuels[6]. Other examples show ICT integrated with scheduling optimization in (i) crude-oil selection or diet; (ii) pathways for hydrocarbon movements[7]; (iii) blendshops, and (iv) regeneration of ion exchange resins in a demineralized water facility for boiler feed water production.

The ongoing project for management of operations in Petrobras refineries will be presented. It is a panel board controlling the operations, gathering real-time information from the plant and communicating to the operators, after the mentioned optimization, what they should execute, where some of them is simply a transmission of radio frequency identification (RFID) to automatic valves.

(1) Thornhill NF, **2015.** In https://workspace.imperial.ac.uk/smartops/Public/ENERGY-SMARTOPSOverview.pdf

(2) Christofides PD, David JF, El-Farra NH, Clark D, Harris KRD and Gipson JN. Smart plant operations: vision, progress and challenges. Aiche Journal, **2007**, 53 (11), 2734–2741.

(3) Davis JF and Edgar TF. Smart Process Manufacturing – A Vision of the Future. In Book: Design for Energy and Environment. Ed. El-Halwagi MM, Linninger AA, **2008**, 150-165.

(4) David JF, Edgar T, Porter J, Bernaden J and Sarli M. Smart manufacturing, manufacturing intelligence and demand-dynamic performance. Comput Chem Eng, **2012**, 47, 145–156.

(5) Industrial Algorithms LLC., **2015**. In http://pt.slideshare.net/alkis1256/ctap-imf.

(6) Kelly JD, Menezes BC, Grossmann IE. Distillation Blending and Cutpoint Temperature Optimization using Monotonic Interpolation. Ind Eng Chem Res, **2014**, 52, 18324-18333.

(7) Magatão SNB, Magatão L, Neves-Jr F, Arruda LVR. Novel MILP Decomposition Approach for Scheduling Product Distribution through a Pipeline Network. Ind Eng Chem Res, **2015**, DOI: 10.1021/ie5046796.

425343 Electrodeposition of Chromium from Chromium(III)-Ionic Liquid Solutions

Abstract:

Electrodeposition of Chromium from Chromium(III)-Ionic Liquid Solutions

Liyuan Sun* and Joan F. Brennecke

Department of Chemical and Biomolecular Engineering, University of Notre Dame

Notre Dame, IN 46556, USA

lsun4@nd.edu

Electrodeposition of chromium (Cr) is indispensable for various industries, such as automotive, aerospace, electronics, optics, and sensors. The most common industrial chromium coatings are usually electrodeposited from hexavalent chromium (Cr(VI)) aqueous solutions, which possesses severe concerns because Cr(VI) is highly toxic. A straightforward alternative is less toxic trivalent chromium (Cr(III)) based solutions. However, the high stability of the hexaaquochromium(III) complex formed in aqueous electrolyte solutions may obstruct or even prevent the electrodeposition of chromium from aqueous solutions. Another considerable drawback of electrodeposition of chromium from aqueous solutions is hydrogen evolution, which results in safety issues and reduced current efficiency. Using ionic liquids (ILs) for Cr(III) electrodeposition is a promising alternative that could address the above issues. So far, only a few limited examples of chromium electrodeposition from ILs have been reported (e.g., using a chromium chloride hexahydrate/choline chloride solution,[1] or chromium chloride hexahydrate/1-buthyl-3-methylimidazolium tetrafluoroborate ([bmim][BF_4]) solution[2]).

In this work, we investigate new Cr(III)-IL solutions as potential candidates for chromium electrodeposition. To optimize the transport properties of the mixtures, different amounts of ultrapure water is added to the Cr(III) salt ¨C IL mixtures. The conductivity, viscosity, and 'ionicity' of the solutions are characterized over the temperature range from 283.15 K to 353.15K. Electrochemical reduction of Cr(III) is demonstrated by cyclic voltammetry and preliminary electrodeposition tests. The physicochemical and electrochemical properties of the solutions depend heavily on the water content and the ion type. The new Cr(III) salt ¨C IL mixtures are excellent candidates for chromium electrodeposition under carefully selected conditions where hydrogen evolution can be avoided or inhibited.

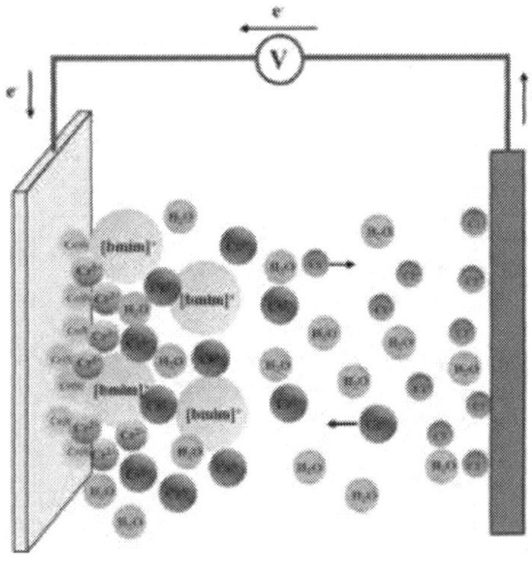

Reference:

1. Abbott, A. P.; Capper, G.; Davies, D. L.; Rasheed, R. K. Ionic liquid analogues formed from hydrated metal salts. Chemistry-a European Journal 2004, 10, 3769-3774.

2. Eugenio, S.; Rangel, C. M.; Vilar, R.; do Rego, A. M. B. Electrodeposition of black chromium spectrally selective coatings from a Cr(III)-ionic liquid solution. Thin Solid Films 2011, 519, 1845-1850.

425635 A Process Engineering Perspective on the Small Scale Conversion of Natural Gas and Biomass to Transportation Fuels

Abstract:

The conversion of natural gas and biomass to liquids (GBTL) at the smaller scale has invited significant research attention, owing to developments in process intensification. Opportunities exist to utilize natural gas in remote and offshore locations, and biomass synergistically in isolated areas [1]. Technologies based on the GBTL concept offer opportunities in optimizing the usage of natural gas along with biomass [2]. It also serves towards one of the options to develop commercially viable biomass technologies to enable biofuel production nationwide, by supporting the goal of the Energy Independence and Security Act for the production of 36 billion gallons per year of renewable transportation fuels by 2022.

The research community has been actively working on concepts related to reducing the scale of associated technology components (gasification, gas conditioning, reforming, Fischer-Tropsch and separation processes) which constitute as some of the key components of a GBTL system. These concepts cover a wide spectrum of process engineering aspects ranging from the effect of feedstock on conversion processes to kinetics associated with chemical reactors (e.g. Reforming, Fischer Tropsch Synthesis). This poster shall offer a reflection on the current understanding of process parameters associated with the small scale co-conversion of natural gas and biomass to transportation fuels, and highlight upon process engineering insights which shall aid in future techno-economic analysis studies.

References:

1. Kaisare et al. (2012), A review on microcombustion: Fundamentals, devices and applications, Progress in Energy and Combustion Science, 38, pp. 321-359

2. Bioenergy Technolgies Office, "The Potential for Natural Gas to Enhance Biomass Technologies," 2014. Available at: http://energy.gov/sites/prod/files/2014/04/f14/february2014_gbtl_webinar.pdf

425778 Comparison Between Traditional and Dynamic Heat Exchangers through Experimental Verification and Aspen HYSYS Simulation

Abstract:

The purpose of this experiment was to develop a heat exchanger (HX) network using dynamic HXs, devices that can decrease or eliminate HX efficiency and performance losses related to transient changes in flow rates. Dynamic HXs can accommodate varying flow rates of ethanol while maintaining constant outlet temperatures throughout the HX network. The main temperature that needs to remain constant is the temperature of the outlet ethanol. Traditional heat exchangers often represent the slowest part of the process to adjust to load changes. Through the dynamic heat exchanger developed in this experiment, the heat exchanger was able to adjust to load change significantly faster. This was accomplished by using pumps, tank system and proportional-integral-derivative controllers (PID controller) to change the flow rate of ethanol. As the flow rate of feed ethanol increased, the flow rate of return ethanol increased, thus maintaining constant outlet temperatures. The scope of this experiment includes using various inlet ethanol flow rates whilst observing transient responses for all units specifically heat exchangers. Aspen HYSYS dyanmics was used to simulate both traditional and dynamic designs. The results of this experiment are verified with simulation's results to indicate that a dynamic heat exchanger is a novel way of decreasing the time needed for a heat exchanger to adjust to load change.

423143 Studying Nonlinear Control System of Atmospheric Distillation Unit Using RGA Method

Abstract:

In this research steady state and dynamic simulation of atmospheric unit of Shiraz oil refining company is carried out. Experimental ASTMD86 curves of different products are compared to those obtained by simulations. Effect of different factors such as feed temperature, product flows and pump-around heat removal variation on quality and quantity of products are studied. Controllers are perfectly tuned in dynamic simulation to keep quantity and quality of products in desirable range. It is observed that control system performs nonlinearly using one of the simplest and accurate methods of system identification, mirror test. The best control pairs are determined and control unit structure is studied. In order to specify control pairs, Relative Gain Array (RGA) is used. The capacity of refinery is improved without any alteration in refinery structure solely by changing control structures.

426326 Mechanistic Understanding of the Mechanical-Activation Enhancement for TiO2 Production Via Acidolysis of Ilmenite

Abstract:

Abstract

The current sulphate process for TiO2 production via ilmenite is vigorously digested with concentrated sulfuric acid, which discharges considerable dilute sulphuric acid waste in industry. This paper describes recent developments of a mechanical-activation process for the intensification of the ilmenite leaching using dilute sulphuric acid (e.g., concentration of 60 wt.%). The used sulphuric acid can be mixed with waste acid and recycled into the leaching process. The unactivated and mechanically activated ilmenites were carefully characterized by using X-ray diffraction, particle size analysis, BET surface area and scanning electron microscope techniques for the understanding the structural changes during the mechanical activation process. The results revealed that mechano-chemical activation significantly affected on the lattice deformation, crystal size specific surface area and the granule size in mechanical activation systems, which led to a significant increase of leaching rate (from 36% to ~73%). The wet milling was found to be more effective than dry milling due to surface regions with larger lattice strain, significantly larger surface area, and more smaller size crystallites. Moreover, the reaction principles of the mechanically activation were discussed and a possible mechanism was proposed. This methodology is economically and scientifically significant for energy-saving and intensifying the leaching process, which could be widely used in solid-liquid reactions.

Keywords: Titanium; Mechanical activation; Sulphuric acid leaching; Wet milling

References

1. Li C, Liang B, Guo L. Dissolution of mechanically activated Panzhihua ilmenites in dilute solutions of sulphuric acid. *Hydrometallurgy*. 2007;89(1-2):1-10.

2. Liang B, Li C, Zhang C, Zhang Y. Leaching kinetics of Panzhihua ilmenite in sulfuric acid. *Hydrometallurgy*. 2005;76(3-4):173-179.

3. Li C, Liang B, Guo L, Wu Z. Effect of mechanical activation on the dissolution of Panzhihua ilmenite. *Minerals Engineering*. 2006;19(14):1430-1438.

4. Li C, Liang B, Guo L. Dissolution of mechanically activated Panzhihua ilmenites in dilute solutions of sulphuric acid. *Hydrometallurgy*. 2007;89(1):1-10.

5. Bin L, Chun L. A New Sulfuric Acid Recycle Routine in TiO_2 Pigment Production. *AIChE 2013 Annual meeting conference proceedings*.

6. Jia L, Liang B, Lü L, et al. Beneficiation of titania by sulfuric acid pressure leaching of Panzhihua ilmenite. *Hydrometallurgy*. 2014;150:92-98.

7. Welham NJ. Mechanical enhancement of the dissolution of ilmenite. Minerals Engineering. 1998;17(9):827-841.

8. Chen Y, Williams J, Campbell S, Wang G. Increased dissolution of ilmenite induced by high-energy ball milling. *Materials Science and Engineering: A*. 1999;271(1):485-490.

9. Sasikumar C, Rao D, Srikanth S, Mukhopadhyay N, Mehrotra S. Dissolution studies of mechanically activated Manavalakurichi ilmenite with HCl and H_2SO_4. *Hydrometallurgy*. 2007;88(1):154-169.

Fischer-Tropsch synthesis under periodic operation in a microstructured reactor

V. Le Courtois[1], L. Guillou[1,2]

1. Unité de Catalyse et de Chimie du Solide, UMR CNRS 8181, Ecole Centrale de Lille, F-59651
2. Fives Proabd, F-68071

Abstract

A micro-structured reactor was used to study the influence of a maintained CO and H_2 periodic flow switching on the performances of the Fischer-Tropsch synthesis. The influences of the duty cycle and of the period of the cycle for high frequency periodic feeds were investigated. Some discrepancies from the behaviour at steady-state were observed: i) the C_{10+} selectivity is influenced by the duty cycle and ii) there is an optimum period length to get an increased CO conversion level for each duty cycle. For the investigated experimental space, up to 3.5 fold C_{10+} yield increase was obtained in respect of a continuously fed reactor at steady state. These differences cannot be explained on the basis of the kinetic rate form available.

Keywords: Fischer-Tropsch synthesis; Periodic operation; Microstructured reactor; Cobalt

1. Introduction

The terms "periodic", "pulsed" or "cyclic operation" describe a similar process control strategy consisting in the modification of a process controlling parameter according to a time repeated protocol. As far as catalytic reactors are concerned periodic operations were defined as engineering tools which can allow to tune or control selectivity and/or conversion (Silveston et al., 1995). It can namely be applied to any control parameter whatever it is composition, temperature (Silveston et al., 2004a) or pressure (Silveston et al., 2004b). Most widespread applications mainly concerned composition cycling so far.

The applications of the Fischer-Tropsch synthesis (FT) under periodic operation were reviewed by Adesina et al. (1995). Their purpose was essentially aiming at modifying the product

[2] corresponding author: loic.guillou@fivesgroup.com
present address: Fives Proabd, 20C rue de Chemnitz – CS11437 – F68071 Mulhouse Cedex France

distribution and the productivity of a given hydrocarbon product range. Other kind of pulsed applications, in order to gather data to validate a reaction pathway or to do mechanistic studies, were reported too (Khodadadi et al., 1998). Ever since only a few other studies were added and might be used to update the latest review: Nikolopoulos & Gangwal (2003) used iron and cobalt modified catalysts and our group studied the influence of periodic operation Co/Al_2O_3 and Co/SiO_2 in catalytic membrane reactors in which both temporal and spatial feed modulation were combined (Léonard, 2001 and Guillou et al., 2004). The application of a forced feed policy showed in all cases some effects upon FT catalytic activity and selectivity. The pulse frequency was the main influencing parameter studied. For Co based catalysts the periodic hydrogen injection was mainly studied either under single step change strategies or as a cyclic modification of P_{H2} (Adesina et al., 1986 & 1995, Khodadadi et al., 1998, Léonard, 2001 & Guillou et al., 2004). A basic assumption was that the chain propagation of FT product distribution could be restricted over narrowed lengths by periodic H_2 injection. In all cases the periodic operation evidenced the existence of an optimum adjustment of the feed parameters to maximize the activity and the product distribution. These feed parameters were the applied frequencies (or periods (T)) and the duty cycles (dc, i.e. the fractional time of exposure to a given reactant in respect of the entire period length)) as defined by Silveston et al. (1995).

In our previous work (Léonard, 2001 & Guillou et al., 2004) we underlined the inner limits of the forced transient policies for FT: the benefit of such strategies is rapidly lost for complex fluid mixing patterns and was therefore found to give satisfying results only in small laboratory sized devices. Scaling-up appeared to be unreasonable, most of the benefits being potentially lost at bigger scales. The application of forced feed composition variation in traditional reactors thus appeared as being unreasonable. Yet it has been established that progresses made in microreactor technology may re-actualize the purpose and meaning of forced transient feed. Rouge et al. (2001) successfully applied a fast periodic operation in a microstructured channel reactor for catalytic dehydratation of isopropanol over γ-alumina. Moreover the design of such microreactors implies that they can be scaled-out through the parallelization of unit cells identical to the laboratory studied reactor, which allows a conservation of laboratory testified performances. The application of unusual process control policies such as periodic feed modulation therefore becomes relevant and scalable.

The scope of this study is to investigate the influence of high frequency pulses with various duty cycles in a microchamber reactor. The range of frequencies is up to one or two order of magnitude higher than previous studies. The effects on the FT performances are expressed as the CO conversion and the associated C_{10+} selectivity. A 20 wt% Co/SiO_2 catalyst is used.

2. Experimental

Catalyst

The catalyst was a composite material with an active phase made out of a 20 wt% Co/SiO_2 catalyst layer. Its preparation required a complex sequence which has been documented elsewhere. The catalyst was grafted on a 5 µm thick silica glass (Guillou, 2005). The latter was obtained thanks to a modified remote enhanced plasma chemical vapour deposition process over an AISI316L stainless steel plate (Guillou et al., 2005). The preparation of the catalyst was based on a sol-gel modified method (Guillou et al., 2006). A classical sol made out of the catalyst and support precursors ($Co(NO_3)_2$ and $Si(C_2H_5O)_4$) acidified with HNO_3 was used. Contrary to the classic solution proposed elsewhere (Ernst, 1997 & Ernst et al., 1999), a binary solvent (water and ethanol) was used. Its composition was set at 42 wt% water. Amounts of tetraethoxyorthosilane and hexahydrate cobalt nitrate to be added were then calculated to get a formulation compatible with the desired catalyst final composition i.e. 20 wt% Co/SiO_2. All reagents and solvents used were commercially available products used without further purification step. The grafting step itself used a sequential spray coating process. This method allowed getting catalysts layers with a thickness varying from 5 to 110 µm. The adhesion was A5 according to the ASTM-D3359-02 test.

In this study a 5µm bonding layer was used. It was coated with a 50 µm thick catalyst layer which corresponded to 44.1 mg of 20 wt% Co/SiO_2 catalyst. Typical specific areas determined by nitrogen adsorption according to the B.E.T. method were set around 260 $m^2\ g^{-1}$. Pore volume was typically 0.17$cm^3\ g^{-1}$ with 50% of microporous contribution (Guillou, 2005).

Reactor

The reactor used for catalytic activity tests consisted in a stacking of a single catalytic plate between two mechanically engineered AISI316L blocks (Figure 1). Upper and lower blocks were equipped with K thermocouple and power cartridges to ensure electrical heating and temperature control. The upper block acted as a gas distributor and collector. Inlet and outlet were located orthogonally to the catalytic plate. The reaction chamber (200 µL) was obtained by engraving the upper block with a 40x20x0.25 mm rectangular groove. A similar groove was engraved in the lower block and received the catalyst coated plates. Resident time distribution determinations (RTD) showed that under the conditions reported in this contribution, the hydrodynamic was relevant of a continuous stirred tank reactor (Guillou, 2005).

Though convenient for rapid assembly and catalyst testing in microstructured environment, such a stacking solution applied to reactor structure limits the range of working pressure achievable. In absence of bonding of the structured foils, the maximum pressure achievable is very limited for metal/metal stacks: for example, specifications for a similar commercially available catalyst testing microreactor are set at 0.3 MPa (IMM, 2003) whereas the design described here was tested as being leakage free up to 0.4 MPa (Guillou, 2005). It corresponds to the pressure gap usually reported for non-bonded stacked microstructures (Commenge, 2005). As a consequence it was decided to limit the working total pressure at 0.1 MPa.

Evaluation of activity and product distribution

Prior to FT, catalysts were in-situ activated under pure H_2 (680 mL h^{-1} STP) for 4 hours at 450 °C according to a protocol presented elsewhere (Balloy, 1997).

After reduction, reactor was cooled down to reaction temperature (180 °C) and inlet feed was preheated at reactor temperature. Gas flow was switched to the proper syngas ($CO+H_2$) composition; a constant N_2 flow, set at 93 NmL h^{-1} STP (Bronkhorst Hi-Tec mass flowmeter regulator), was used in the syngas flow. Total flow rate was 600 Nml h^{-1} STP. N_2 was used as an internal standard for CO conversion (X_{CO}) and product selectivity calculation purposes. Catalyst was held under steady state feed conditions for 72 h prior to begin the transient experiments.

Outlet stream was heated at 250 °C to avoid product condensation. Gaseous products were analyzed on-line by gas chromatography using serial 13X molecular sieve and DHES packed columns connected to a thermal conductivity detector and a TDX-DB1 flash chromatography capillary column connected to a flame ionization detector. This set up allowed to achieve accurate information on CO conversion (X_{CO}) and product distribution (C_1-C_4, C_5-C_9, C_{10+}) within very short analysis times (1 to 15 minutes maximum for the complete gaseous products range); however within such analytic conditions, the C_1-C_4 cut is represented by a single composite peak that cannot be deconvoluted and as such the C_1 (methane) selectivity is not independently reachable. Thus selectivities must be expressed as hydrocarbon cuts. Special attention was dedicated to the C_{10+} hydrocarbon cut which represents the economically attractive product of choice for synfuel applications.

Periodic injection of reactant was controlled by switching CO and H_2 flows delivered by two mass flow meters (Brooks) in serial with two 2/3 solenoid valves. Valve commutations were controlled with a Zelio programmable automat (Siemens-Télémécanique) to selectively shunt reactant flows in a time controlled way.

Right after reactor outlet vent a static mixer was used to continuously gather the educts. Its design allowed a perfect mixing of the collected flows over a 600 s time gap at 750 Nml h^{-1} STP total flow rate. Thus the flow composition was mechanically averaged. It was checked that the corresponding mean composition was comparable to a calculated one from integration of instantaneous composition values. Less than 3% of difference between mean measured values after static mixer and integrated instantaneous values was found. It validated the use of the mean composition taken at the exhaust of the static mixer to calculate the averaged outlet CO conversion (X_{CO}) and mean product distribution (C_1-C_4, C_5-C_9, C_{10+}). These were used to get various yield (Y) values expressed as $Y_i = X_{CO} \times C_i$ for a given hydrocarbon cut i with selectivity C_i taken from C_1-C_4, C_5-C_9 or C_{10+}.

Under such conditions it was ensured that CO conversion and product distribution based on the analysis of gas phase remained constant over the 30-72 h time gap. All mean measurements reported were done after 72 h under fixed reaction conditions for constant and periodic feed policies in order to ensure that steady state was potentially attained.

Steady state experiments

Steady state experiments under constant feed flow conditions were done on the same catalyst sample over a range of total inlet flow rates ($<Q_t>$) from 150 to 688 Nml h^{-1}. Experimental conditions are gathered in Table 1.

N_2 flow was set at 93 Nml h^{-1} while CO and H_2 flows were selected to ensure an inlet reactant partial pressure ratio $P_{H2}°/P_{CO}°=2$. All experiments were done before the application of the various reported periodic forced feed policies except runs at $<Q_t>=600$ Nml h^{-1}. These were done respectively before and after the application of forced transient feed strategies and were used to check out the stability of the catalyst used.

Pulsing strategy

During forced periodic feed the microchamber reactor was kept under a constant flow of N_2 set at 93 NmL h^{-1}. H_2 and CO flows were cyclically injected in the microchamber reactor which caused step changes of contact times. This periodic switch was kept applied over 72 h for all the conditions studied. The cycle was defined over a single period as shown on Figure 2. Various duty cycles (dc) and period lengths (T) were applied.

The periodic forced feed policies were developed while keeping a constant value of 2 for the integral ratio of injected reactant molar flows defined under isotherm and isobar conditions as follows:

$$\frac{\int_0^T \overset{\circ}{n}_{H_2}(t)\times dt}{\int_0^T \overset{\circ}{n}_{CO}(t)\times dt} = 2 \qquad (1)$$

For all experiments CO flow (Q_{CO}) was kept constant at 198 Nml h^{-1} while H_2 flow was tuned to a flow value permitting to get the imposed constant integral ratio of inlet partial pressures in respect of selected T and dc values. The latter parameters covered the 60-480 s and 0.2-0.67 ranges respectively. Detailed T and dc conditions are gathered in Table 2. The corresponding mean total flows ($<Q_t>$), was calculated as:

$$<Q_t> = dc\times(Q_{CO}+Q_{N_2})+(1-dc)\times(Q_{H_2}+Q_{N_2}) \quad (2)$$

The monitoring of the composition of the gaseous educts was used to determine eventual mixing between the periodically switched flows. Composition profiles were reconstructed from the TCD analysis of sequentially trapped vent gas and compared with corresponding compositions of various N_2+H_2+CO mixes.

3. Results and discussion

Continuous feed and validation of periodic policies

Steady state FT performances over a 50 µm thick 20 wt% Co/SiO_2 catalyst obtained after 72 h under a continuous syngas feed of given composition for various inlet total flow rates are given in Table 1.

It appears that conversion increased with the residence time within the microchamber reactor, X_{CO} reaching 30.7% at 150Nml/h. It corresponds to a highly diluted reactant feed as F_{CO+H2}/F_{N2}= 0.61.

But even under these conditions it was still possible to reach a high conversion in reason of the significant corresponding contact time of 4.8 s. At maximum flow, which corresponded to a minimal dilution (F_{CO+H2}/F_{N2}=6.4) and a minimal contact time ($F_{CO+H2+N2}/V$= 1 s), CO conversion dropped to 13.9%. The product distribution appeared as being sensitive to the conversion level reached, the higher the conversion, the higher the C_1-C_4 selectivity. The maximum light gas selectivity achieved 79.3% at 150 Nml h^{-1} total flow rate. Similarly maximum C_{5+} production reached 47.2% for a corresponding 688 Nml h^{-1} total flow rate. These results were coherent with the well-known behaviour of Co based catalyst (Steynberg & Dry, 2004).

An experiment 600 Nml h^{-1} was done before the continuous and the periodic forced feed experiments, as a reference set point, and a second one (star marked) after, ending the whole experimental campaign. Thus they can be used to estimate the catalyst stability. The FT performances after the 21 experiments under various T and dc values appeared to be fully recovered. After 504 hours under continuous feed reaction conditions and 1512 hours under periodic feed reaction conditions, the catalyst layers exhibited conversion levels and product distributions similar to the ones obtained after the early first 72 hours of the first run. Thus it was considered that the periodic experiments done between the two identical continuous feed conditions could be considered as being meaningful.

Conversion and C_{10+} yield surface responses

FT performance evaluation under forced periodic feed gave the X_{CO} and product distributions reported in Table 2.

The monitoring of vent gas composition profile showed negligible mixing effects in all cases. Thus reactor atmosphere substitutions times (s.t.), calculated under the assumption of a continuously stirred tank, can be considered as being relevant of real transient delays which correspond to mixed composition of N_2+CO+H_2. These s.t. values can be evaluated at values of 2.5 s for a switch from H_2 to CO flow whatever T and dc were. For switches from CO to H_2 flow, s.t. values are functions of both T and dc. Corresponding numeric values are reported in Table 2.

Conversion and selectivity of the various hydrocarbon cuts appeared to be highly dependant on the feed policy applied, namely T and dc. These two parameters are correlated and must be simultaneously considered for result representation. A convenient format is a surface response plot.

Such graphics are proposed for X_{CO} and Y_{C10+} in Figure 3 and Figure 4 respectively. They represent some isolevel curves for CO conversion and C_{10+} yield respectively; they plot X_{CO} or Y_{C10+} vs. the H_2 and CO exposure times. Exposure times are calculated from the T and dc value according to their definition, i.e.:

$$CO_{\text{exposure time}} = dc \times T \text{ and } H_{2\,\text{exposure time}} = (1\text{-}dc) \times T \qquad (3)$$

The level curves proposed were interpolated from the experimental results restricted to the 60-240 s time gaps from table 2 for H_2 and CO. A routine fitting protocol from Matlab7.1 software package with a cubic interpolation function was applied. Such interpolated figures must be considered for general trends identification purpose only.

Some maxima were experimentally found for the CO conversion and C_{10+} yield. They appeared as being period and duty cycle dependant. Experimentally a maximum conversion was achieved for T=120 s and dc=0.5 (Table 2). This experimental maximum corresponded to a maximal interpolated conversion zone which could reach up to 26% for the shortest H_2 and CO periods (Figure 3). It can be predicted from Figure 3 that a maximum conversion area could be found to be around an optimum parameter setting defined for symmetric injections (i.e. dc=0.5 which corresponded to $CO_{\text{exposure time}} = H_{2\,\text{exposure time}}$ in (3)) and below T≈280 s. A second local maximum area is found for T=300s and dc=0.8.Experimentally it corresponded to X_{CO}=24.3%. But contrary to the former experimental maximum it is mainly only accompanied by light gases formation with C_1-C_4=81.8% (Table 2) whereas the first previously identified maximum could be associated with C_{10+} circa 17% (Table 2).

Maximum interpolated C_{10+} yield area was predicted as corresponding to the longest H_2 exposures for short CO exposures (Figure 4). Maximum Y_{C10+} area can be defined for a time frame centred on the 60-120 s and 140-240 s exposure time gaps for CO and H_2 respectively. These conditions appeared as being relevant of low H_2 flows and low $\langle Q_t \rangle$ values.

The maximum value experimentally obtained for C_{10+} yield was 6.9% for (H_2; CO) = (240 s; 60 s) exposures times (Table 2).

Influence of dc

Figure 5 gathers the results of the various periodic policies used under a single representation for C_{10+} selectivity and CO conversion. For a given functioning point (T=300 s dc=0.2)s, up to Y_{C10+}=7% was obtained. The points associated to the different feed strategies were located in very different places from the continuous feed ones (Figure 5, full diamonds). Their location appeared to be scaled according to the dc value applied. The shorter dc was, the higher the C_{10+} selectivity was. Thus for high dc values (0.6 and 0.8), testifying of very short exposure time to H_2, performances were found to be below the continuous feed. Reciprocally, longer exposure times to H_2 shifted the C_{10+} selectivity to higher levels, thus controlling a vertical translation of performance points. It corresponded to situations where the H_2 flow is quite low which means that a longer time (Table 2) is required to fully replace the gaseous atmosphere of the microchamber reactor. It was favourable to more consequent mixing time of H_2 and CO within the micro chamber, whereas low dc values tend to favour larger composition segregation within the microsystem, gaseous atmosphere substitution delays being shortened (Table 2).

Influence of T

Period (or frequency if 1/T is considered) appeared to influence the maximum conversion achieved for a given dc. This finding was similar to some previously reported results over different catalysts or for longer periods (Adesina et al., 1995 & Nikolopoulos & Gangwal, 2003). Here, for example at dc=0.33, maximum conversion (17.1%) was obtained for T=180 s; for longer period (T=360 s), conversion dropped to 1.8%. Similarly a shorter period gave a lower conversion at only 14.8% for T=90 s. Similar trends could have been extracted for other fixed dc values (Table 2).

Combination of dc and T respective effects on the location of the functioning point let appear that T defined mainly the CO conversion level and dc the C_{10+} selectivity in Figure 5.

Other studies underlined a similar influence of T but found some optimum settings for larger parameter values such as T=40 min or did not considered the influence of dc (Adesina et al., 1995 & Nikolopoulos & Gangwal, 2003). These results were obtained from switches between a syngas feed and H_2 pulses mainly. Disparities might be the indication of some possible extra influencing parameters over the location of optimum functioning point such as for example the H_2 content in the CO flow. Thus for cycled switches between a syngas flow and a pure H_2 flow, maximum attainable yields might get a different position from the one obtained with diluted, yet segregated reactant cycled switches. It may indicate one or many metastable functioning points for a catalyst which surface properties were shown to rely on syngas composition and exposure duration for a model Co(0001) catalyst (Wilson & de Groot, 1995).

Comparative performances with continuously fed conditions

The FT performance obtained for various feed policies gave some surface responses that appeared to be highly T and dc correlated. Yet one could wonder of the potential benefit of such a complex feed management policy in regard of a basic continuous feed.

$<Q_t>$ might have been selected for comparison purpose between steady state experiments under constant feed flow conditions and forced transient feed experiments. It would however not necessary be a most pertinent comparison criterion with all the possible forced transient feed conditions especially for higher T values (Léonard, 2001) or dc set away from 0.5. Indeed it appeared from data gathered in Table 2 that under similar mean $<Q_t>$ values X_{CO} could vary on a range as wide as 1.8 to 17.1%. Being so obviously not comparable between them, the pertinence of comparison of forced transient feed with continuous mixed flow strategy at similar $<Q_t>$ appears as being meaningless and another representation is required.

Figure 5 represented the C_{10+} selectivity associated to the corresponding X_{CO}. These coordinates defined a location that can be compared to the isoyield level curves and are a most convenient format to gather the various tested feed policy under a single plot. Experimental results associated to continuous feed conditions showed that for most cases the steady state C_{10+} production gave yield values tending toward 2.5% or even below for higher conversions. Even under higher flow rates, although favourable to an increased C_{10+} selectivity, the associated conversion drop hindered the possible yield benefices. This distribution is comparable to the one observed in the same reactor and for a similar catalyst under close operating conditions (Guillou et al., 2005). The selection of proper values for parameters dc and T allowed to tune the location of the functioning point; it let appeared that it was possible to reach some performances up to Y_{C10+}=7.9% for T= 300 s and dc=0.2. This yield level is 3.5 fold higher than the maximum achievable one under steady state continuously fed conditions as reported here (Figure 5) or elsewhere (Guillou, 2005).

Contrary to what was obtained by Léonard (2001) and Guillou et al. (2004) with membraneous H_2 distributors, the shortest periods (T=60 and 90 s) did not necessary give results tending toward the performances of continuous feed of mixed reactants. Within the inert membrane packed bed reactor used in a fast succession of H_2 and CO pulses meant a possible increased degree of mixing between the two gases imposed compositions because of a variable axial and radial dispersive influences of the packing under cycled flows. In the present case the T and dc policies still appeared as having an influence over FT performance. This effect was attributed to the combination of the distributor location and the peculiar reactor design. The former was as close as possible form the reactor inlet that insured minimal mixing between injected flows. The design of the latter allowed, as far as it was concerned, some atmosphere substitution times varying from 0.4 to 3.7 s for switch from CO to H_2 and 2.5 s for switches from H_2 to CO: these insured a complete substitution of the chamber atmosphere even under the longest H_2 exposure times (corresponding to the smallest flow rates and higher s.t. according to Table 2). Such conditions are readily hardly achievable in classical apparatus in laboratory and were made possible only because of the use of the microchamber reactor.

Catalyst stability

Another remark that can be addressed on the general behaviour of the catalyst is its stability: over 1512 hours of periodic feed under short periods, which defined aggressive conditions according to Nikolopouklos & Gangwal. (2003), the catalyst did not appeared as having suffered from the imposed cyclic modulation of reactant feed. More interestingly, the periodic switch between CO and H_2 feeds created some situations with high or low P_{H2}/P_{CO} ratios. On the first hand, high P_{H2}/P_{CO} ratios can potentially favour excessive water formation that is supposed to be responsible for deactivation of the catalysts (Storsæter et al, 2005). The P_{H2O}/P_{H2} ratio actually potentially controls the relative importance of reversible over irreversible deactivation (Bremaud et al., 2005). On the other hand, low P_{H2}/P_{CO} ratios are known to favour carbon deposition through Boudouard reaction which causes another deactivation mechanism through coking. Here no significant deactivation was found between the beginning and the end of the periodic fed experimental set.

4. Conclusion

The influence of periodic operation on FT performance was tested in a microstructured reactor. Policies were defined using two parameters, T and dc, set at conditions of low period value, and variable period geometry. Such conditions complete the scope of already explored transient feed policies while focusing upon lowly or un-reported parameter ranges. It exhibited promising C_{10+} yield increase which appeared as being inaccessible under classical continuous state feed condition. Up to 3.5 fold Y_{C10+} was obtained for a precise tuning of T and dc values as both appeared as having an important influence over hydrocarbon cut selectivity and CO conversion levels respectively.

Additionally the aggressive switching did not seem to have a detectable influence over catalyst durability and performances.

The choice of a microstructured reactor destined to be scaled out allows envisaging keeping the optimal feed policy even for devices designed to get a higher productivity level: conservation of unit reaction cell design can allow a possible deployment of the most potent unconventional feed policy.

However the scope of parameters explored was deliberately restricted by the initially defined working hypotheses:

$$Q_{CO}=\text{constant} \ \& \ \frac{\int_0^T \overset{\circ}{n}_{H_2}(t) \times dt}{\int_0^T \overset{\circ}{n}_{CO}(t) \times dt} = 2$$

Thus experimental space remains to be expanded to other conditions especially higher total pressure or replacement of pure CO by various syngas compositions.

Acknowledgments

Financial support of French Ministère de l'Education Nationale, de l'Enseignement Supérieur et de la Recherche is greatly acknowledged. Dr. Guillou personally thanks Ph. Kerleau (L.G.P.C.) for fruitful discussions.

Miscellaneous

Original research work dated 2005 and previously unpublished.

Matlab7.1 software package (L.G.P.C., licence N°72329) was used for representation of surface response data

References

Adesina A.A., Hudgins R.R., Silveston P.L. (1986). Can. J. Chem. Eng., **64**, 447-454

Adesina A.A., Hudgins R.R., Silveston P.L. (1995). Catal. Today, **25**, 127-144

Balloy D. (1997). Ph. D. thesis, U. Sc. Tech. Lille, Villeneuve d'Ascq, France

Bremaud M., Fongarland P., Anfray J., Jallais S., Schweich D., Khodakov A.Y. (2005). Catal. Today, **106**, 137-142

Commenge J.-M. (2005). 1st thematic school on Reactor Engineering, Frejus, France

Ernst B. (1997). Ph. D. thesis, U. L. Pasteur, Strasbourg, France

Ernst B., Libs S., Chaumette P., Kiennemann A. (1999), Appl. Catal. A, **186,** 145-168

Guillou L., Léonard S., Le Courtois V., Vanhove D. (2004). 6th International Conference on Catalysis in Membrane Reactors, Lahnstein, Germany

Guillou L. (2005). Ph. D. thesis, U. Tech.Compiègne, Compiègne, France

Guillou L., Le Courtois V., Supiot Ph. (2005). Mat. & Tech., **93,** 335-345

Guillou L.,Balloy D., Supiot Ph., Le Courtois V. (2006), Appl. Catal. A, submitted

Khodadadi A.A., Hudgins R.R., Silveston P.L. (1998). Can. J. Chem. Eng., **76,** 239-246

Léonard S. (2001). PhD thesis, U. Tech. Compiègne, Compiègne, France

Nikolopoulos A.A., Gangwal S.K. (2003). Chain limiting of FT reactor, Final report, US-DOE contract N°DE-FG26-99FT40680, available on www.fischer-tropsch.org

Process Technology of Tomorrow (2003). part II, IMM catalogue

Rouge A., Spoetzl B., Gebauer K., Schenk R., Renken A. (2001). Chem. Eng. Sci., **56,** 1419-1427

Silveston P.L., Hudgins R.R., Renken A (1995). Catal. Today, **25**, 91-112

Silveston P.L., Hudgins R.R. (2004a). Chem. Eng. Sci., **59**, 4043-4054

Silveston P.L., Hudgins R.R. (2004b). Chem. Eng. Sci., **59,** 4055-4064

Steynberg A.P., Dry M.E. (Ed.) (2004). In "Fischer-Tropsch Technology", Studies Surf. Sci. Catal., **152,** Elsevier, Amsterdam

Storsæter S., Borg Ø., Blekkan E.A., Holmen A. (2005). J. Catal., **231,** 405-419

Wilson J., de Groot C. (1995). J. Phys. Chem., **99**, 7860-7866

<Q_t> (Nml/h)	X_{CO} (%)	Selectivity (%)		
		C_1-C_4	C_5-C_9	C_{10+}
150	30.7	79.3	18.1	2.6
250	27.7	68.2	28.0	3.8
350	25.3	69.2	23.5	7.3
450	20.7	62.1	26.4	11.4
600	14.2	58.8	28.1	13.1
600*	14.3	59.9	27.2	12.9
688	13.9	52.8	26.4	20.8

Table 1: CO conversion and product distribution under various total flow rates for continuous feed at fixed $P°_{H2}/P°_{CO}$=2, T=180 °C and P=0.1 MPa - * represents the validation run under continuous feed done after the 21 different periodic feed policies

Experimental conditions				Results			
T (s)	dc	H_2 s.t. (s)	$<Q_t>$ (Nml/h)	X_{CO} (%)	Selectivity (%)		
					C_1-C_4	C_5-C_9	C_{10+}
60	0.50	1,5	390.5	16.2	59.6	26.3	14.2
90	0.33	2,4	294.3	14.8	47.3	27.8	24.9
90	0.67	2,4	490.0	18.9	70.8	19.3	10.0
120	0.50	1,5	390.5	28.5	59.7	25.0	15.4
150	0.20	3,7	212.6	13.4	35.9	29.7	34.4
150	0.80	0,4	561.0	8.2	87.1	7.4	5.5
180	0.33	2,4	294.3	17.1	48.9	27.6	23.5
180	0.67	0,8	490.0	18.4	68.3	19.4	12.4
240	0.25	3,2	241.5	10.8	37.0	30.0	33.0
240	0.50	1,5	390.5	25.8	59.9	23.0	17.5
240	0.75	0,6	538.3	15.8	68.9	19.9	11.2
300	0.20	3,7	212.6	20.7	36.7	29.9	33.3
300	0.40	2,0	330.6	21.5	41.6	30.7	27.7
300	0.60	1,0	449.8	8.7	58.2	25.2	16.7
300	0.80	0,4	568.6	24.3	81.8	11.6	6.7
360	0.33	2,4	294.3	1.8	49.1	26.2	24.7
360	0.50	1,5	390.5	20.0	60.9	21.6	17.5
360	0.67	0,8	490.0	10.7	70.5	17.2	12.3
420	0.43	1,8	348.7	10.4	44.5	30.6	24.9
420	0.57	1,2	433.3	6.8	57.2	24.3	18.5
480	0.50	1,5	390.5	5.5	48.9	27.9	23.3

Table 2: Experimental conditions and corresponding FT performances for various periodic feed policies (T=180 °C – P=0.1 MPa)

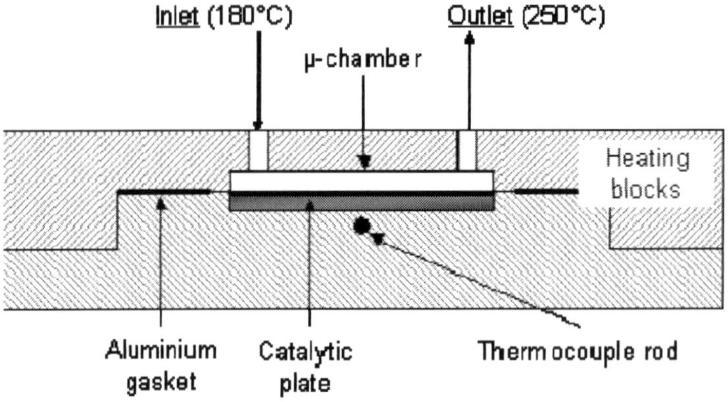

Figure 1: Microchamber test reactor for AISI316L plate supported catalysts

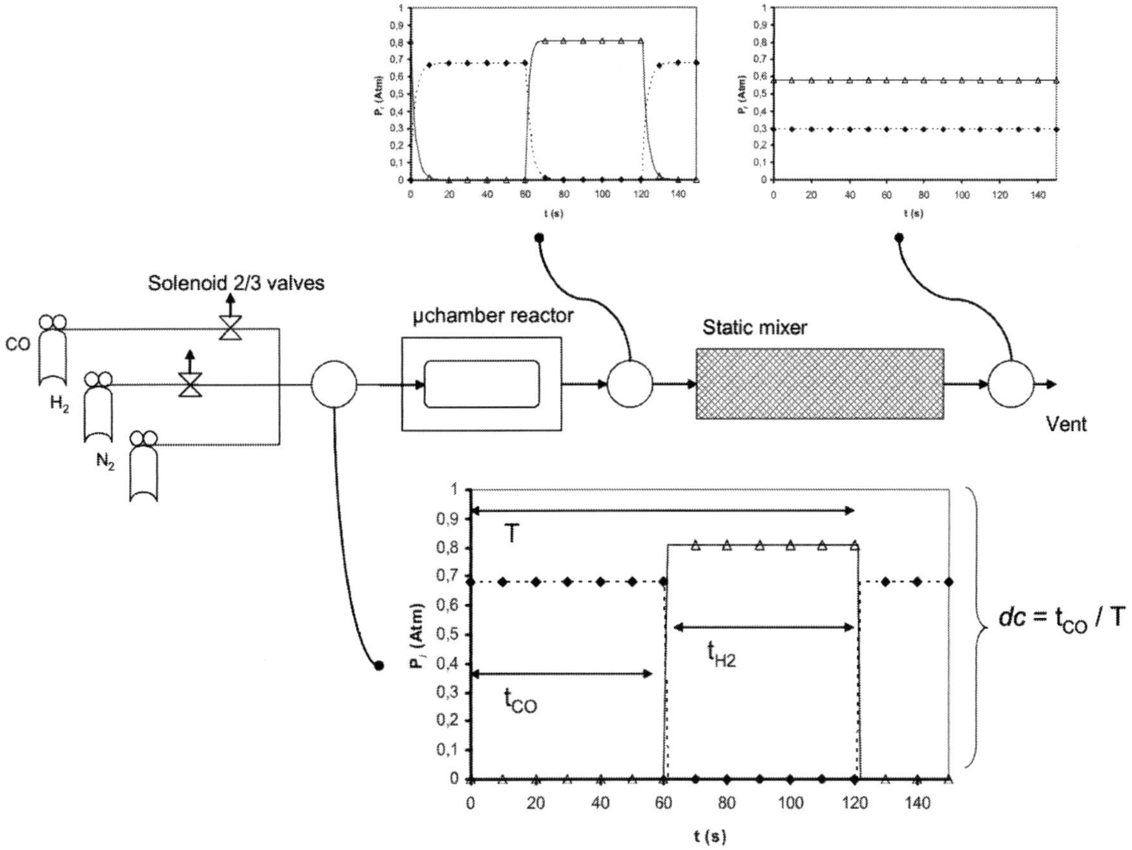

Figure 2: Schematic view of reactor setup and P_i=f(t) at reactor inlet and outlet and after static mixer

(Theoretical visualisations of P_i=f(t) for T=120 s and dc=0.5 at X_{CO}=0 in a 200 µL differential reactor)

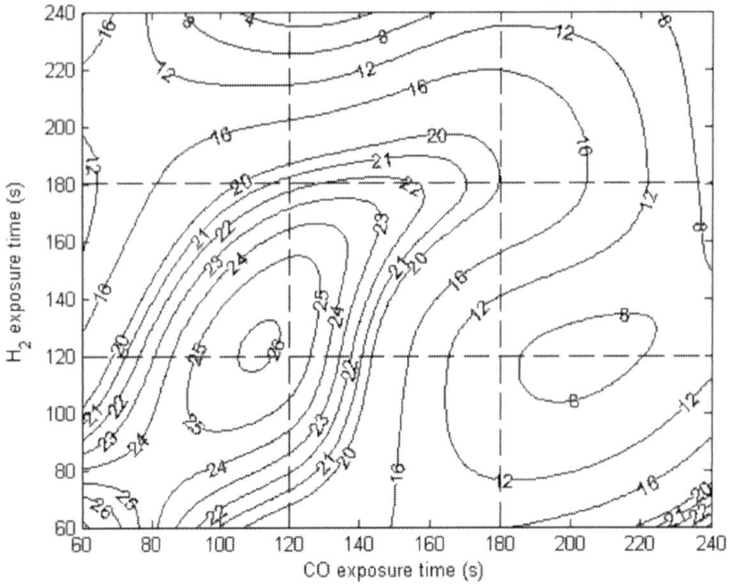

Figure 3: Interpolated CO conversion levels for various CO and H_2 cycled exposure times

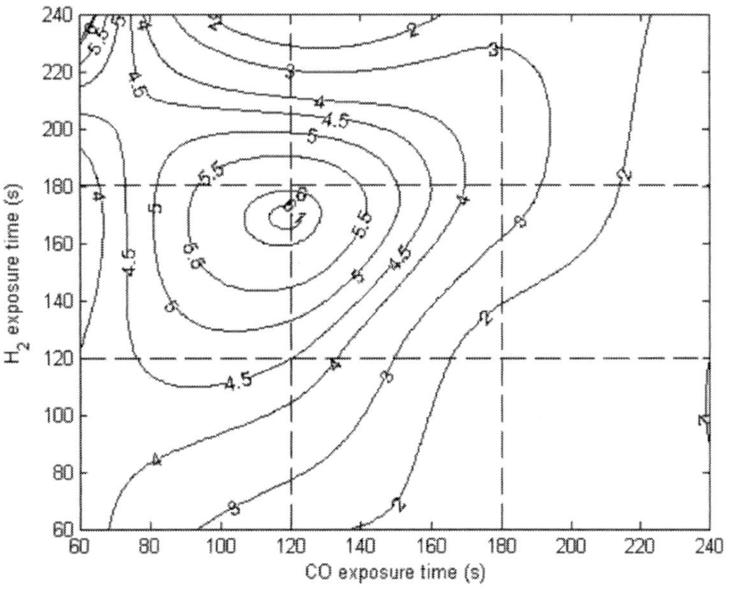

Figure 4: Interpolated Y_{C10+} levels for various CO and H_2 cycled exposure times

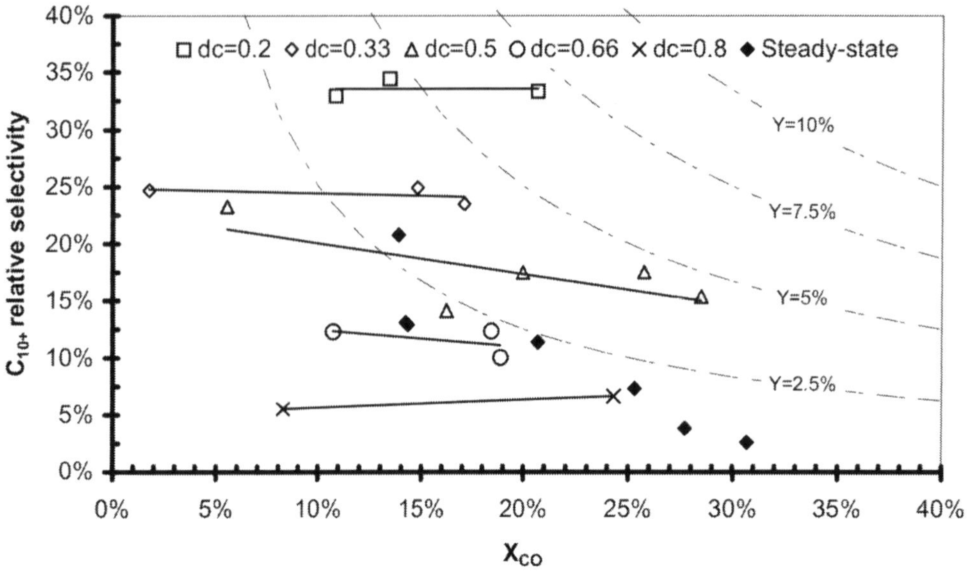

Figure 5: C_{10+} selectivity dependency over X_{CO} for continuous and periodic feed policies and influence of dc – isoyield level curves indicatively shown

428119 Simultaneous Oxidation and 2-Octanol Extraction of Iron from Simulated Ilmenite Hydrochloric Acid Leachate

Abstract:

Abstract

Separation of ferrous iron from titanium (IV) in a simulated ilmenite hydrochloric acid leachate by simultaneous oxidation and 2-Octanol extraction was investigated. The effects of extraction time, oxygen flow rate, O/A phase ratio, temperature, and hydrochloric acid concentration were studied. The results showed that with pure oxygen as oxidant and 100% (v/v) 2-Octanol as extractant, the oxidation of ferrous and extraction of iron were up to 99% and 86%, respectively while no titanium was co-extracted under O/A 1:3, hydrochloric acid concentration 9.35 mol/L, oxygen flow rate 1000 ml/min, reaction temperature 35 °C and reaction time 30 min. With two stages of cross-current operation, the total iron extraction amounted over 99.5% and still no titanium was co-extracted. The composition of extraction complex was $HFeCl_4 \cdot 5$(2-Octonol). 100% of iron stripping was obtained using distilled water as a stripping liquor at an O/A ratio of 1:1 through a single contact. Thus, an almost complete separation of iron (II) and titanium (IV) from the simulated leachate by simultaneous oxidation and 2-Octanol extraction was feasible. The process could by-produce Fe_2O_3 with high purity. The kinetics of both the extraction and stripping of iron (III) were very fast while the oxidation of iron (II) was comparatively slow. The oxidation could be accelerated significantly by increasing hydrochloric acid concentration and reaction temperature. Enhancement of the oxidation through improving the structure of gas-liquid-liquid multiphase reactor was also expected. An improved hydrochloric acid process for pigment production was proposed.

Keywords: solvent extraction; iron; 2-Octanol; oxidation; hydrochloric acid process

[1] R. K. Biswas, M. A. Habib, N. C. Dafedar, 1992, A study on the recovery of titanium from hydrofluoric acid leach solution of ilmenite sand Hydrometallurgy, 28 (1), 119-126.

[2] Wang, X. M., Liang, B., Lü,L., Wu, P., Li, C., Ma, J. Simultaneous oxidation and extraction of iron from simulated ilmenite hydrochloric acid leachate [J]. Hydrometallurgy. 129-130 (2012) 105–110.

[3]Saji, J.K., Rao, T.P., Iyer, C.S.P., Reddy, M.L.P., 1998. Extraction of iron(III) from acidic chloride solutions by Cyanex 923. Hydrometallurgy 49, 289–296.

[4] Saji, J.K., Saji, J., Reddy, M.L.P., Ramamohan, T.R., Rao, T.P., 1999b. Solvent extraction of titanium(IV) from acidic chloride solutions by Cyanex 923. Hydrometallurgy 51(1), 9–18

[5] Thomas, J., Surender, G.D., Reddy, M.L.P., 2002. Liquid–liquid extraction of tetravalent titanium from acidic chloride solutions by trialkylphosphine oxide. Proceedings of International Symposium on Solvent Extraction (ISSE), September, 2002, Bhubaneswar, India, pp. 159–168.

[6]Sato, T., 2003. Liquid–liquid extraction of titanium(IV) from hydrochloric acid solutions by di-(2-ethylhexyl) phosphoric acid. Sigen-to-Suzai 119, 175–181.

[7] Liang, B., Li, C., Zhang, C.G., Zhang, Y.K., 2005. Leaching kinetics of Panzhihua ilmenite in sulfuric acid. Hydrometallurgy 76 (3-4), 173-179.

[8] Li, C., Liang, B., Guo, L.H., 2007. Dissolution of mechanically activated Panzhihua ilmenites in dilute solutions of sulfuric acid. Hydrometallurgy 89 (1-2), 1-10.

[9] Li, C., Liang, B., Wang, H.Y., 2008. Preparation of synthetic rutile by hydrochloric acid leaching of mechanically activated Panzhihua ilmenite. Hydrometallurgy 91 (1-4), 121-129.

429646 Recovery of Natural Gas from Marginal Wells with a Deep Well Reactor

Abstract:

With the recent surge in availability, natural gas (methane) has become an important feedstock for the energy and chemicals market. Yet the low energy density and high compression cost necessitate high flow rates close to market for economic viability. Subsequently, a significant quantity of gas remains in the ground in 'abandoned wells,' once the well has hit the economic threshold. Furthermore, 'remote wells,' cannot be tapped because the cost of liquefaction and transportation are cost prohibitive. Similarly, oil-associated gas is flared in preference to the more desirable oil, wasting energy and opportunity. Technologies with low capital cost and low complexity could be leveraged to monetize low flow rate and low pressure natural gas. Currently, all gas-to-liquid technologies (e.g. Fischer-Tropsch Synthesis) are incredibly capital-intensive and only economically viable at very large scales. A new low-capital paradigm is required to tap the gas available only at low flow rate and pressure. Here we present a mathematical model of one such possibility that we term the 'Deep Well Reactor,' which utilizes pre-existing structures for the biological conversion of methane to liquid fuels and chemicals. One of the major issues is cost-effective mass transfer of the methane to the reactive phase. We show that this technology can effectively overcome these limitations without any extra complexity. Furthermore, we stipulate minimum conversion rates that would be required for a feasible process with biological or inorganic catalysts.

434341 Modeling and Optimization of Upgrading Reaction System in Gas to Liquid Process

Abstract:

Upgrading of heavy hydrocarbons that were produced from Fischer-Tropsch process in GTL complex is one of the important technologies to obtain high quality fuels. In this work, we developed a mathematical modeling of upgrading reaction system, and optimized operating condition to maximize the conversion of heavy hydrocarbons to distillate products. To estimate the product distribution, all feasible reaction mechanisms were identified, and their reaction kinetics were applied to the modeling of reaction system. The study results showed the estimated figures such as product distribution and conversion from the modeling were in good agreement with experimental data. Finally, the reactor operating conditions such as temperature, pressure and space velocity were optimized to maximize the reactor performance.

This work was supported by Korea Institute of Science and Technology and funded by Ministry of Trade, Industry and Energy (Project No. 20142010102790)

This work was supported by a Special Education Program for Offshore Plant by the Ministry of Trade, Industry and Energy Affairs (MOTIE).

435246 Conceptual Design of Distillation Columns for Azeotropic Systems Using a Rigorous Method

Abstract:

Conceptual Design of Distillation Columns for Azeotropic Systems Using a Rigorous Method

Elsa Fernández-Martínez,[a] Ángel Castro-Agüero,[a]

[a] Facultad de Ciencias Básicas Ingeniería y Tecnología, Universidad Autónoma de Tlaxcala, Apizaco, Tlaxcala, México, 90300.

Azeotropic mixtures show non idealities, which significantly complicates the separation of components; but also, the knowledge about the thermodynamics of system can help to obtain the design of distillation column. Azeotropic and extractive distillations are essential, because there are numerous processes by which different mixtures are separated by this kind of distillations. A number of important petrochemicals from of hydrocarbons mixtures can be separated by azeotropic and extractive distillations starting the biomass (Van Dongen and Doherty, 1985).

The mixture in study is Ethanol/Water and Ethylene Glycol is added like entrainer. Liquid solution models (Wilson and NRTL) were used to predict the vapor-liquid equilibrium of binary and ternary systems. The parameters of models were obtained from Kamihama, N, et al (2012). The Ethanol/Water mixture has been studied for many researchers and several entrainers have been used by breaking azeotrope. The homogeneous azeotropic distillation is generated with Ethylen Glycol like entrainer, (Kiva, et al, 2003). Julka, et al studied the best election of entrainers using residue curve maps, due to that entrainer determines the separation sequence.

The main purpose of this paper is calculating minimum reflux ratio for a single column, because this structural variable is important for conceptual design. Subsequently the rest of the other variables like feed location, theoretical stages, among others are determined. The proposal design model is based on the generation of a vector space by the vectorial operations and the solution of systems of linear and nonlinear equations that represent the balances of mass and energy, as well as the equilibrium relationship. This approach yields the structural variables of the conceptual design of distillation column for azeotropic mixtures. Mathematical procedure is based on concepts of Ponchon-Savarit method. Doherty and Malone (2001) show a methodology to obtain conceptual design for azeotropic mixtures.

References

Doherty, M. F. y Malone, M. F. "Conceptual Design of Distillation Systems", Editorial Mc Graw Hill, primer edición, (2001).

Julka, V., Chiplunkar, M. and O'Young, L., "Selecting Entrainers for Azeotropic Distillation", *AIChE CEP*, *Reactions and Separations*, p. 47, 53, (2009).

Kamihama, N., Matsuda, H., Kurihara, K., Tochigi, K and Oba, S., "Isobaric Vapor-Liquid Equilibria for Ethanol+Water+Ethylene Glycol and Its Constituent Three Binary Systems", *Journal of Chem. and Eng. Data*, **57**, p.339-344, (2012).

Kiva, V. N., Hilmen, E. K. and Skogestad, S., "Azeotropic phase Equilibrium Diagrams: a Survey", *Chem. Eng. Sci*, **58**, 1903-1953, (2003).

426884 Design of Reactive Dividing Wall Column Using Difference Cascade Points in Composition Space

Abstract:

Using a geometric design approach, stage calculations of a reactive dividing wall column with the ternary isomolar reaction were carried out under the given operating conditions. Before specifying the operating conditions, the minimum values of them were estimated by assuming the feed-pinch where the liquid molar composition of a feed stage is equal to that of above or below feed stage. For given operating conditions, the required number of stages, liquid/vapor molar composition, and the extent of reaction on each reactive stage in terms of both reaction equilibrium and reaction kinetic constraints were determined. This geometric stage calculation could provide a potential on integrating reaction and separation and evaluating the feasibility of the ternary isomolar reaction in a highly complex system of reactive dividing wall column.

424097 Towards Tailored Monolithic Sponges for Highly Exothermic Catalytic Processes in Chemical Energy Storage

Abstract:

Towards Tailored Monolithic Sponges for Highly Exothermic Catalytic Processes in Chemical Energy Storage

Lars Kiewidt (kiewidt@uni-bremen.de) and Jorg Thšming (thoeming@uni-bremen.de)

Center for Environmental Research and Sustainable Technology (UFT), University of Bremen, Leobener Straμe, 28359 Bremen, Germany

Monolithic ceramic and metallic sponges, also known as open-cell foams, have been generating substantial interest as catalyst support for fixed-bed reactors in current and recent years. In contrast to conventional packed beds composed of pellets or extrudates, they combine low pressure drops and large volumetric surface areas with superior heat and mass transport properties, and are thus well-suited as catalyst support for highly endo- or exothermic processes. Three prominent examples of *exothermic* processes, which are currently discussed as promising options for effective storage of excess renewable energy, are the Sabatier process (Power to Gas, PtG), the Fischer-Tropsch process (Power to Liquid, PtL), and the synthesis of methanol (PtL) from green syngas. The performance of these processes, however, is vitally dependent on the effective removal of heat from the reaction zone to avoid hot spots that induce thermodynamic limitations on conversion and yield, promote undesirable side reactions, and decrease catalyst lifetime by thermal sintering. In addition, thermal runaway has to be prevented for safe operation. Consequently, catalyst supports with enhanced and adjustable heat transport properties are required to enable efficient small-scale units for decentralized Chemical Energy Storage as proposed in our former study [1].

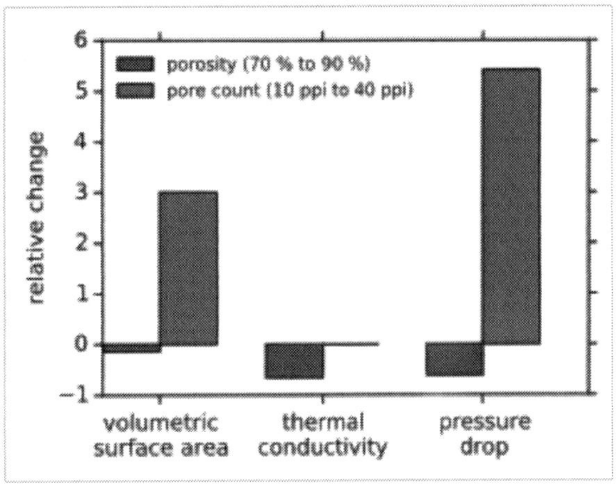

Figure 1: Sensitivity of the volumetric surface area, the effective thermal conductivity, and the pressure drop of monolithic sponges on variations of porosity and pore count.

In this study we demonstrate how the structure of monolithic sponges, represented by their pore count and open porosity, and further their *spatial distribution* can be utilized to tune the heat transport properties locally, and thus intensify heat transport where necessary to avoid hot spots. In order to show the potential of tuning the volumetric surface area, the effective thermal conductivity, and the pressure drop by adjusting the sponge structure, we conducted a sensitivity analysis at typical reaction conditions (300 °C, 10 bar, 4:1 H_2/CO_2-mixture) using recently published correlations [2–4]. Figure 1 shows that the volumetric surface area and the pressure drop of monolithic sponges can effectively be tuned by adjusting the pore size, whereas the effective thermal conductivity can be altered by changing the porosity. Consequently, the heat transport properties of monolithic sponges and the release of heat can be tuned locally by acutely adjusting the structure of the sponge.

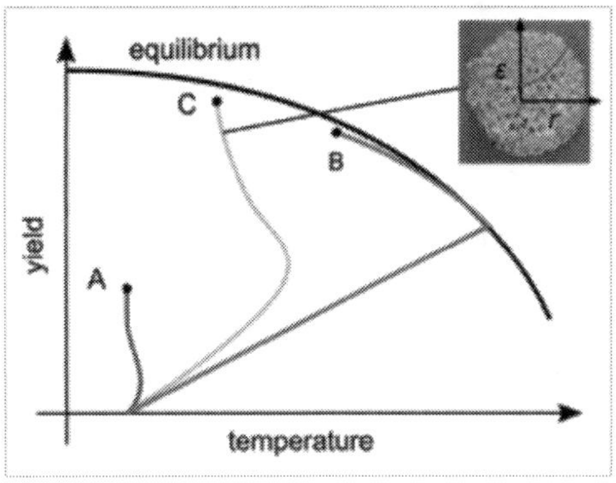

Figure 2: Qualitative temperature profiles for graded (A), uniform (B), and tailored (C) monolithic sponge catalyst supports in the yield-temperature plane.

Therefore, we propose *tailored catalyst supports* based on monolithic sponges with a *graded* structure to balance high heat transport, low pressure drops, and high space-time yields (see Fig. 2). As a first proof of principle we utilized a 2-d fixed-bed reactor model to simulate the methanation of CO_2 over a 20 wt.-% Ni/Al_2O_3 catalyst. Figure 3 shows the temperature distributions in a 2.5 m long and 25 mm in diameter single-tube reactor with external cooling and a 4:1 H_2/CO_2-mixture at 325 °C and 5 bar(a) as feed, for a uniform sponge (Fig. 1a; 40 ppi, 80 % porosity), and a graded sponge with linear radial porosity profile (Fig. 1b; 40 ppi, center: 75 % porosity; wall: 90 % porosity). The axial temperature profiles are qualitatively depicted in Fig. 2 (profiles A and B). The hot spot close to the inlet can effectively be reduced from 625 °C to 360 °C using the graded sponge. Because of the lower temperature, the CH_4-yield drops simultaneously from 80 % to 30 % for the same reactor length. Therefore, we use the developed model to find the optimal porosity profile that leads to the optimal temperature profile (Fig. 2, profile C) [1] and provides high yields, tolerable hot spots, and low pressure drops in order to intensify the catalytic process. Further, we give an outlook on the experimental realization of graded sponges and their application in CO_2-methanation.

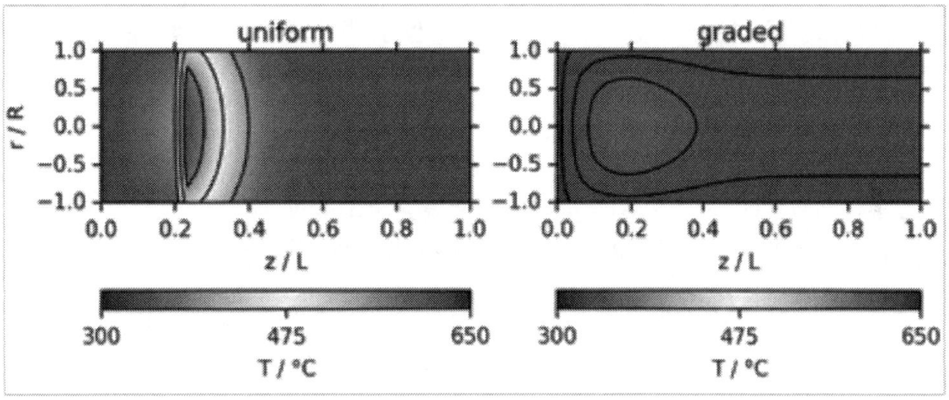

Figure 3: Temperature distribution in a sinlge-tube methanation reactor for (a) a uniform sponge (80 % porosity, 40 ppi), and (b) a graded sponge (40 ppi) with linear porosity profile (axis: 75 % porosity; wall 90 % porosity) at 325 °C inlet temperature and 5 bar(a) (isobaric). The tube is 25 mm in diameter and 2.5 m long.

References

[1] L. Kiewidt, J. Thšming, Predicting optimal temperature profiles in single-stage fixed-bed reactors for CO2-methanation, Chem. Eng. Sci. (2015). doi:10.1016/j.ces.2015.03.068.

[2] A. Inayat, H. Freund, T. Zeiser, W. Schwieger, Determining the specific surface area of ceramic foams: The tetrakaidecahedra model revisited, Chem. Eng. Sci. 66 (2011) 1179–1188. doi:10.1016/j.ces.2010.12.031.

[3] E. Bianchi, T. Heidig, C. Visconti, G. Groppi, H. Freund, E. Tronconi, An appraisal of the heat transfer properties of metallic open-cell foams for strongly exo-/endo-thermic catalytic processes in tubular reactors, Chem. Eng. J. 198-199 (2012) 512–528. doi:10.1016/j.cej.2012.05.045.

[4] E. Bianchi, T. Heidig, C.G. Visconti, G. Groppi, H. Freund, E. Tronconi, Heat transfer properties of metal foam supports for structured catalysts: Wall heat transfer coefficient, Catal. Today. 216 (2013) 121–134. doi:10.1016/j.cattod.2013.06.019.

427150 Numerical Hydrodynamics Study of Gas Solid Vortex Reactor for Process Intensification in Fluidization Technology

Tuesday, November 10, 2015: 8:55 AM
Salon I (Salt Lake Marriott Downtown at City Creek)

Kaustav Niyogi, Maria N. Pantzali, Geraldine J. Heynderickx and Guy B. Marin, Laboratory for Chemical Technology, Ghent University, Ghent, Belgium

The Gas Solid Vortex Reactor (GSVR) is a novel type of reactor, which surpasses major limitations of conventional gravitational Fluidized Bed (FB) reactors by replacing the gravitational field with a *centrifugal field*[1, 2]. In the GSVR, the process gas flows through tangentially inlet slots positioned along the circumferential wall of a static cylindrical chamber. The gas is forced to leave the chamber via the centrally placed exhaust. The solid particles, which are fed into the chamber via a solids inlet, are entrained by the gas and set to rotation. When the centrifugal force exerted on the rotating particles outweighs the drag force exerted by the flowing gas, a rotating solids bed is formed near the circumferential wall of the GSVR (Fig. 1). Since the centrifugal force can exceed gravitational force significantly, *high gas-solid sli*p velocities are developed. Hence, the GSVR can achieve higher heat and mass transfer as compared to conventional FB, leading to *Process Intensification* (PI)[3]. Moreover, the GSVR differs from Rotating Fluidized Bed reactors (RFB), as it enables centrifugal fluidization in a *static* geometry, thus minimizing mechanical abrasion caused by moving parts. Among the several possible applications of the GSVR, drying of biomass[4], effluent SO_2/NO_x adsorption[5], even nuclear rocket fuel propulsion[6] have been investigated in literature.

In the Laboratory for Chemical Technology, an experimental semi-batch GSVR setup has been constructed and tested under various operational conditions[7-9]. However, the need for non-intrusive measurement techniques[8], has limited the information that the experiments alone can provide. This necessitates the use of Computational Fluid Dynamics (CFD) simulation of the hydrodynamics inside the GSVR.

In the present work, the commercial CFD software, FLUENT® 14.0 is used to numerically study the detailed three-dimensional (3D) cold flow hydrodynamics in the GSVR. Transient CFD calculations are performed on $1/9^{th}$ section of the whole geometry to reduce computational cost. The multiphase flow is simulated in an Eulerian-Eulerian framework, where the two phases, air and polymer particles, are treated as interpenetrating continua. Kinetic Theory of Granular Flow (KTGF) modeling is used to close the solid phase equations. As the GSVR can give rise to the formation of dense beds, separate turbulence modeling is considered per phase, to account for four-way coupling of phase momentum equations. Since the main interest is focused on extracting global fluidization behavior inside the GSVR, only time-averaged data are considered.

First, a detailed sensitivity study is carried out to identify the key numerical parameters influencing the simulation results. The simulation data is found to be highly sensitive to the values of the particle-particle restitution coefficient and the particle-wall specularity coefficient. Particle-particle friction modeling is also found to become important at higher solids capacity, for accurate prediction of maximum solids content that the GSVR can hold at a given gas flow rate. Optimal parameter values have been determined by comparing the simulation data with their experimental counterpart[9] over a wide range of operating conditions (gas flow rate, solids capacity) and particle properties (density, size). The validated model is then used to investigate channeling, slugging inside the GSVR. This helps identify the full range of operability of the reactor. Next the degree of fluidization is investigated. The solids are found to exhibit a complex pattern of bed behavior, ranging from almost packed to fluidized regime. The simulation results are used to construct a fluidization regime map for the GSVR (Fig. 2). This fluidization regime map can be applied to compare the GSVR performance with the traditional FB and RFB technologies and assess the extent of PI achieved in the GSVR.

Acknowledgment

The computational work was carried out using the STEVIN Supercomputer Infrastructure at Ghent University, funded by Ghent University, the Flemish Supercomputer Center (VSC), the Hercules Foundation and the Flemish Government – department EWI. The authors acknowledge the financial support from the European Research Council under the European Union's Seventh Framework Programme FP7/2007-2013/ERC grant agreement n° 290793.

References

1. Ekatpure RP, Suryawanshi VU, Heynderickx GJ, de Broqueville A, Marin GB. Experimental investigation of a gas–solid rotating bed reactor with static geometry. Chem. Eng. Process. Process Intensif. 2011;50(1):77-84.

2. De Wilde J, de Broqueville A. Experimental investigation of a rotating fluidized bed in a static geometry. Powder Technol. 2008;183(3):426-435.

3. Ashcraft RW, Heynderickx GJ, Marin GB. Modeling fast biomass pyrolysis in a gas–solid vortex reactor. Chem. Eng. J. 2012;207–208(0):195-208.

4. Volchkov EP, Dvornikov NA, Yadykin AN. Characteristic features of heat and mass transfer in a fluidized bed in a vortex chamber. Heat Transf. Res. 2003;34:486-498.

5. R.W. Ashcraft, J. Kovacevic, G.J. Heynderickx, G.B. Marin. Assessment of a gas–solid vortex reactor for SO2/NOx adsorption from flue gas. Ind. Eng. Chem. Res. 52 (2013) 861–875.

6. Lewellen W.S., Stickler D.B. Two-Phase Vortex Investigation Related to the Colloidal Core Nuclear Reactor. ARL TR 72-0037, Aerospace Research Laboratory, Wright Patterson Air Force Base, OH, USA, (1972).

7. Dutta A, Ekatpure RP, Heynderickx GJ, de Broqueville A, Marin GB. Rotating fluidized bed with a static geometry: Guidelines for design and operating conditions. *Chem.Eng.Sci.* 2010;65(5):1678-1693.

8. Kovacevic JZ, Pantzali MN, Heynderickx GJ, Marin GB. Bed stability and maximum solids capacity in a Gas–Solid Vortex Reactor: Experimental study. *Chem.Eng.Sci.* 2014;106:293-303.

9. Kovacevic JZ, Pantzali MN, Niyogi K, Deen NG, Heynderickx GJ, Marin GB. Solids velocity fields in a cold-flow Gas–Solid Vortex Reactor. *Chem.Eng.Sci.* 2015;123:220-230.

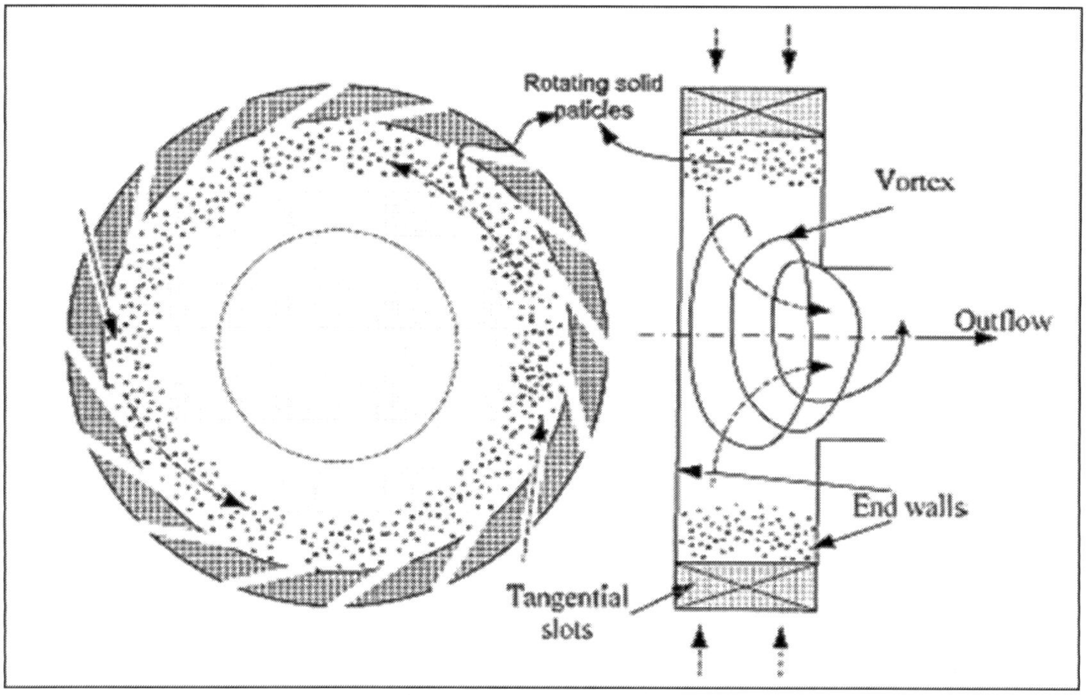

Fig. 1. A schematic representation of the Gas-Solid Vortex Reactor

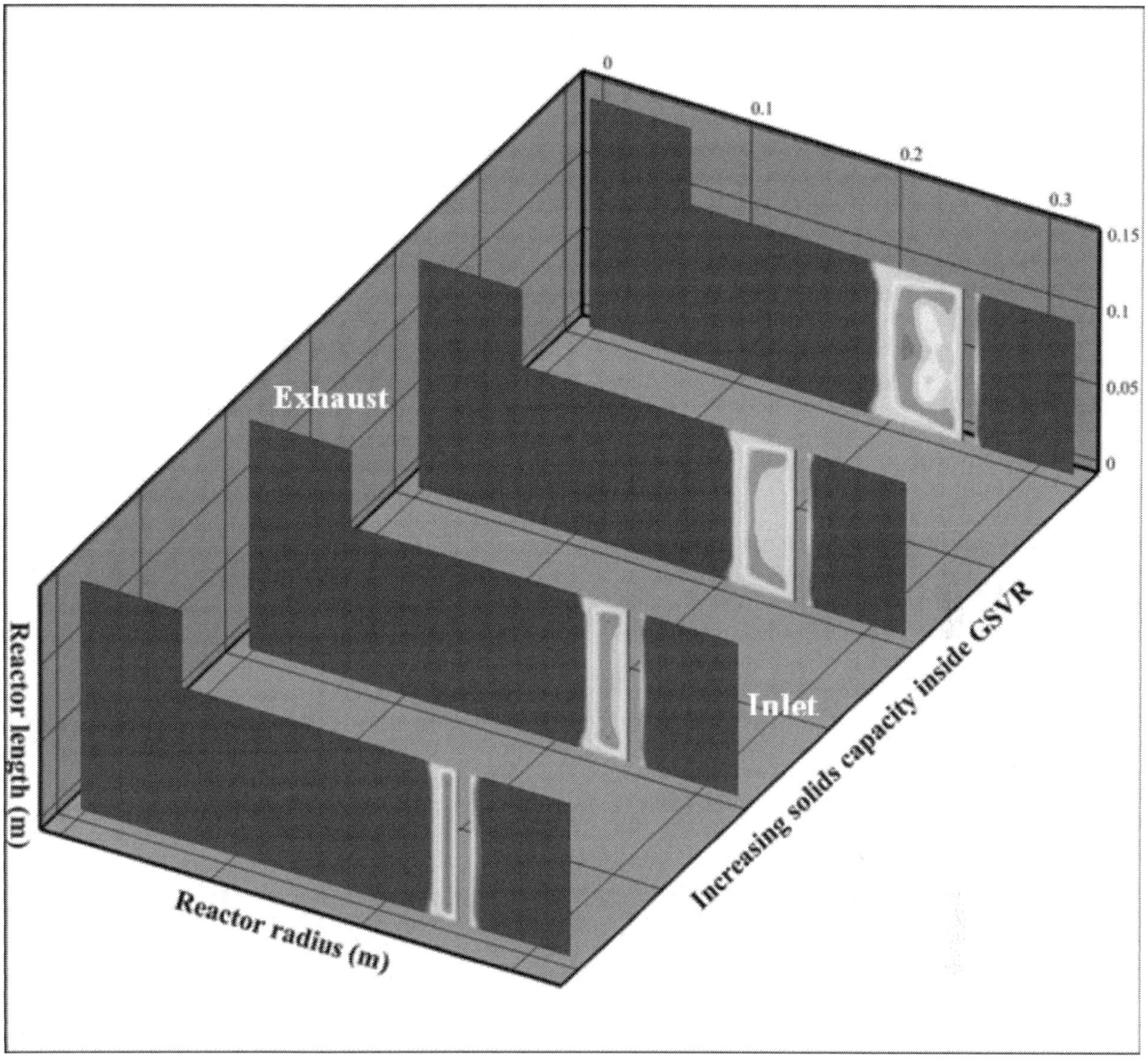

Fig. 2. Field of solids volume fraction in an azimuthal GSVR plane. Fluidization regimes with increasing solids capacity

415830 Twin-Screw Extruders for Solvent Devolatization from Polymers

Abstract:

Wiped Film Evaporators (WFE's) are used extensively in the chemical and polymer industry for bulk as well as trace devolatization of solvents from the final product. The energy from the externally heated walls, along with the shear heat and surface renewal from the internal rotating blades is used to remove the solvents. However, in case of polymer products with high viscosity which require a high operating temperature, the WFE's can result in degradation and/or discoloration of the product due to the high residence time and temperature.

To overcome these disadvantages a twin-screw extruder was used for the trace devolatization of solvent from a high viscosity polymer. The presentation covers the theoretical comparison of WFE's with extruders in terms of residence time, shear rate, surface renewal and vacuum capability. An overview of the pilot testing that shows the effect of key control variables such as feed rate, extruder screw RPM, barrel temperature, and vacuum on vent ports on the response variables such as specific mechanical energy, molecular weight build, color build and residual solvent in the final product is provided. The switch from WFE's to a twin-screw extruder for trace solvent devolatization resulted in higher production rates as well as improved quality in terms of color, residual solvent content, particulate count and overall product consistency. The key learnings could be applied to similar high viscosity polymeric systems.

429126 Model-Based Optimization of Reaction and Process Conditions for the Reactive Absorption of Carbon Dioxide

Abstract:

With the rise of process intensification and innovative reactor technologies traditional reactor design methods, which predefine reactor types based on heuristics and optimize the parameters of the chosen reactor afterwards, possess limited capability especially when dealing with complex reaction systems. In this work, we propose a new method to identify the best reaction and process conditions for multiphase systems with severe mass transfer limitation. The method allows for the design of innovative and tailor-made multiphase reactor concepts including different intensification options, which is exemplified by the reactive absorption of carbon dioxide.

As basis for the optimization, a recently proposed design methodology [1] based on the concept of elementary process functions [2] is applied, and appropriately extended in order to include the rigorous treatment of multiple phases as described in detail below. The idea is to track fluid elements which travel through the process and change their states within, and to continuously provide them with the optimal fluxes according to the states, in order to achieve optimal reaction and process conditions. The optimal fluxes can be calculated by solving a dynamic optimization problem, which is formulated based on the apparatus-independent model equations.

In this contribution, we present a significant extension of the design methodology in order to rigorously consider multiple phases simultaneously, and to deal with the large concentration gradient in the phase. Considering the gas phase as a homogeneous fluid element and the liquid phase as an inhomogeneous fluid element, balance equations are applied to each phase and coupled on the length scale by the interfacial transfer, with the diffusion term accounting for the limiting mass transfer in the liquid phase and Henry's law accounting for the limiting mass transfer between two phases. Reactor characteristics such as liquid hold-up and specific gas-liquid interfacial area are initially assigned values in a typical range, while the influence of these quantities is investigated in the further optimization by sensitivity analysis.

The optimal inlet conditions and the optimal temperature profile are obtained to maximize efficiency. Sensitivity analysis with simultaneous optimization reveals maximum efficiency improvement resulting from better reactor characteristics. These results serve as the basis of the intensified reaction concept and detailed reactor design.

This extension of the optimization methodology is not limited to the specific example presented here, but is generally applicable also to other multiphase reaction systems. In this regard, it presents an important contribution in the field of process intensification in multiphase reactor design.

References

[1] A. Peschel, H. Freund, K. Sundmacher, Methodology for the design of optimal chemical reactors based on the concept of elementary process functions, Ind. Eng. Chem. Res., 49 (2010) 10535-10548.

[2] H. Freund, K. Sundmacher, Towards a methodology for the systematic analysis and design of efficient chemical processes. Part 1. From unit operations to elementary process functions, Chem. Eng. Process., 47 (2008) 2051-2060.

425759 Transient Natural Gas Liquefaction Process Comparison- Dynamic Heat Exchanger Under Transient Changes in Flow

Abstract:

In this study two methods for transient modeling of natural gas liquefaction model have been done using Aspen HYSYS Dynamics. Traditional heat exchanger design (THED) and dynamic heat exchanger design (DHED) are investigated. In THED, Single mixed refrigerant (SMR) process has been simulated and optimized with Aspen HYSYS Dynamics. For DHED, new types of heat exchanger design and optimization method; Model predicted control theory (MPCT) have been presented and transient responses have been compared with traditional design. SMR model has been modified and two methanol streams are added to system to regulate flow rate and temperature profiles during transitions. Operating conditions, MR compositions and heat exchanger's sizing are the same for both designs. Temperatures, pressures and other operating conditions are perfectly controlled. K-values and U values technique are used to make flowrates and temperature of heat exchanger's streams steady. LNG and NG flowrates variation have been shown in time for all cases. Natural gas (NG) consumption and Liquefied natural gas (LNG) production have been compared to show the optimality of new design. Temperature profiles are compared and Last but not least, the efficiency graph for both methods are given and compared to show the adjustability of heat exchanger with NG flow rate change in transient mode.

426469 Multi-Objective Optimization for High Heat Removal and Productivity in Fischer-Tropsch Synthesis Microchannel Reactor Using Discrete Catalyst Dilution Zone

Abstract:

For the fabrication of effective, robust, and safe Fischer-Tropsch Synthesis (FTS) microchannel reactor, it is required to meet both high heat removal and C_5^+ productivity. In this research, axial symmetric thermal non-equilibrium porous media reactor model was developed in computational fluid dynamics (CFD) and C++ user defined function (UDF) for implanting given reaction kinetics. To manipulate the local temperature profile, which represents high heat removal, catalyst was diluted, i.e., mixed with various amount of inert bead to pack the reactor channel with discrete zone. Meanwhile, the C_5^+ productivity was simultaneously optimized to satisfy the economic feasibility, as well as the safety. For optimizing several objectives containing discontinuities and inconsistent function simultaneously, machine learning regression of CFD data set was performed to get differentiable function set. Out results suggested the optimized multiple catalyst dilution zones that decreased maximum temperature less then limited temperature (~5°C) and maximized the C_5^+ yield set as pareto efficient curve.

An Introduction to:

犹他-青海联盟 Utah-Qinghai EcoPartnership

US-China Sub-National EcoPartnerships

犹他-青海联盟 Utah-Qinghai EcoPartnership

MISSION

TO DEVELOP RELATIONSHIPS OF TRUST THROUGH JOINT COOPERATION IN AREAS OF COMMON INTEREST

犹他-青海联盟 Utah-Qinghai EcoPartnership

MISSION

JOINT RESEARCH, PROTECTION AND TRANSFER OF INTELLECTUAL PROPERTY

犹他-青海联盟 Utah-Qinghai EcoPartnership

MISSION

IMPROVING HUMAN HEALTH AND WELFARE, WITH A FOCUS ON THE ENVIRONMENT, FOOD PRODUCTION AND CLEAN ENERGY

犹他-青海联盟 Utah-Qinghai EcoPartnership

MISSION

TO FACILITATE THE DEVELOPMENT AND IMPLEMENTATION OF MARKETABLE TECHNOLOGIES FOR ECONOMIC DEVELOPMENT

犹他-青海联盟 Utah-Qinghai EcoPartnership

BACKGROUND

Utah-Qinghai EcoPartnership - 2011

ECOPARTNERSHIP signing ceremony in the Treaty Room at the U.S. Department of State with then secretary of State Hillary Clinton. U.S.-China sub-national partnerships provide cooperation in both the private and public sectors on environmentally important issues

犹他-青海联盟 Utah-Qinghai EcoPartnership

BACKGROUND

Utah-Qinghai EcoPartnership - 2011

Inspired by a common vision, then Governor of Qinghai Province, Luo Huining, and Utah Governor Gary Herbert, developed a partnership aimed at joint innovation for environmental sustainability.

Utah-Qinghai EcoPartnership - Overview

Regular Conferences (annual or bi-annual)

Discuss common needs, make technology presentations and progress reports, plan joint initiatives

(Attendees: Government groups, universities, technology providers, potential funders, UQEP)

Technology Implementation and Commercialization

IP management, investment, new or existing companies, local economic development

UQEP

Not for profit organization sponsored by the U.S. State Department and the Chinese National development and Reform Commission

Technology Identification and Joint Research

Universities, start-ups, private firms

Initiate Demonstration Projects

Determine objectives, set time-table, secure funding, publish findings

---

犹他-青海联盟 Utah-Qinghai EcoPartnership

ACCOMPLISHMENTS

Renewable Energy Project and Bio-waste Remediation

This anaerobic digestion facility represents the successful transfer of technology to efficiently process bio-waste pollutants, converting them into renewable energy in the form of natural gas while reducing greenhouse gas emissions and waterway

犹他-青海联盟 Utah-Qinghai EcoPartnership

ACCOMPLISHMENTS

Soil Remediation Technology Demonstration Project

Technology from the University of Utah was used to clean organic chemicals from the soil of a contaminated industrial site. The soil remediation process called Heightened Ozonation Treatment (HOT), was used in a demonstration project to clean up a vacated industrial site on the shores Wuxi City's Lake Taihu.

犹他-青海联盟 Utah-Qinghai EcoPartnership

ACCOMPLISHMENTS

Heavy Metals Detection Technology

New Technology Research from the University of Utah is behind new mercury detection technology which will help to manage heavy metals in Qinghai and elsewhere.

犹他-青海联盟 Utah-Qinghai EcoPartnership

ACCOMPLISHMENTS

Sand Filtration Process to Clean Drinking W

Completed the Purdue-Utah village water project in two small villages located nearly two hours from the provincial capital, Xining. Purdue University scientists, supported by the Purdue-led U.S.-China Eco-Partnership for Environmental Sustainability implemented slow sand filter technology.

犹他-青海联盟 Utah-Qinghai EcoPartnership

ACCOMPLISHMENTS

Utah-Qinghai-Purdue Joint Water Resource Lab (in planning)

Joint effort between Purdue University and Qinghai Normal University led by the Utah-Qinghai EcoPartnership for ongoing efforts to improve water availability and quality for villagers who rely on drinking water directly from rainwater collection.

犹他-青海联盟 Utah-Qinghai EcoPartnership

ACCOMPLISHMENTS

Utah-Qinghai Joint High Altitude Medical Lab

Conceived in cooperation with the University of Utah and Qinghai University in 2013, completed in 2015. The 65,000 square foot, $5M facility has been sanctioned by the Chinese Ministry of Science and Technology. The lab will contribute to research and cooperation on high altitude medicine between the two

犹他-青海联盟 Utah-Qinghai EcoPartnership

NEXT STEPS

CONTINUE THE UQEP'S MISSION AS THE

UTAH SUSTAINABILITY COMMISSION

- PROMOTE COOPERATION AND TRUST WITHIN GOVERNMENT AND BUSINESS TO ADDRESSES COMMON INTERESTS AND ENVIRONMENTAL ISSUES
- SPONSOR INTERNATIONAL EXCHANGES AND CONFERENCES
- PROMOTE TECHNOLOGY DEVELOPMENT, PROTECTION, APPLICATION AND COMMERCIALIZATION
- SPONSOR TECHNOLOGY DEMONSTRATIONS PROJECTS

416159 Patent Protection Across Borders: Tips for Obtaining International IP

Abstract:

Patent protection across borders requires special attention from inventors and companies and special efforts from legal counsel. Many businesses have lost significant opportunities by failure to properly consider the challenges of international IP protection. We will discuss best practices and guidelines that can help you and your employer be more successful and cost effective in protecting your business in multiple nations.

421131 Innovation with Asian Teams: Overcoming Stereotypes and Cultural Gaps

Abstract:

Innovation with Asian teams can be extremely successful, but often poses problems for Western companies that lack an understanding of the cultural and language barriers that can get in the way. Managing innovation across borders with Asian partners can be most effective when Western managers or team members have been properly prepared. Here we share experiences from our innovation work with Asian teams with an emphasis on recommendations for effective innovation in China and India.

421158 International Innovation, Licensing, and R&D: Panel Discussion

Abstract:

In this panel discussion, speakers will share their extensive international, IP, and innovation experience to provide guidance on how to best manage international opportunities in business, R&D, and IP. Topics covered include how to protect trade secrets and IP (including licensing), how to deal with regulatory burdens, how to identify and overcome cultural barriers, and how recognize when there is trouble hindering innovation.

419257 Design and Control of Processes Producing Derivatives of Furfural: Furan, Furfuryl Alcohol and 2-Methyfuran

Abstract:

Furfural is an important platform for second-generation, lignocellulosic biomass utilization, and about 60-70 percent of furfural produced worldwide is converted to furfuryl alcohol (Hoydonckx et al., 2000; Zeitsch, 2000). In this work, a process for the sustainable simultaneous manufacture of furfural alcohol (FAL) and 2-methylfuran (2-MF) is designed and optimized. Energy consumption is reduced by employing a feed-effluent heat exchanger and employing a 3-phase distillate receiver at the top of the distillation column and refluxing only the organic phase. Control studies suggest that a simple decentralized control structure without controller parameter scheduling can reject disturbances including changes in the production rate and feed composition and can also transition smoothly between product distributions.

Furan is also one of the major products produced from furfural, and it is an important starting material in chemical synthesis. Therefore in this work the design of a process to produce furan from furfural is also considered. The process design is based on one presented by Zeitsch (2000). The fresh furfural feed stream enters an absorption tower to recover furan vapor from the carbon monoxide product stream. The liquid out stream from the absorption tower is fed to the reboiler of a reactive distillation column. The reaction occurs only in the reboiler. At the top of the column, a condenser is cooled to minus 30 degree Celsius to recover most of the furan as a liquid. The liquid product stream from the reactive distillation column contains a significant amount of carbon monoxide. It enters a short stripping column in order to separate furan and carbon monoxide. As the temperature is increased, the solubility decreases, so furan nearly free of carbon monoxide can be collected from the sump. The vapor streams from the top of the reactive column and the top of the stripping column are combined and fed to the absorption tower in order to remove furan from the carbon monoxide. For this process also, key design variables are identified and adjusted to determine the optimal design. Control studies also suggest that this process can also be effectively controlled with a decentralized PID control structure.

References

1. Hoydonckx, H. E.; Van Rhijn, W. M.; Van Rhijn, W.; De Vos, D. E.; Jacobs, P. A., Furfural and Derivatives. In *Ullmann's Encyclopedia of Industrial Chemistry*, Wiley-VCH Verlag GmbH & Co. KGaA: 2000.
2. Zeitsch, K. J., The Chemistry and Technology of Furfural and its Many By-Products; *Elsevier*: London, **281**; pp156-158, 2000

424277 Efficient Hydrogen Power Plant Design for Continuous Power Supply from Renewable Hydrogen

Abstract:

Development and implementation of renewable energy conversion processes are essential to reduce the dependence on limited fossil fuel reserves and get atmospheric greenhouse gas (GHG) under control emissions. Electricity sector is the largest source of primary energy consumption with the highest growth rate and is currently responsible for greater than 40% of the global CO_2 emissions. Thus, the increase in renewable power production capacity is especially important. Yet intermittencies and geographical limitations for renewable energy collection need to be overcome by developing efficient renewable energy conversion processes and synergistically integrated energy storage options.

Among the energy storage options, storing energy in chemical bonds through synthesis of chemicals prompted interest for large scale applications due to its competitive storage efficiencies, convenience of usage, and high energy densities [1]. Hydrogen has been proposed to be the enabling energy career that will be a key player in a sustainable economy. Hydrogen is the major carbon-free energy carrier and a key chemical that finds primary use in production and upgrading of fuels and chemicals. In 2012, US consumed more than 1.07 EJ energy equivalent hydrogen [2]. Despite the fact that current hydrogen production is predominantly from natural gas reforming, advancements in renewable hydrogen production techniques such as electrolytic or thermal water splitting, will eventually replace the fossil fuel based hydrogen production capacity.

Regardless of the hydrogen production method, efficient electricity generation from hydrogen is critical. Here, we introduce a novel turbine based hydrogen power cycle, which can achieve ~70% hydrogen-to-electricity efficiency at high operating temperatures. In addition to the very high hydrogen-to-electricity efficiency, when integrated with concentrated solar power plants [3], the process ensures the continuous operation of power generation equipment (e.g. turbines, condensers). The proposed hydrogen power cycle minimizes exergetic losses and increases production efficiency. The novel features of the hydrogen power cycle and case studies for integration with solar thermal hydrogen production and natural gas reforming are presented.

[1] Al-musleh EI, Mallapragada DS, Agrawal R. Continuous power supply from a baseload renewable power plant. Applied Energy, 122, 83-93, 2014.

[2] Joseck F. DOE Hydrogen and Fuel Cells Program Record. ed Satyapal S (U.S. Department of Energy), 2012.

[3] Gençer E, Tawarmalani M, Agrawal R, Integrated solar thermal hydrogen and power coproduction process for continuous power supply and production of chemicals, Computer Aided Chemical Engineering, 37, 2291-2296, 2015.

426810 Self-Heat Recuperation Technology for Sustainable Process Design

Abstract:

The reduction of carbon dioxide (CO2) emission has become a major target in efforts to suppress global warming. The combustion of fossil fuel for heating produces a large amount of CO2. Thus, the reduction in energy consumption is one of the most concerned issues. So far, to reduce the energy consumption, heat recovery technology, which exchanges heat between the hot and cold streams in a process, has been applied to thermal processes.

Recently, authors have developed a self-heat recuperation technology based on exergy recuperation to reduce energy consumption of chemical processes. This technology consists of the followings; (i) a process unit is divided into functions to balance the heating and cooling loads; (ii) the cooling load is recuperated by compressors and exchanged with the heating load to recirculate the self heat in the process. Thus, applying the self-heat recuperation technology to chemical processes, not only the latent heat but also the sensible heat of the process stream can be circulated into the processes without any heat addition. As a result, the energy consumption and exergy destruction of a process can be greatly reduced in the steady state.

In this research, authors summarize the design of chemical processes based on the self-heat recuperation technology, and introduce several applications and energy saving potentials by using the self-heat recuperation technology.

429696 Multi-Objective Decision Making Under Uncertainty for Process Sustainability Enhancement

Abstract:

Process sustainability enhancement is a focal point in industrial sustainability. In chemical process systems, energy and material efficiency, waste generation, process safety, heath impact, and certainty product quality, are among key areas when assessing process sustainability. There are a number of sustainability metric systems adopted by the chemical and allied industries, such as those developed by AIChE and IChemE. Sustainability performance assessed using the selected sets of metrics should provide valuable information for decision makers. However, such information could be frequently difficult to use in decision making for two reasons. First, the assessment results are grouped in three categories reflecting economic, environmental and social sustainability performance, and in each category, there could be much information derived using many specific sustainability indicators. There is a lack of effective methods for effective use of multi-dimensional assessment results. Second, it is very likely that the assessment results contain much uncertain information, mainly because the data used for assessment are incomplete, imprecise, and uncertain. In addition to the challenges in assessment data use, decision makers may consider a number of scenarios based on corporation's different goals in different development stages. Apparently, this is a very challenging multi-objective decision-making problem under uncertainly.

In this paper, we introduce a sustainability performance improvement methodology by resorting to optimization and uncertainty theories. In this methodology, the sustainability assessment results are all expressed by intervals, if the data used for assessment are uncertain. We then formulate sustainability performance improvement task as a number of optimization problems with different objectives over a period of development phases, such as those for minimum investment, maximum sustainability performance improvement, maximum investment efficiency, and minimum time. A variety of economic, environmental, and social development expectations and technical feasibilities on process, product and technologies are formulated as constraints associated with the optimization objectives. In solution derivation, we use a genetic algorithm technique, while the data uncertainty problem is dealt with by a Monte Carlo simulation technique. To demonstrate methodological applicability, we present a case study on biodiesel manufacturing, where sustainability performance improvement objectives are comprehensively studied and the solutions derived for different development goals are analyzed. It is shown that the introduced methodology could be a useful decision-support tool to assist industrial decision makers in analyzing system's sustainability and identifying promising decisions for future development.

430537 Sustainability Indicators in Computer-Aided Process Evaluation and Design

Abstract:

Sustainability indicators provide an approach to evaluating process sustainability that integrates with computer-aided process methods. As processes are developed, their design determines input and output flows responsible for resource use and releases (i.e., a process life cycle inventory). These flows are the bases for process indicators, which can be evaluated with the U.S. Environmental Protection Agency's GREENSCOPE methodology and tool. GREENSCOPE indicators, placed in context with best-target and worst case limits, can be used in the evaluation and design of chemical processes, representing nearly 140 indicators in four basis areas: environment, economics, energy, and (mass) efficiency. Computer-aided process methods are one avenue for using these indicators. In this work, GREENSCOPE indicators for the oxidation of toluene are calculated, and an optimization is performed across a range of conversions. Intermediate conversions realize the maximum total utility, as lower conversions increase toluene emissions and higher conversions increase byproduct formation. Tradeoffs are resolved using marginal rates of substitution, allowing for the calculation of total utility in terms of only GREENSCOPE indicator scores, best-target / worst-case ranges for the indicators, and marginal rates of substitution. With the indicator ranges being constant and marginal rates of substitution set by user choices, the results depend on indicator scores, providing an intuitive method for connecting process attributes to the optimized results.

433502 Designing Synergies Between Technological and Ecological Systems for Chemical Manufacturing

Abstract:

Design of new industrial processes has typically relied on identifying an optimal design for a system from a set of alternative pathways. Design objectives for environmentally friendly processes usually include economic goals such as maximizing profit and net present value while simultaneously reducing impact from processes by choosing raw materials that are least toxic and identifying pathways that result in minimum impact. A more holistic and system wide approach towards design considers systems at the supply chain and life cycle scale and by combining principles of life cycle assessment and multi-objective optimization results in designs that are optimal from the cradle to grave of products [1] . Industries are also working towards reducing the impact of their current activities by improving process efficiency or by developing new designs from more 'sustainable' materials. However, efforts to design new processes or enhancing system efficiency for existing processes focus primarily on the technological sphere.

Establishment of closed loop production systems with energy- and mass- integration and encouragement of cyclic material flow are more advanced steps in the direction of sustainable systems development resulting in a symbiotic system [2]. Yet, design and development are still confined only to the technological sphere while ignoring larger aspects of interactions with systems in the ecological sphere.

On the other hand, pressure from governmental policies and society has resulted in advances in the field of ecological engineering. Ecologists have focused on the reclamation of wastelands and restoration of ecological systems to address the issue of ecological degradation. In parallel to this, industries are realizing the importance of ecosystem services in supporting human activities and business decisions are slowly incorporating the value of natural capital into their planning. Recent study by Kroeger et.al 2014 [3] and Nowak et.al 2006 [4] have highlighted the economic and environmental benefits of reforestation as an air quality compliance strategy. Several companies have implemented wetlands on manufacturing sites to meet water regulatory issues while saving costs compared to conventional water treatment systems. Such systems may also be environmentally friendlier and provide various services to society such as flood regulation and carbon sequestration [5]. While such efforts help in moving towards the goal of sustainable development, they treat ecological systems as end-of-pipe solutions to treating pollutants. Most of these advances are discrete, lacking integration with systematic design procedures resulting in suboptimal solutions with very little effort to bridge the gap between design efforts in technological and ecological spheres. Besides, these methods do not account for the fact that currently engineering and human activities demand significantly more goods and services from ecological systems than what can be supplied sustainably.

This presentation introduces a framework for designing engineering processes alongside designing the landscape around industrial facilities. The resulting network of technological and ecological systems establishes synergies between both spheres since the ecosystems provide services demanded by technological systems, while technological systems provide nutrients to ecosystems. Ideally, such a design would result in situations where operation of industrial system is within the ecological carrying capacity and the ecological systems can generate enough ecosystem services to support plant operation.

This approach towards design relies on integration of models from various disciplines. The TES system comprises of process models of manufacturing systems that are integrated with pollutant transport models from point sources and of trees that take up pollutants such as particulate matter, nitrogen oxides and sulfur oxides. Processes are represented by detailed engineering models that capture the non-linear behavior of production systems while pollutant transport is captured by means of Gaussian plume models for a selected region. Gaussian models predict the maximum diffusivity of particles from point sources and are used to identify the most optimal location for restoring land around industrial facilities. Trees and forest ecosystems have the ability to take up pollutants from the atmosphere, significantly reducing atmospheric pollutant concentration and can store carbon in their biomass for decades. Besides, they also play a vital role in increasing groundwater filtration capacity in addition to reducing energy usage in buildings. Pollutant uptake by trees is modeled as a function of air pollutant concentration and deposition velocity of particles on trees species which varies spatially with changes in meteorological conditions. A mixed forest ecosystem model having multiple trees species is setup along with wetland models that are analogous to a simple CSTR systems. The index of sustainability for the integrated system is based on the TES metrics developed by Bakshi et.al 2015 [6] calculated as the overshoot in demand for ecosystem services over the supply, per unit demand of ecosystem service. The entire network is setup as a nonlinear program and the model is optimized to maximize the net present value and the index of sustainability for the system. To bring in time dynamics into the framework and to account for the development of ecosystem services through the operating life of a facility, each period of interest is modeled as discrete time periods accounting for a change in key parameters that change with time. Thus the framework considers designing a forest ecosystem and wetland ecological system analogous to unit operations along with process models so as to increase the overall supply of ecosystems services to offset most of the demand created by processes.

This TES design is applied to a biofuel production plant with an on-site utility generation systems in mid-western US. A biodiesel production plant in Ohio with soybean as its main feedstock with a coal combined heat and power plant to meet the utility demand is considered. Most of the land around this facility is almost barren and plume models identify a range of diffusivity distances for particles from the stack. A horizontal subsurface flow wetland is designed to treat the wastewater from the production plant to reusable standards while a mixed forest with the most common tree species in Ohio is designed to take up most of the pollutants generated by the process.

Preliminary results indicate that the TES design is more cost effective compared to conventional systems in that, to be able to reach the goal of zero emissions and pollution neutrality, ecological systems are more cost effective options than conventional pollutant control equipment. In addition, it was observed that within a period of less than 15 years of establishment of the ecological systems, the forest ecosystems have the ability to completely offset the demand for criteria air pollutants with a positive return on investment and shorter payback time compared to conventional systems design.

Thus the TES approach towards integrated design supports the development and restoration of local ecological systems to ensure continuous provisioning of ecosystem services that support functioning of human activities while simultaneously reducing the impact created by industrial processes resulting in a more sustainable techno-ecological system that is also more economical.

References

[1] Azapagic, A., & Clift, R. (1999). The application of life cycle assessment to process optimisation. Computers & Chemical Engineering, 23(10), 1509–1526. doi:10.1016/S0098-1354(99)00308-7

[2] Jacobsen, N. B. (2006). Industrial symbiosis in Kalundborg, Denmark: a quantitative assessment of economic and environmental aspects. Journal of industrial ecology, 10(1 2), 239-255.

[3] Kroeger, T., Escobedo, F. J., Hernandez, J. L., Varela, S., Delphin, S., Fisher, J. R., & Waldron, J. (2014). Reforestation as a novel abatement and compliance measure for ground-level ozone. Proceedings of the National Academy of Sciences, 111(40), E4204-E4213.

[4] Nowak, D. J., Crane, D. E., & Stevens, J. C. (2006). Air pollution removal by urban trees and shrubs in the United States. Urban forestry & urban greening, 4(3), 115-123.

[5] DiMuro, J. L., Guertin, F. M., Helling, R. K., Perkins, J. L., & Romer, S. (2014). A financial and environmental analysis of constructed wetlands for industrial wastewater treatment. Journal of Industrial Ecology, 18(5), 631-640.

[6] Bakshi, B. R., Ziv, G., & Lepech, M. D. (2015). Techno-Ecological Synergy: A Framework for Sustainable Engineering. Environmental science & technology.

414784 Strategies for H2 Production By Steam Reforming of Ethanol with Pressure Swing Adsorptive Reactor

Abstract:

Strategies for H_2 Production by Ethanol Steam Reforming with Pressure Swing Adsorptive Reactor

Yi-Jiang Wu[1], Ping Li[1], Jian-Guo Yu[1] and Alirio E. Rodrigues[2]*

1 School of Chemical Engineering, East China University of Science and Technology, Shanghai 200237, China

2 Laboratory of Separation and Reaction Engineering, Associated Laboratory LSRE/LCM, Department of Chemical Engineering, Faculty of Engineering, University of Porto, Rua Dr. Roberto Frias s/n, Porto 4200-465, Portugal

Tel. 0086-021-64252171 Email: wuyijiang@ecust.edu.cn

Ethanol, which can be produced from renewable biomass resources, is an ideal feedstock for H_2 production, and the overall reaction of ethanol steam reforming can be described as:

$CH_3CH_2OH(g)+3H_2O(g)\leftrightarrow 6H_2(g)+2CO_2(g)$ (ΔH^0_{298K} = +173.3 kJ·mol^{-1}) (1)

However, the product stream always contains a large amount of CO_2, and other undesired products from side reactions. By using an adsorptive reactor, where steam reforming reaction with *in-situ* CO_2 adsorption taking place, the thermodynamic equilibrium can be shifted towards the product side, which is known as sorption-enhanced reaction process (SERP) (Hufton et al., 1999). Most researches have focused on the reaction stage where SERP was performed with different catalysts and adsorbents, while the cyclic operation process for H_2 production from SERP with ethanol as feedstock is yet to be developed and improved.

In this work, a two-dimensional reactor model developed in our previous work (Wu et al., 2014a) has been employed, since the temperature gradient in radial direction is found able to affect the H_2 production performance comparing with the one-dimensional reactor model (Wu et al., 2014b) where the effect of radial temperature difference is ignored. A model considering multi-compound and overall mass balance, Ergun relation for pressure drop, energy balance for the bed-volume element, and nonlinear adsorption equilibrium isotherm coupled with reactions to describe coupled mass, momentum and heat transport phenomena within the adsorptive reactor. Numerical solution of model equations for the cyclic process was obtained by orthogonal collocation with finite elements method. Besides, Ni-based hydrotalcite (Wu et al., 2013b) has been used as the reforming catalyst and the high temperature CO_2 adsorbent employed is the K-promoted hydrotalcite (Wu et al., 2013a), as shown in Fig.1.

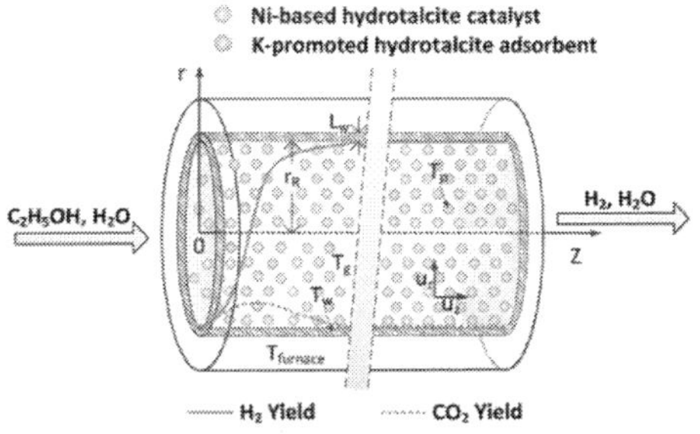

Figure 1 2D reactor model used in this work

The effect of different operating strategies on the performance of cyclic pressure swing sorption-enhanced ethanol steam reforming for high-purity H_2 production has been investigated. The feasibility and effectiveness of conventional pressure swing SERP operating procedure and the use of reactive regeneration (with 10% of H_2 in the feed during the regeneration step) instead of direct steam purge developed by (Xiu et al., 2002) have been investigated. A schematic diagram is illustrated in the following Figure 2.

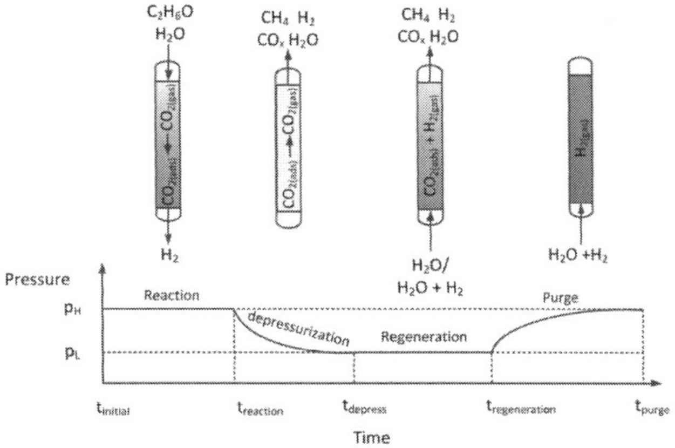

Figure 2 Schematic diagram of the cyclic operation for sorption-enhanced ethanol steam reforming

The simulation of a cyclic pressure swing process has been constructed accordingly with the following steps:

1 Reaction (co-currently to feed). Sorption-enhanced reaction process at p_H;

1 Depressurization (counter-currently to feed). Pressure of the column is reduced to atmospheric pressure (p_L);

1 Regeneration (counter-currently to feed). Regenerating the CO_2 sorbent by steam (y_{H2O} = 100%) or steam with hydrogen (y_{H2O} = 90% and y_{H2} = 10%) at p_L;

1 Purge (counter-currently to feed). Purging the column with H_2 and steam gas-mixture (y_{H2} = y_{H2O} = 50%) with a pressure increase to p_H before the next cycle.

Finally, the effects of operating conditions (reaction temperatures, pressures, length of each step, feeding flow rate as well as the used of reactive regeneration) on the hydrogen purity, productivity and energy efficiency have been investigated by numerical simulation. High purity H_2 product (> 99 mol%, dry basis) with traces of CO content (< 30 ppm) can be produced directly from the pressure swing adsorptive reactor.

References

Hufton, J.R., Mayorga, S., Sircar, S., 1999. Sorption-enhanced reaction process for hydrogen production. AIChE J. 45, 248-256.

Rohland, B., Plzak, V., 1999. The PEMFC-integrated CO oxidation ºª a novel method of simplifying the fuel cell plant. J. Power Sources 84, 183-186.

Wu, Y.-J., Li, P., Yu, J.-G., Cunha, A.F., Rodrigues, A.E., 2013a. K-Promoted Hydrotalcites for CO2 Capture in Sorption Enhanced Reactions. Chem. Eng. Technol. 36, 567¨C574.

Wu, Y.-J., Li, P., Yu, J.-G., Cunha, A.F., Rodrigues, A.E., 2013b. Sorption-enhanced steam reforming of ethanol on NiMgAl multifunctional materials: experimental and numerical investigation. Chem. Eng. J. 231, 36-48.

Wu, Y.-J., Li, P., Yu, J.-G., Cunha, A.F., Rodrigues, A.E., 2014a. Sorption-enhanced steam reforming of ethanol For Continuous High-Purity Hydrogen Production: 2D Adsorptive Reactor Dynamics and Process Design. Chem. Eng. Sci. 118, 83-93.

Wu, Y.-J., Li, P., Yu, J., Cunha, A.F., Rodrigues, A.E., 2014b. High-Purity Hydrogen Production by Sorption-Enhanced Steam Reforming of Ethanol: A Cyclic Operation Simulation Study. Ind. Eng. Chem. Res. 53, 8515¨C8527.

Xiu, G.-h., Li, P., E. Rodrigues, A., 2002. Sorption-enhanced reaction process with reactive regeneration. Chem. Eng. Sci. 57, 3893-3908.

Esterification and Transesterification of Propylene Glycol Methyl Ether Using Heterogeneous Catalyst in Simulated Moving Bed Reactor

Jungmin Oh[1], Shan Tie[1], Gaurav Agrawal[1], Balamurali Sreedhar[2], Megan E. Donaldson[2], Alfred Schultz[2], Timothy C. Frank[2], Andreas S. Bommarius[1] ,Yoshiaki Kawajiri[1]
School of Chemical and Biomolecular Engineering,
Georgia Institute of Technology[1]
The Dow Chemical Company[2]

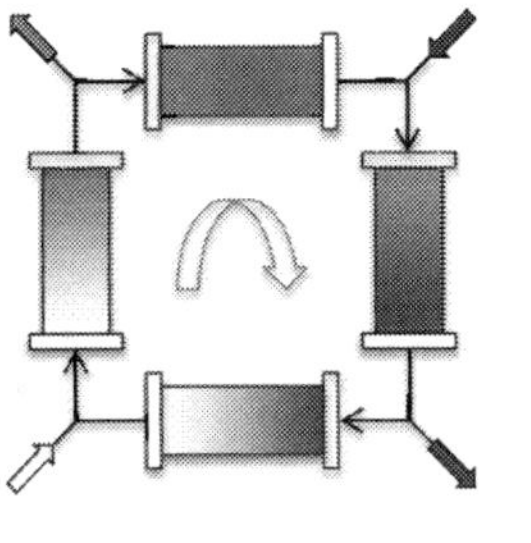

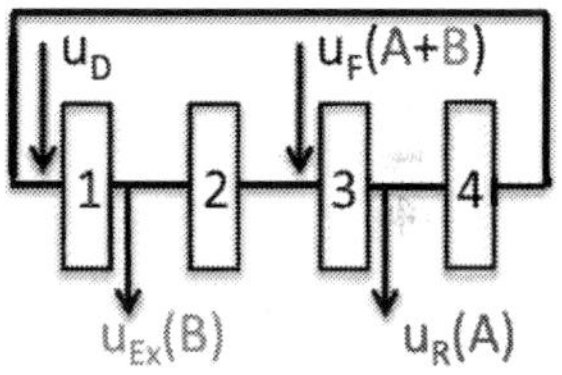

Dow

Georgia Tech | School of Chemical & Biomolecular Engineering

2

Motivation : DOWANOL™ PMA

- Propylene glycol methyl ether acetate (DOWANOL™ PMA)
- The U.S demand for PMA in 2007 was ~117million lbs.[1]

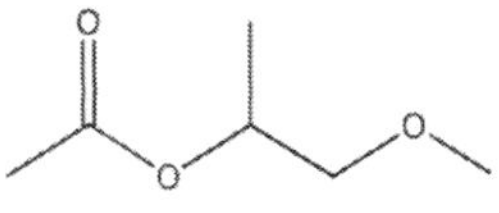

[1] The Dow Chemical Company, Product safety assessment **2008**
TM Trademark of The Dow Chemical Company "Dow" or an affiliated company of Dow

Dow

Georgia Tech | Chemical & Biomolecular Engineering

Potential options for production process of PMA

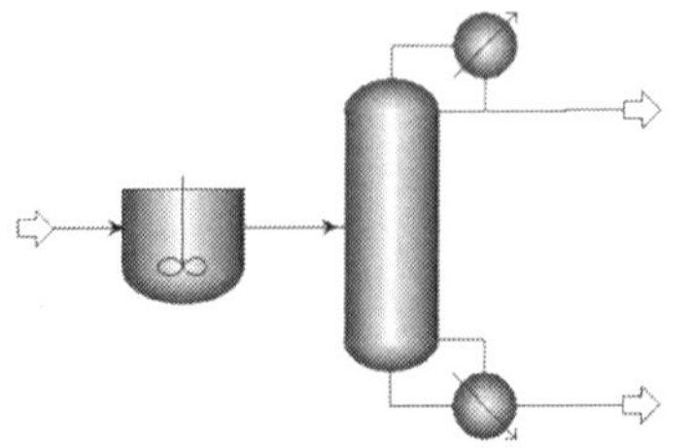

Reaction-then-separation multistep

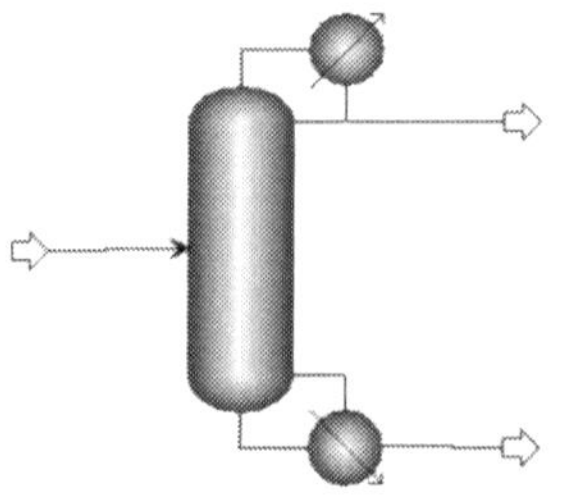

Reactive-distillation one-step

- Limitations
  - Separation : Azeotrope
  - Reaction : Thermodynamic equilibrium limit

Target : Production of PMA through Reactive Chromatography

A + B ⇌ C + D   low conversion

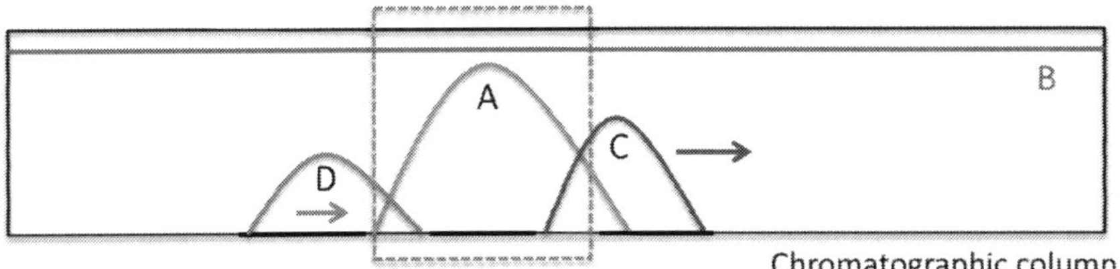

Reaction zone

Higher conversion

Chromatographic column
Packed with Catalyst (Adsorbent)

Esterification:

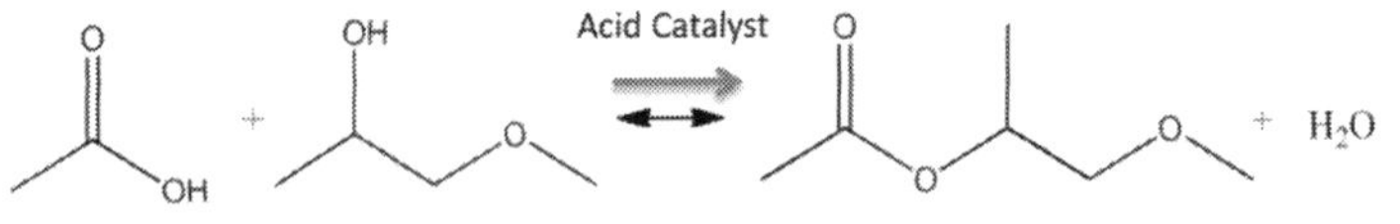

Acetic acid   1-methoxy-2-propanol (PM)   1-methoxy-2-propanol acetate (PMA)   Water

Target : Production of PMA through Reactive Chromatography

A + B ⇌ C + D   low conversion

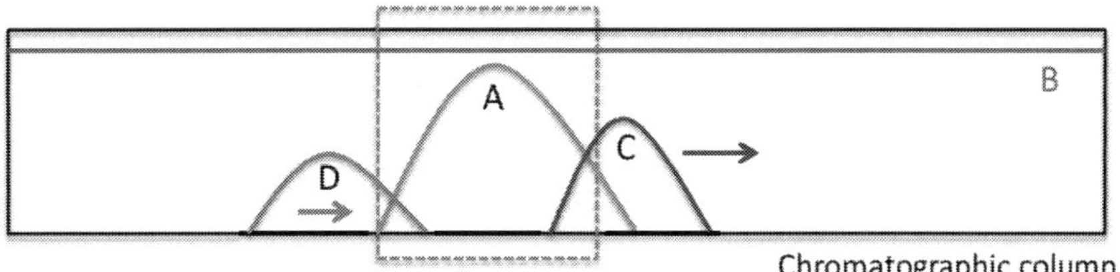

Reaction zone
Higher conversion

Chromatographic column
Packed with Catalyst (Adsorbent)

Transesterification:

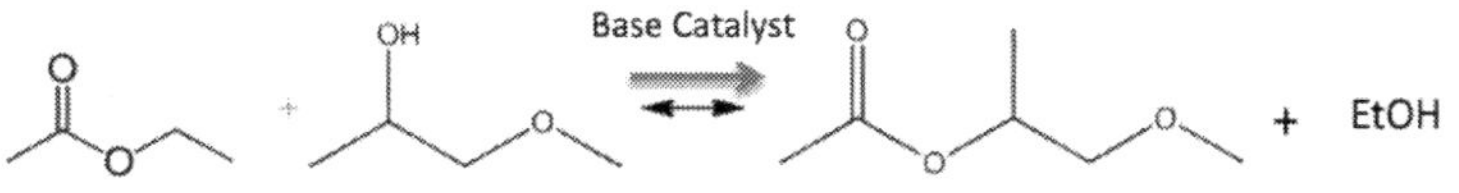

Ethyl acetate   1-methoxy-2-propanol (PM)   1-methoxy-2-propanol acetate (PMA)   Ethanol

Target : Production of PMA through Reactive Chromatography

A + B ⇌ C + D   low conversion

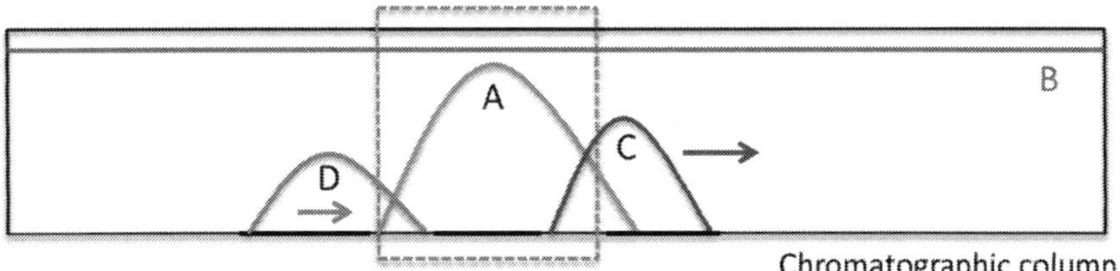

Reaction zone

Chromatographic column
Packed with Catalyst (Adsorbent)

Batch Process → Continuous Process

7

Simulated Moving Bed Reactor (SMBR)

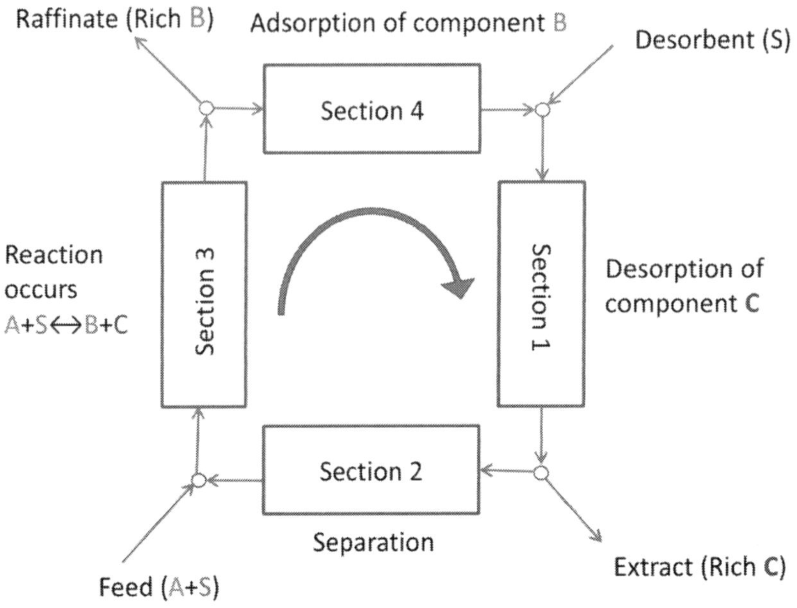

8

Simulated Moving Bed Reactor (SMBR)

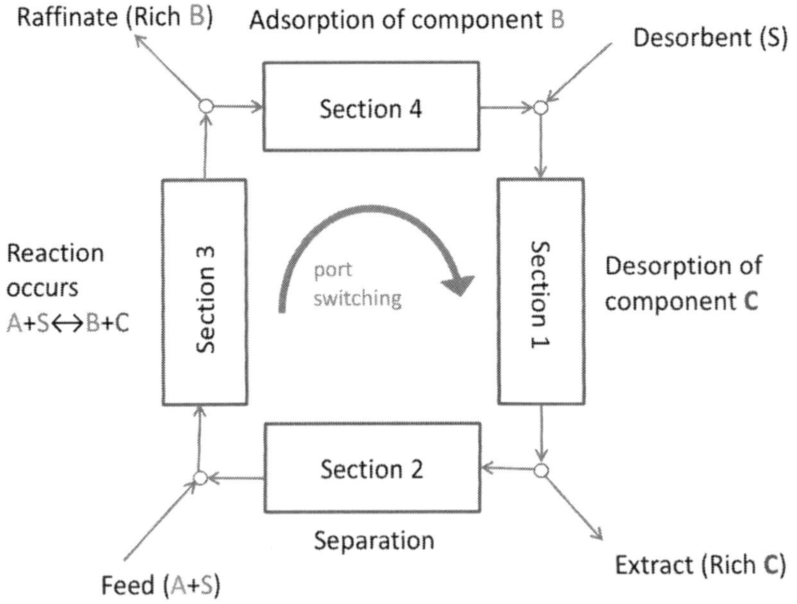

Simulated Moving Bed Reactor (SMBR)

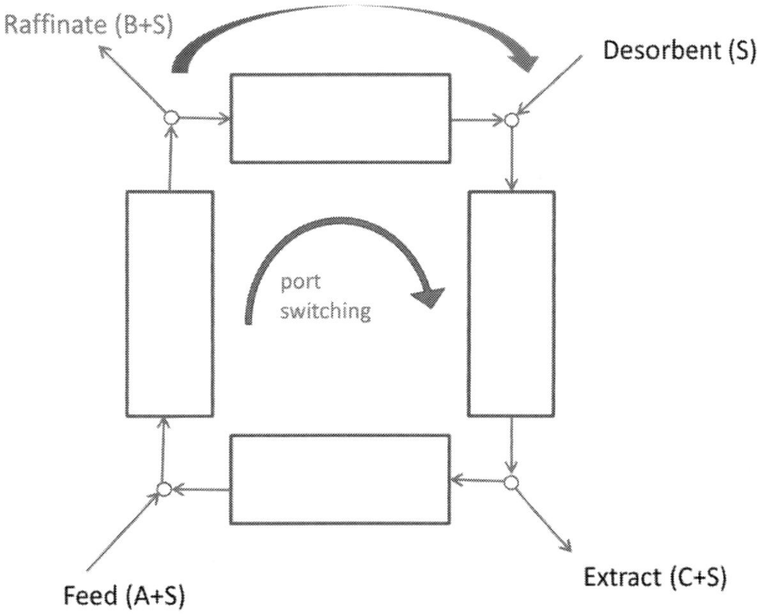

Simulated Moving Bed Reactor (SMBR)

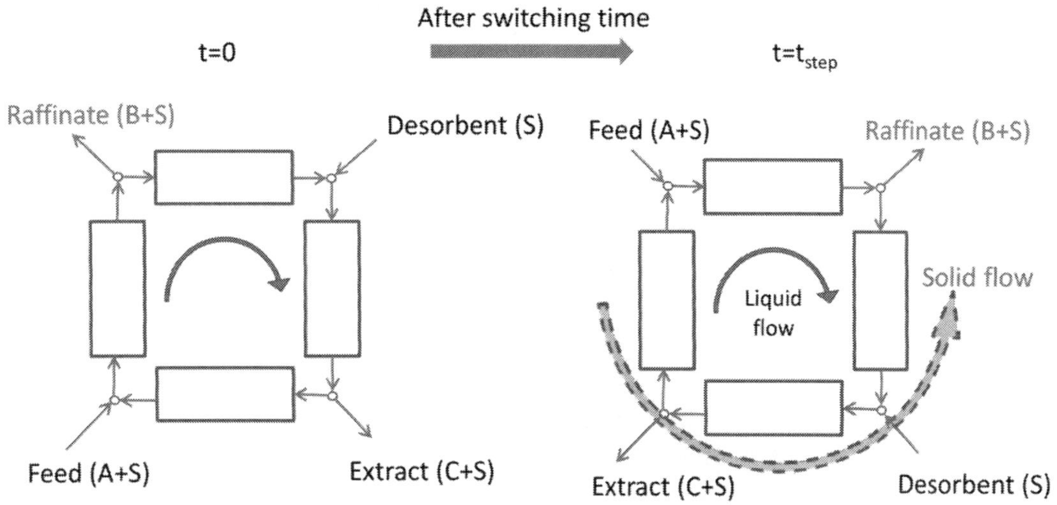

"Simulated" counter-current separation

11

Simulated Moving Bed Reactor (SMBR)

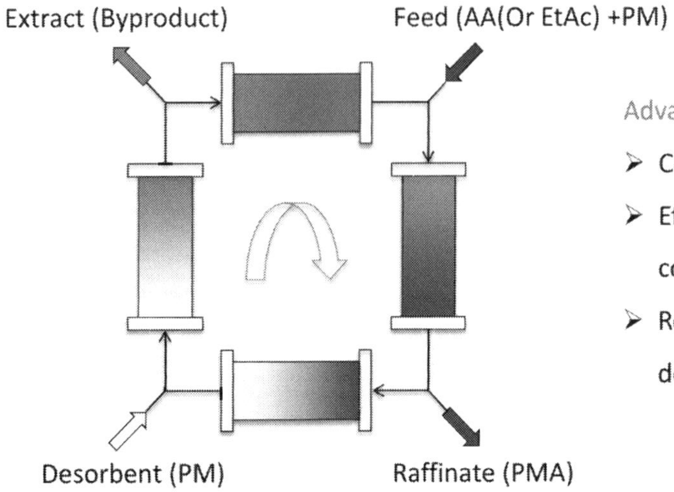

Advantages of SMBR

- Continuous operation
- Efficient use of adsorbent by counter current flow
- Recycle of reactants and desorbents

Georgia Tech | Chemical & Biomolecular Engineering

12

Target: an industrial-scale SMBR process

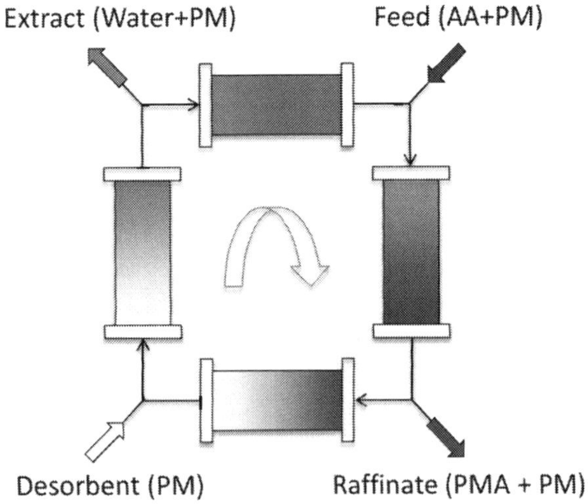

- Novelty of the project
  - Very few previous study of transesterification through reactive chromatography
  - SMBR process that initiated with industrial need

Georgia Tech | Chemical & Biomolecular Engineering

Overview

1. Selection of Catalyst
2. Study of reaction kinetics
3. Single column experiments
4. Parameter estimation
5. SMBR optimization
6. Prep-scale SMBR experiment
7. Modification of SMBR

Georgia Tech | Chemical & Biomolecular Engineering

Selection of Catalyst – Esterification[2]

- AMBERLYST™ 15
  - Swelling ratio test

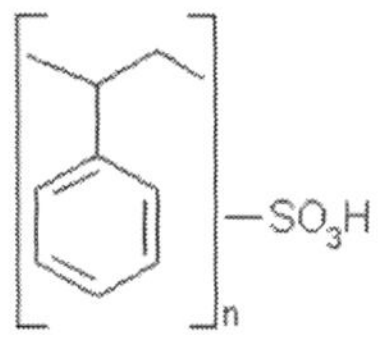

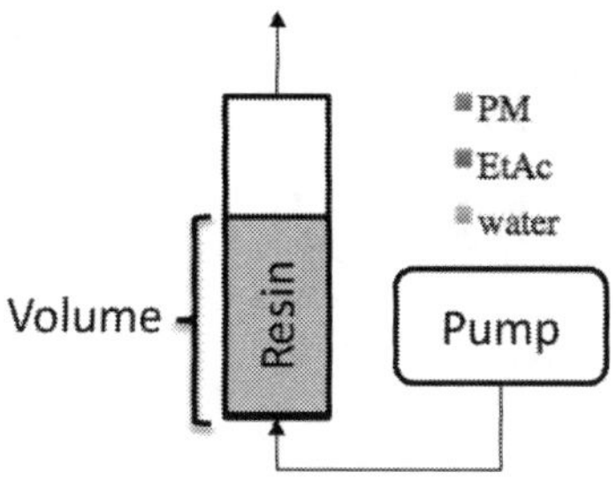

[2] J. Oh, G. Agrawal, B. Sreedhar, M.E. Donaldson, A.K. Schultz, T.C. Frank, A.S. Bommarius, Y. Kawajiri, *Conversion Improvement for Catalytic Synthesis of Propylene Glycol Methyl Ether Acetate by Reactive Chromatography: Experiments and Parameter Estimation*, **Chemical Engineering Journal** 2015 (397-409)

Georgia Tech | Chemical & Biomolecular Engineering

15

Selection of Catalyst – Esterification[2]

- AMBERLYST™ 15
  - Swelling ratio test

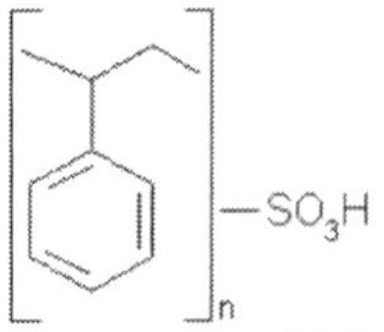

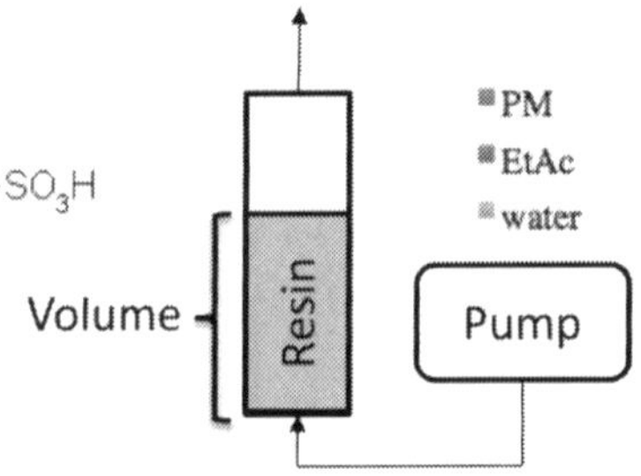

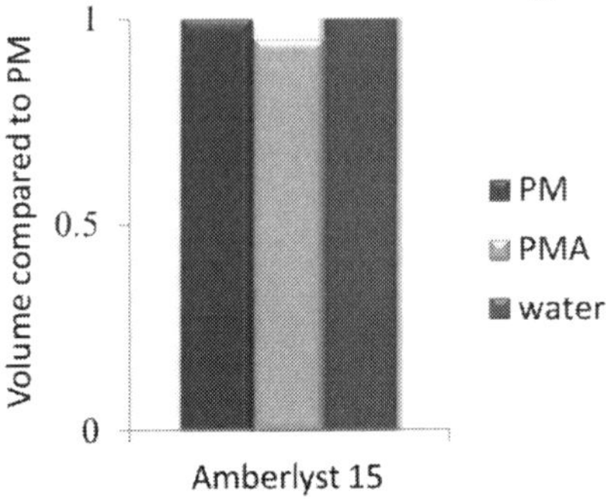

Good packing material for reactive chromatography

[2] J. Oh, G. Agrawal, B. Sreedhar, M.E. Donaldson, A.K. Schultz, T.C. Frank, A.S. Bommarius, Y. Kawajiri, *Conversion Improvement for Catalytic Synthesis of Propylene Glycol Methyl Ether Acetate by Reactive Chromatography: Experiments and Parameter Estimation*, **Chemical Engineering Journal** 2015 (397-409)

Dow

Georgia Tech | Chemical & Biomolecular Engineering

16

Selection of Catalyst – Transesterification

- Swelling ratio test
  - Which resin gives the best swelling ratio for the column packing ?

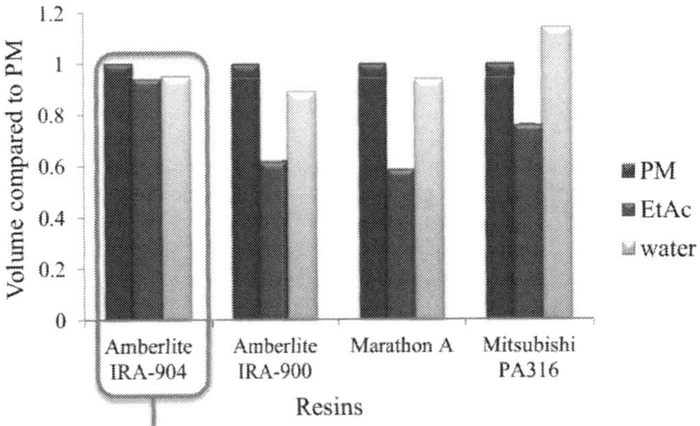

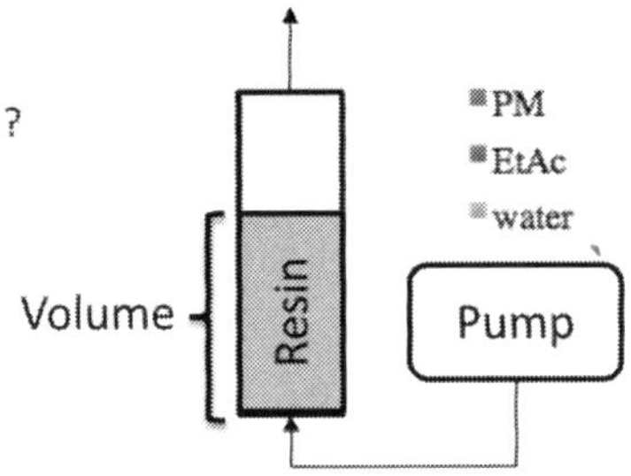

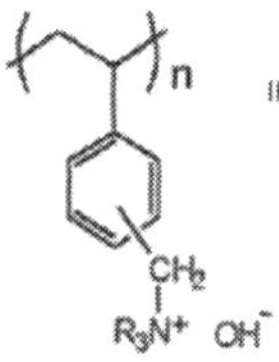

IRA-904

Minimum difference in swelling ratio

IRA904 Selected

Dow

Georgia Tech | Chemical & Biomolecular Engineering

Overview

1. Selection of Catalyst
2. Study of reaction kinetics
3. Single column experiments
4. Parameter estimation
5. SMBR optimization
6. Prep-scale SMBR experiment
7. Modification of SMBR

Batch Experiment Results : Esterification

➢ Higher temperature, higher reaction rate
- Small change in final conversion

Temperature (°C)	50	60	70	90

➢ Higher initial reactant mole ratio, higher the reaction rate
- Increases conversion

Acetic acid : PM	1:1	1:1.25	1.8:1	2.5:1

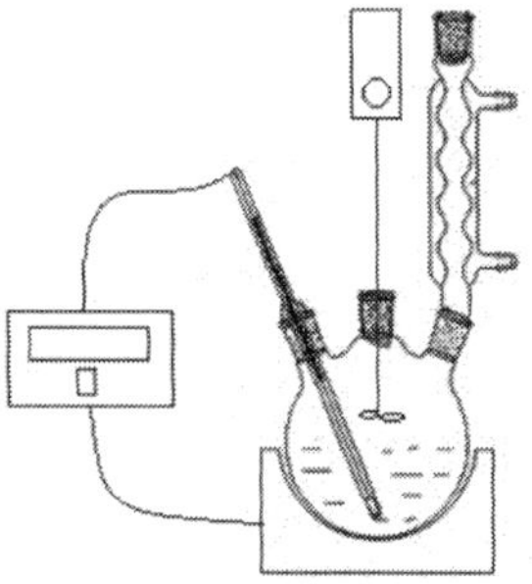

➢ More catalyst, higher reaction rate
- No significant effect on final conversion

Catalyst (g/mL reactant)	0.1	0.2

Final conversion strongly limited by equilibrium

Reaction Rate Model

- No external or internal mass transfer resistance

Stirring speed	600rpm	800rpm
Catalyst particle size	d<707um	707um<d

- Heterogeneous reaction rate models

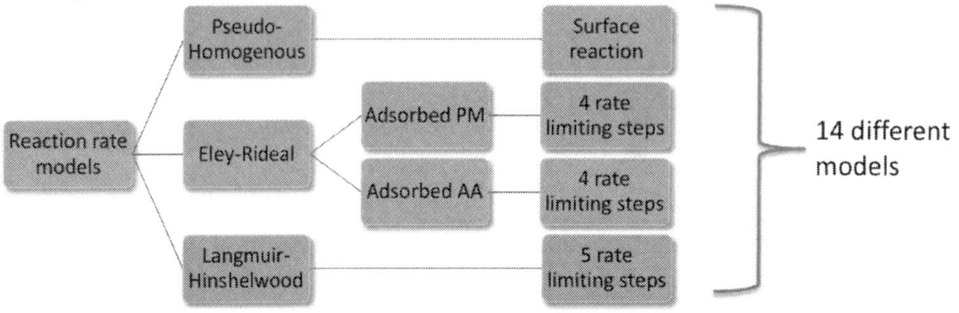

- Rate limiting steps
  - Surface reaction
  - Adsorption of reactants
  - Desorption of products

Reaction Rate Model

- No external or internal mass transfer resistance

Stirring speed	600rpm	800rpm
Catalyst particle size	d<707um	707um<d

- Heterogeneous reaction rate models

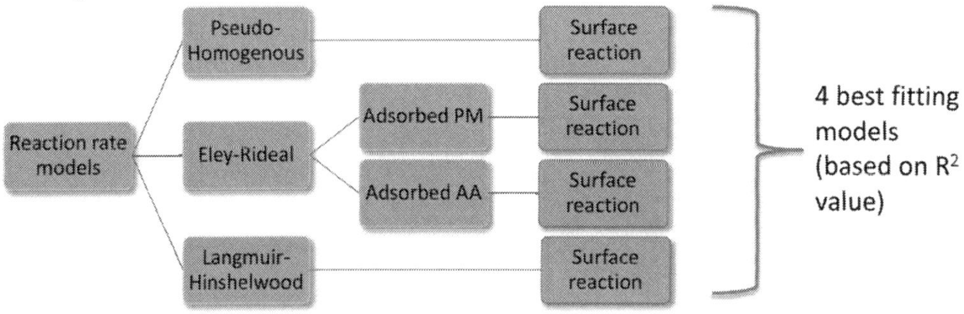

- Surface reaction is rate limiting step
- Adsorption based models (ER and LH) not suitable
  - ✓ Large confidence interval of adsorption constants

→ PH model selected

Reaction Rate Model : Pseudo-Homogeneous model

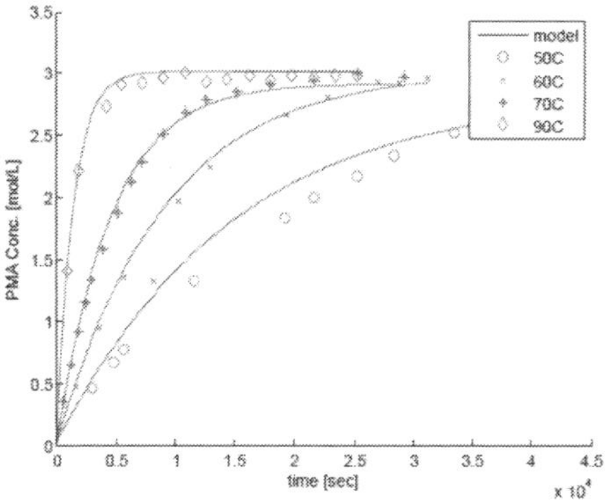

Model	PH
$k_0 * 10^2$ [L^2/(g·sec·mol)]	3.608 ± 0.86
Ea [kJ/(K·mol)]	61.76 ± 0.94
rmse	0.211
R^2	0.99
SRS	2.74

$$r = w_{cat} k_0 exp^{\frac{-E_a}{RT}} [C_{AA} C_{PM} - \frac{C_{PMA} C_{water}}{K_{eq}}]$$

Overview

1. Selection of Catalyst
2. Study of reaction kinetics
3. Single column experiments
4. Parameter estimation
5. SMBR optimization
6. Prep-scale SMBR experiment
7. Modification of SMBR

Experimental Setup

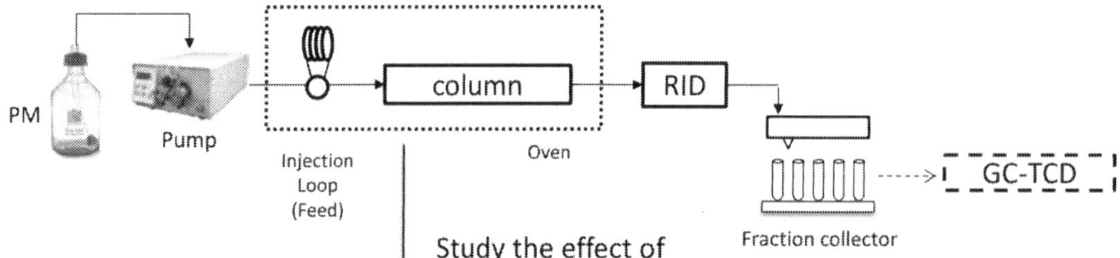

Study the effect of
- Temperature

	Esterification	Transesterification
Feed	Acetic acid	Ethyl Acetate
Resin	AMBERLYST™ 15	IRA 904

Effect of Temperature

Esterification[2]

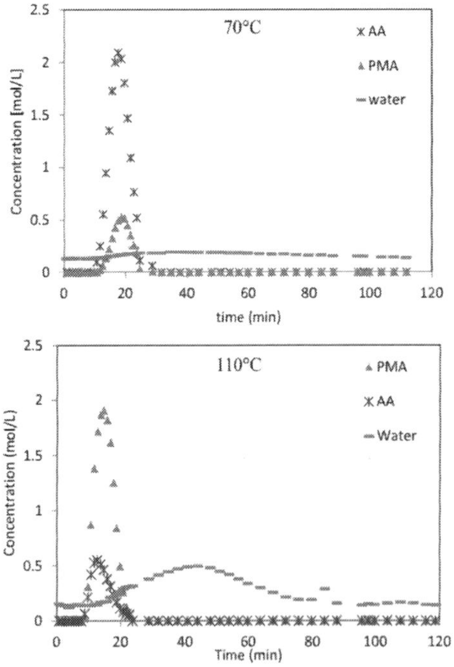

Transesterification

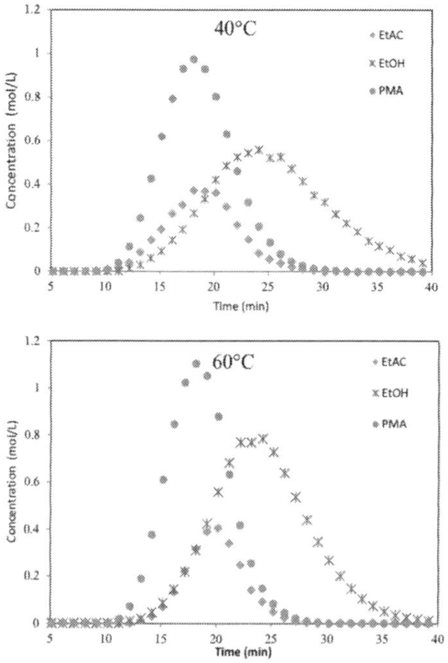

50% EtAC 1mL injection ; flow 0.5ml/min; Column 25×0.8

Effect of Temperature

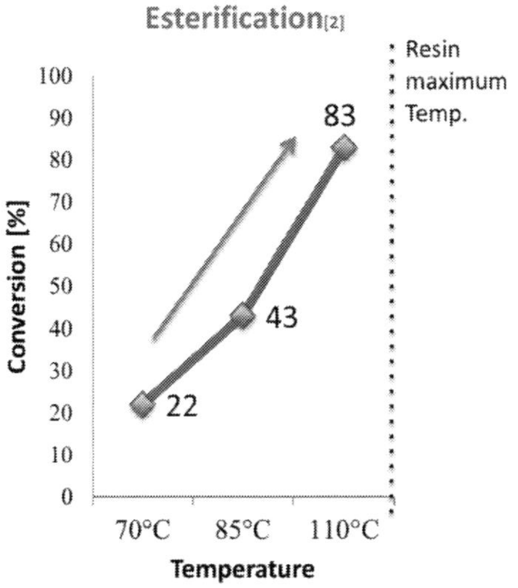

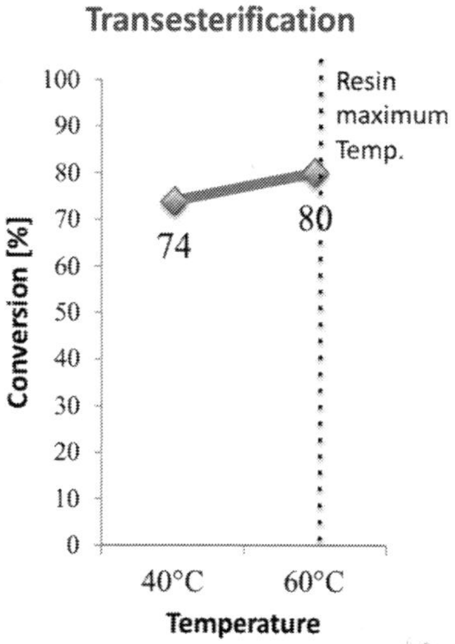

⇨ Conversion increases with temperature

⇨ No significant effect of temperature on conversion

[2] J. Oh, G. Agrawal, B. Sreedhar, M.E. Donaldson, A.K. Schultz, T.C. Frank, A.S. Bommarius, Y. Kawajiri, *Conversion Improvement for Catalytic Synthesis of Propylene Glycol Methyl Ether Acetate by Reactive Chromatography: Experiments and Parameter Estimation*, **Chemical Engineering Journal** 2015 (397-409)

DOW

Georgia Tech | Chemical & Biomolecular Engineering

Experimental Setup

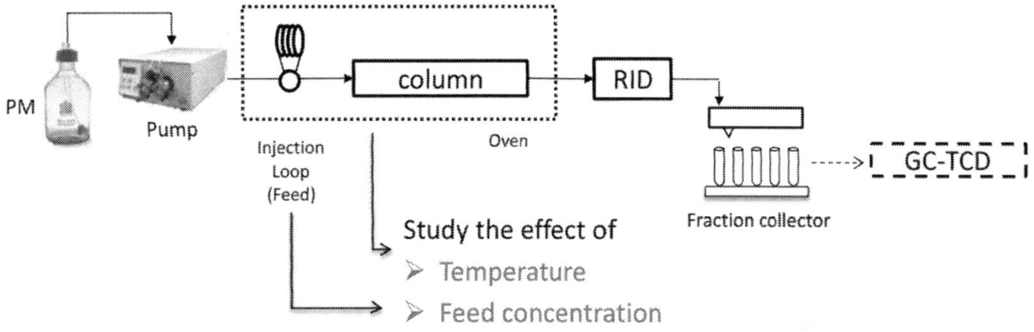

	Esterification	Transesterification
Feed	Acetic acid	Ethyl Acetate
Resin	AMBERLYST™ 15	IRA 904

DOW

Georgia Tech | Chemical & Biomolecular Engineering

Effect of Feed Concentration

Esterification[2] **Transesterification**

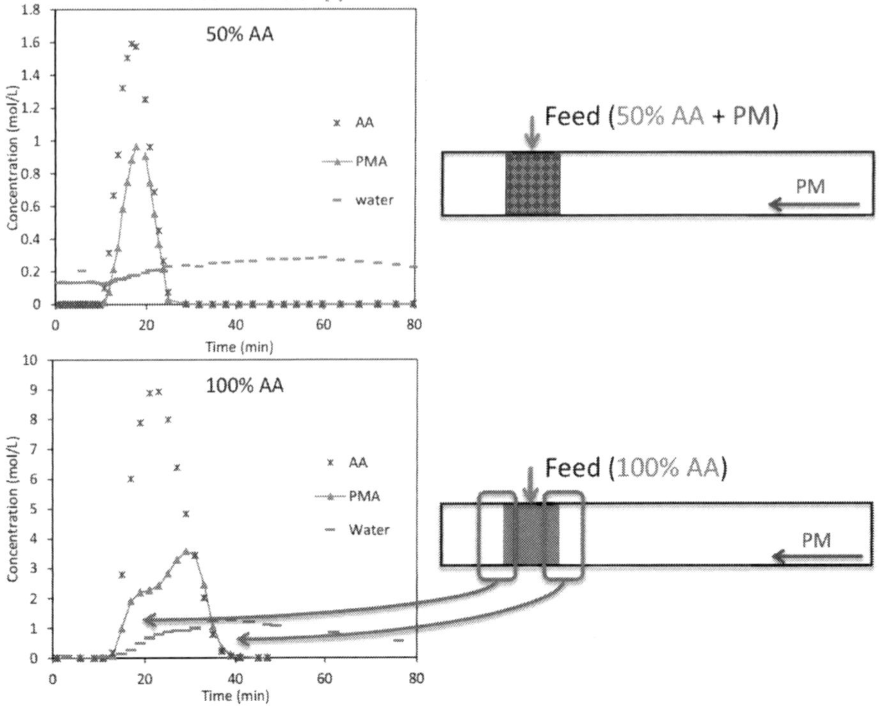

Effect of Feed Concentration

Transesterification

Esterification[2]

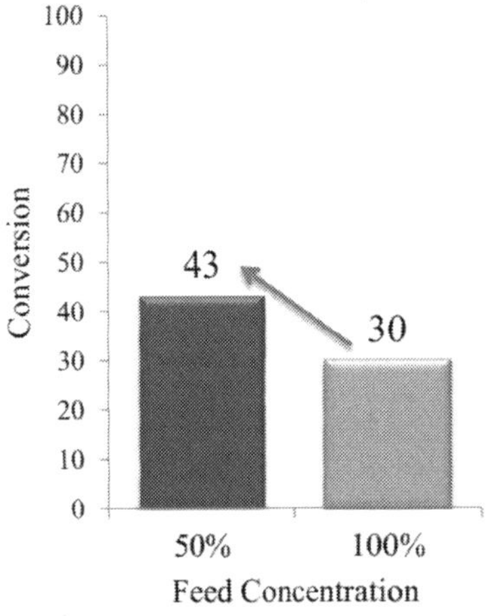

Feed concentration affects separation and conversion

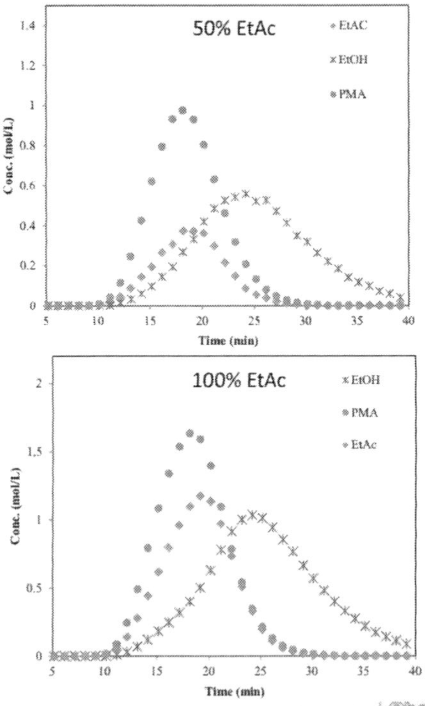

Effect of Feed Concentration

29

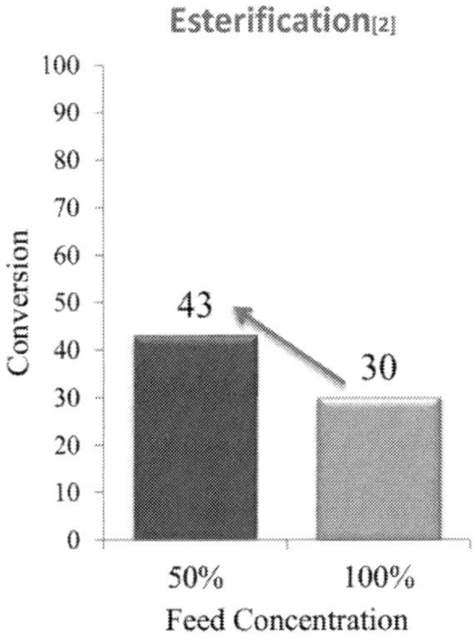

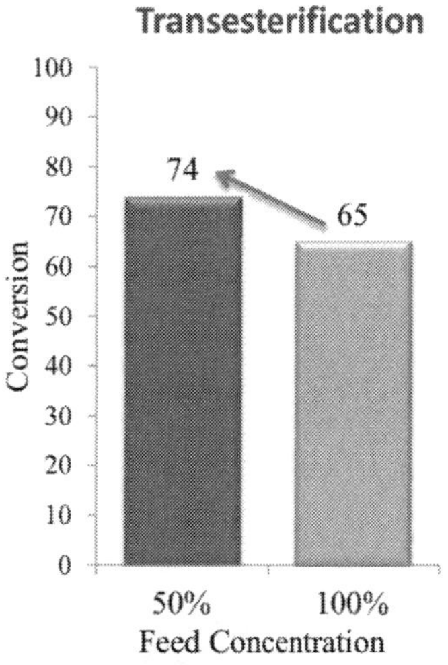

Feed concentration affects separation and conversion

Feed concentration affects conversion

[2] J. Oh, G. Agrawal, B. Sreedhar, M.E. Donaldson, A.K. Schultz, T.C. Frank, A.S. Bommarius, Y. Kawajiri, *Conversion Improvement for Catalytic Synthesis of Propylene Glycol Methyl Ether Acetate by Reactive Chromatography: Experiments and Parameter Estimation*, **Chemical Engineering Journal** 2015 (397-409)

Georgia Tech Chemical & Biomolecular Engineering

30

Esterification vs. Transesterification for PMA production

	Esterification	Transesterification
Applicability to SMBR	Yes	Yes
Catalyst	Acidic	Basic
Advantage	Constant catalytic activity	Low operating temperature
Disadvantage	Large desorbent consumption	Catalytic activity loss
Highest conversion reached from single column	83% (110°C)	80% (60°C)

Georgia Tech Chemical & Biomolecular Engineering

Overview

1. Selection of Catalyst
2. Study of reaction kinetics
3. Single column experiments
4. Parameter estimation
5. SMBR optimization
6. Prep-scale SMBR experiment
7. Modification of SMBR

Modeling and Parameter Estimation – Esterification[3]

$$\frac{\partial C_i}{\partial t} + \frac{1-\epsilon}{\epsilon} k_{m,i}(q_i^{eq} - q_i) + u\frac{\partial C_i}{\partial x} = D_{ax}\frac{\partial^2 C_i}{\partial x^2}$$ Liquid phase mass balance

$$\frac{\partial q_i}{\partial t} = k_{m,i}(q_i^{eq} - q_i) + \nu_i r \qquad q_i^{eq} = H_i C_i$$ Solid phase mass balance

$$r = k_1\left(q_{AA}\, q_{PM} - \frac{1}{K_{eq}} q_{PMA}\, q_{Water}\right)$$ Reaction rate

$$\min_{H_i, K_{m,i}, D_{ax}, k_1, K_{eq}, e_b} \sum_{k=0}^{N_{exp}} \sum_{i=1}^{N_{Comp}} (C_i^{k,mod} - C_i^{k,exp})^2 + \rho \sum_{m=1}^{N_{reg}} (\theta_{reg,m}^{mod} - \theta_{reg,m}^{exp})$$ Objective function

Parameters

k_1, K_{eq}, H_i, D_{ax}

$k_{m,1}, k_{m,2}, k_{m,3}, k_{m,4}$

50% AA, 5 ml injection, 0.5 ml/min, 110°C

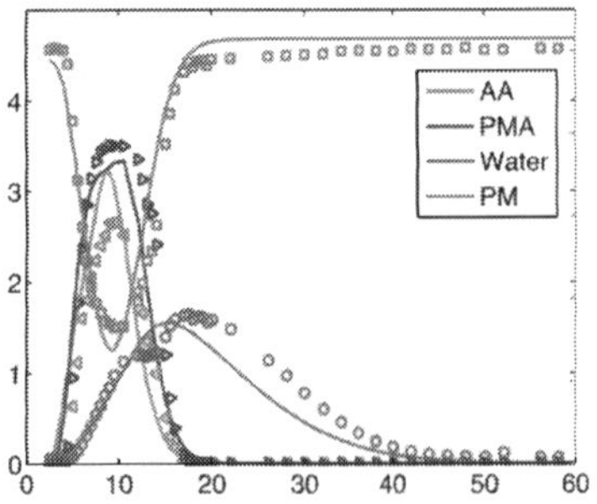

75% AA, 5 ml injection, 0.5 ml/min, 110°C

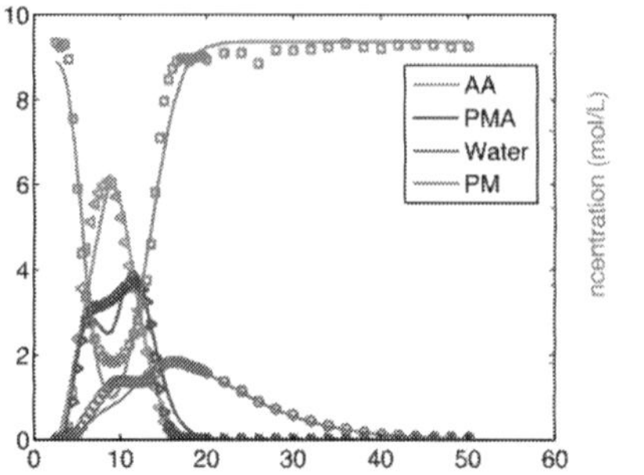

Reliable model and parameters

[3] G. Agrawal, J. Oh, B. Sreedhar, S. Tie, M. Donaldson, T. Frank, A. Schultz, A. Bommarius, Y. Kawajiri. *Optimization of reactive simulated moving bed systems with modulation of feed concentration for production of glycol ether ester*, **Journal of Chromatography A**, 2014 (196-208)

Modeling and Parameter Estimation - Transesterification

$\frac{\partial C_i}{\partial t} + \frac{1-\epsilon}{\epsilon} k_{m,i}(q_i^{eq} - q_i) + u\frac{\partial C_i}{\partial x} = 0$	Liquid phase mass balance
$\frac{\partial q_i}{\partial t} = k_{m,i}(q_i^{eq} - q_i) + \nu_i r \quad q_i^{eq} = H_i C_i$	Solid phase mass balance
$r = k_1(q_{EtAc} q_{PM} - \frac{1}{K_{eq}} q_{PMA} q_{EtOH})$	Reaction rate
$\min_{H_i, k_{m,i}} \sum_{k=1}^{N_{exp}} \sum_{i=1}^{N_{Comp}} \left(C_i^{k,mod} - C_i^{k,exp}\right)^2$	Objective function

Parameter estimation
$k_1, K_{eq}, H_1, H_2, H_3, H_4$
$k_{m,1}, k_{m,2}, k_{m,3}, k_{m,4}$

40°C, 50% EtAc — 60°C, 50% EtAc — 40°C, 100% EtAc

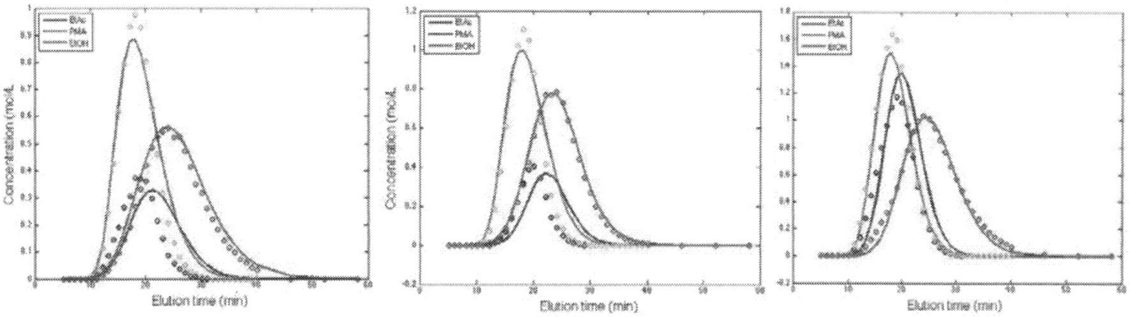

⇨ Reliable model and parameters

Dow

Georgia Tech | Chemical & Biomolecular Engineering

Overview

1. Selection of Catalyst
2. Study of reaction kinetics
3. Single column experiments
4. Parameter estimation
5. SMBR optimization
6. Prep-scale SMBR experiment
7. Modification of SMBR

Dow

Georgia Tech | Chemical & Biomolecular Engineering

SMBR Optimization

Problem formulation

Max. PMA production rate (in raffinate outlet)

Max. Conversion of Reactant

Constraints :

Parameters	Value
Water content in raffinate outlet	≤ 1 wt%
PMA recovery in raffinate outlet	≥ 90%

$$\begin{cases} \frac{\partial C_i}{\partial t} + \frac{1-\epsilon}{\epsilon} k_{m,i}(q_i^{eq} - q_i) + u\frac{\partial C_i}{\partial x} = D_{ax}\frac{\partial^2 C_i}{\partial x^2} \\ \frac{\partial q_i}{\partial t} = k_{m,i}(q_i^{eq} - q_i) + \nu_i r \\ q_i^{eq} = H_i C_i \\ r = k_1(q_{AA}\, q_{PM} - \frac{1}{K_{eq}} q_{PMA}\, q_{Water}) \end{cases}$$

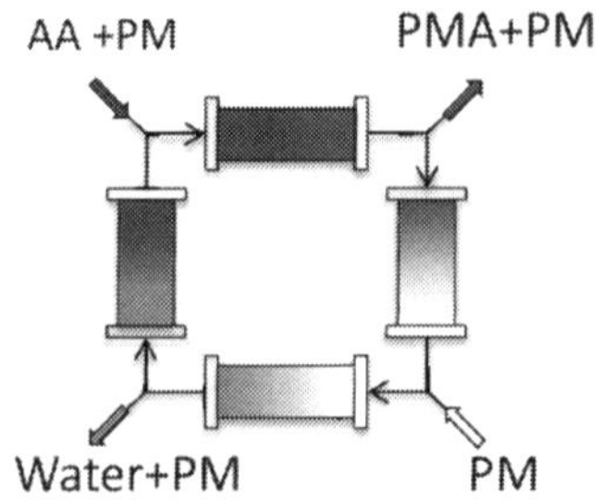

Optimal operating parameters:

$u_{zone1}{}^{opt}$
$u_F{}^{opt}$
$u_{Ex}{}^{opt}$
$u_D{}^{opt}$
$t_{switch}{}^{opt}$

Dow

Georgia Tech Chemical & Biomolecular Engineering

SMBR Optimization Results

Esterification : 95% Conversion Target

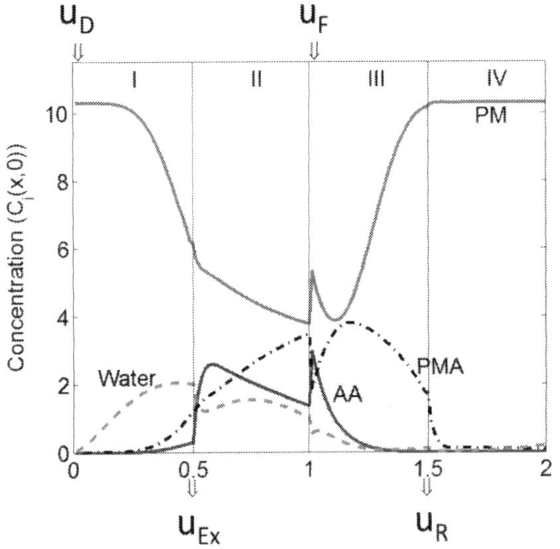

PARAMETERS	CONDITIONS
Resin	AMBERLYST™ 15
Oven Temperature	110 °C
Desorbent Composition	100% PM
Feed composition (%)	25% AA; 75% PM
Desorbent flow (mL/min)	1.6
Extract flow (mL/min)	1.6
Feed flow (mL/min)	0.38
Recycle flow (mL/min)	0.2
Switch time (min)	51.3

Dow

Georgia Tech Chemical & Biomolecular Engineering

Overview

1. Selection of Catalyst
2. Study of reaction kinetics
3. Single column experiments
4. Parameter estimation
5. SMBR optimization
6. Prep-scale SMBR experiment
7. Modification of SMBR

SMBR Unit

- Max Temp :120°C
- Max Pressure: 500 psi
- Max flow rates : 100ml/min

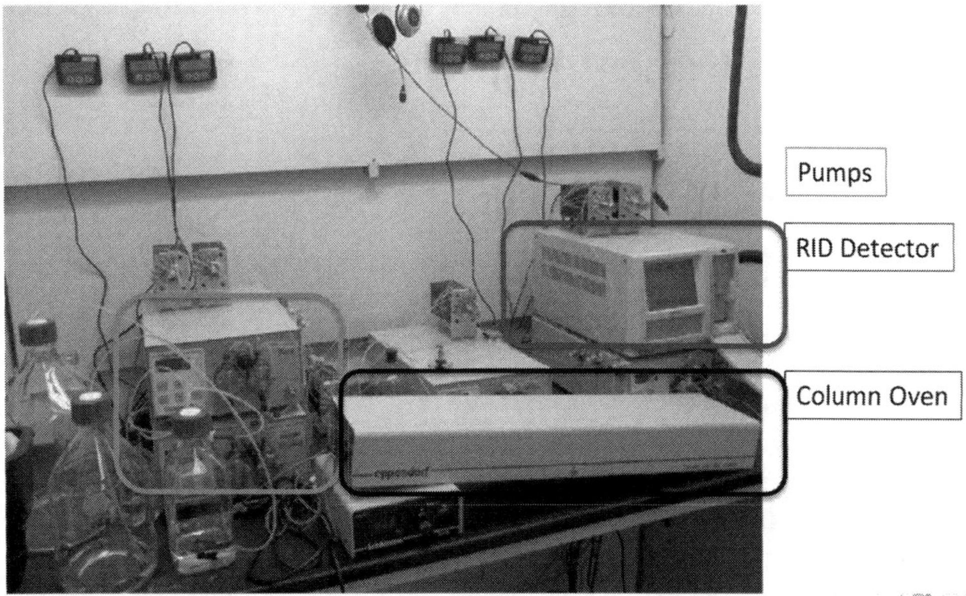

SMBR Model Prediction vs. Experimental Results

Esterification

Conversion

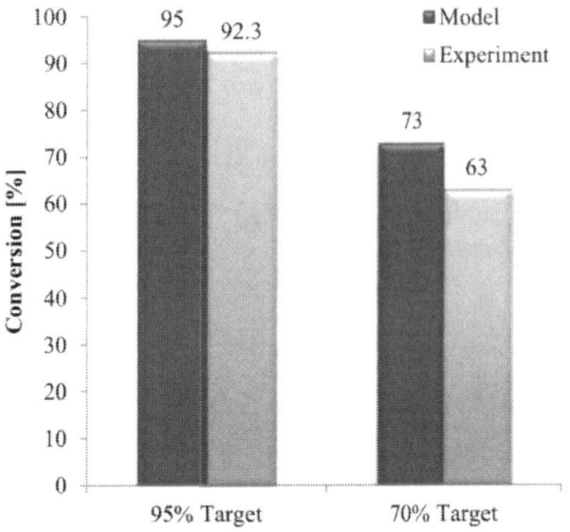

Production Rate

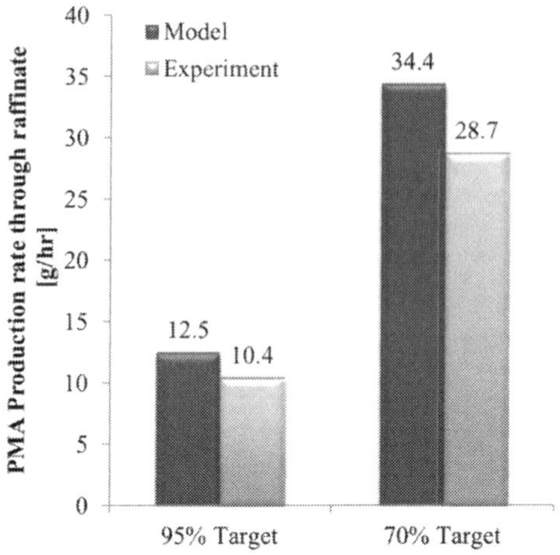

Successful prediction of experimental results with single column parameters

Overview

1. Selection of Catalyst
2. Study of reaction kinetics
3. Single column experiments
4. Parameter estimation
5. SMBR optimization
6. Prep-scale SMBR experiment
7. Modification of SMBR

SMBR, reaction catalyzed only by homogeneous catalyst

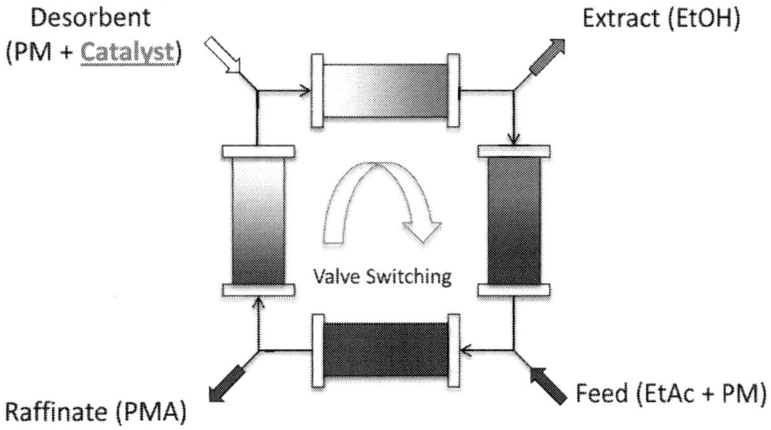

- Use resin as separation medium, but not as catalyst.
- Reaction is catalyzed only by homogeneous catalyst (NaOH or Sodium ethoxide, etc.)

Georgia Tech | Chemical & Biomolecular Engineering

Conclusions

- SMBR process that has a high potential for commercialization (PMA synthesis)
- Novel application of transesterification reaction to the SMBR process
- Screened potential catalysts for the SMBR application
- Conducted single column experiments and analyzed the influence of temperature and feed concentration.
- Developed a model for SMBR process and validated it for esterification reaction
- Future work : Extend the SMBR study to the transesterification reaction
- Potential for the addition of homogeneous catalyst into the system

Georgia Tech | Chemical & Biomolecular Engineering

Acknowledgments

- Dr. Kawajiri group
- Dr. Bommarius group
- The Dow Chemical Company
  - University Partnership Program

DOW®

Georgia Tech

PROCESS REVITALIZATION

FALL 2015 AICHE PRESENTATION

J. CROLL AND G. HEMMER

CHEMISTRY THAT MATTERS™

Greg Hemmer, Chief Scientist

SABIC's Innovative Plastics Business – Technology & Innovation, High Performance Polymer, Global Process Technology

Josh Croll, Senior Manager

SABIC's Innovative Plastics Business – Technology & Innovation, High Performance Polymer, Global Process Technology

Evelyn Pearson, Senior Manager

SABIC's Innovative Plastics Business – Technology & Innovation, Core Polymer, Global Process Technology

No. 1

ASSET DESIGN

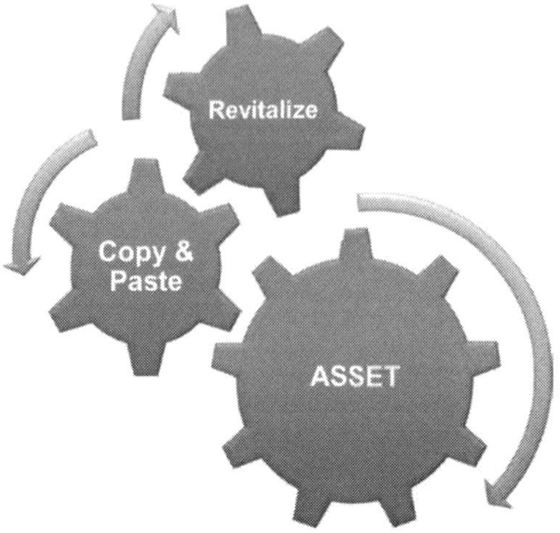

COPY AND PASTE APPROACH

+ SPEED
+ CERTAINTY OF RESULT
- HIGHER CAPEX & OPEX
- COMPETITIVE ADVANTAGE

REVITALIZATION

- SPEED
- CERTAINTY OF RESULT
+ HIGHER CAPEX & OPEX
+ COMPETITIVE ADVANTAGE

PROCESS REVITALIZATION DEPENDS ON TIME REQUIRED TO DEVELOP AND ASSESS CHANGE AND THE ACHIEVABLE / TANGIBLE BENEFITS

No. 2

PROCESS IMPROVEMENT ROADMAP

TIME

LEVEL 1
Optimize

Minimal changes in existing chemistries

Simplify and right sizing / scaling of existing unit operations

External manufacturing as a source of supply vs in-house synthesis

LEVEL 2
Modernize

Modest changes in existing chemistries

Scaling and intensification of existing unit operations

Batch to continuous transformations

External manufacturing as a development partner

LEVEL 3
Next Gen

Alternative reaction pathways / chemistries and solvents

Alternative unit operations and advanced intensifications

Extensive pilot and semiworks required

External manufacturing as semiworks

IDEATION PROCESSES

FOUNDATIONAL INFORMATION

No. 3

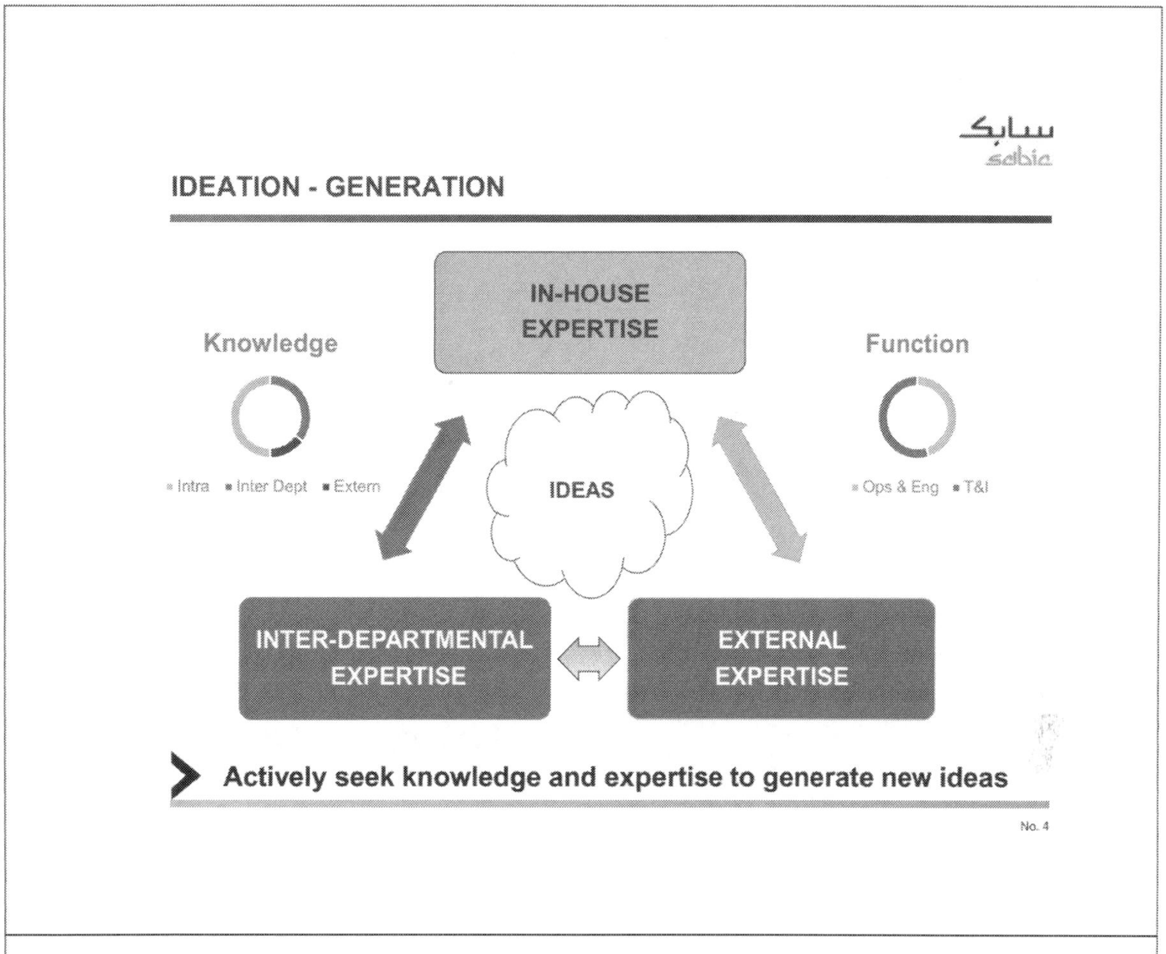
سابك
sabic
IDEATION - GENERATION
IN-HOUSE EXPERTISE
Knowledge
Function
Intra
Inter Dept
Extern
IDEAS
Ops & Eng
T&I
INTER-DEPARTMENTAL EXPERTISE
EXTERNAL EXPERTISE
Actively seek knowledge and expertise to generate new ideas
No. 4

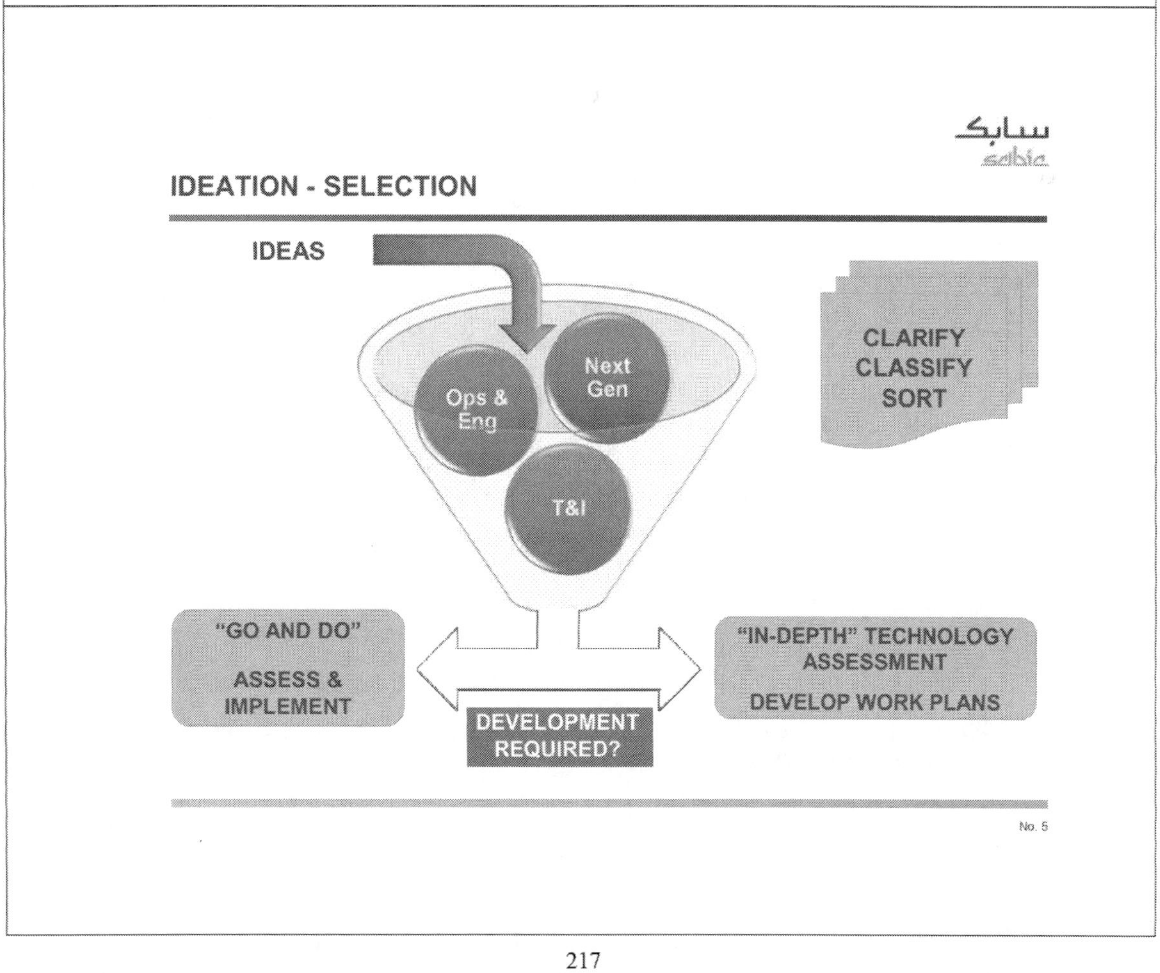
سابك
sabic
IDEATION - SELECTION
IDEAS
Next Gen
Ops & Eng
T&I
CLARIFY
CLASSIFY
SORT
"GO AND DO"
ASSESS & IMPLEMENT
"IN-DEPTH" TECHNOLOGY ASSESSMENT
DEVELOP WORK PLANS
DEVELOPMENT REQUIRED?
No. 5

FOUNDATIONAL INFORMATION

Documentation

- Process narrative
- Equipment & process details
- Operating strategies
- Critical to design parameters
- Material of construction

Properties & Kinetics

- Time vs temperature sensitivity data
- Liquid-liquid partition coefficients
- Crystallization kinetics
- Chemical reaction kinetics

Modeling

- Deeper insight regarding ex. unit ops
- Model-centric process development allows virtual experimentation

Benefits

Questioning the Status Quo
Uncovering inconsistencies
Identifying

- Competency needs
- Equipment design
- Operational protocols
- Control logic

Discovery of hidden factories

⇩

Build firm foundation for future change efforts

Identify no-cost opportunities to increase capacity in the existing process

No. 6

LEVEL 1 (OPTIMIZATION) EXAMPLE #1

Two-step to One-Step Acid Recovery

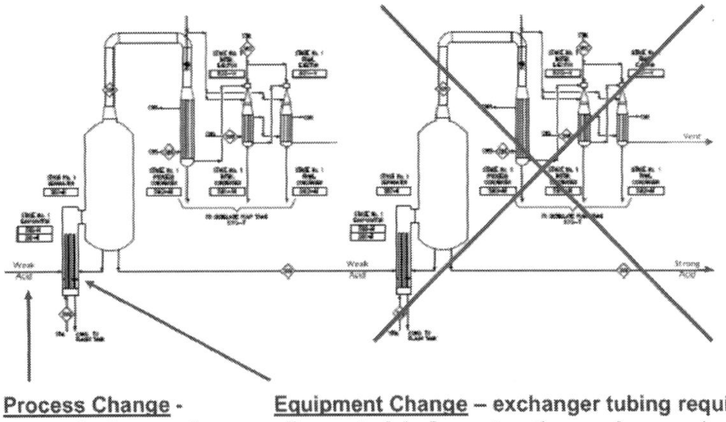

Process Change - concentration can be increased by various upstream modifications

Equipment Change – exchanger tubing requires high alloy material of construction – advances in fabrication allow scaling versus original installation

No. 7

LEVEL 1 (OPTIMIZATION) EXAMPLE #2

Two-step to One-Step Monomer Dissolution Processes

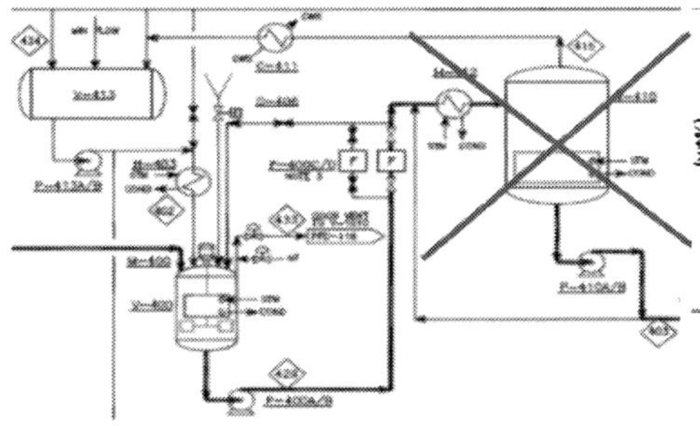

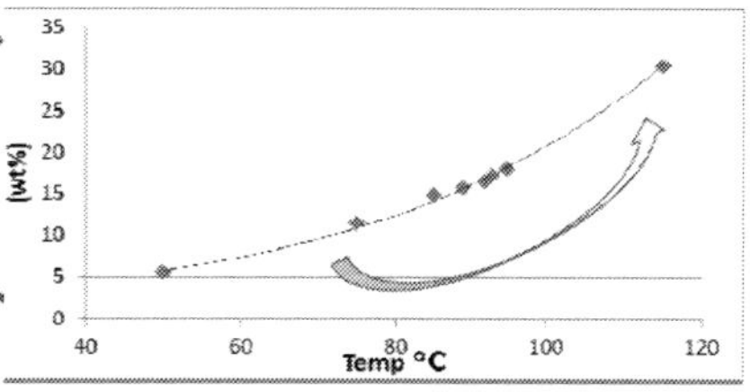

Improving foundational information and questioning status quo drives intensification

No. 8

LEVEL 2 (MODERNIZATION) EXAMPLE #1

Acid Concentration Column Intensification

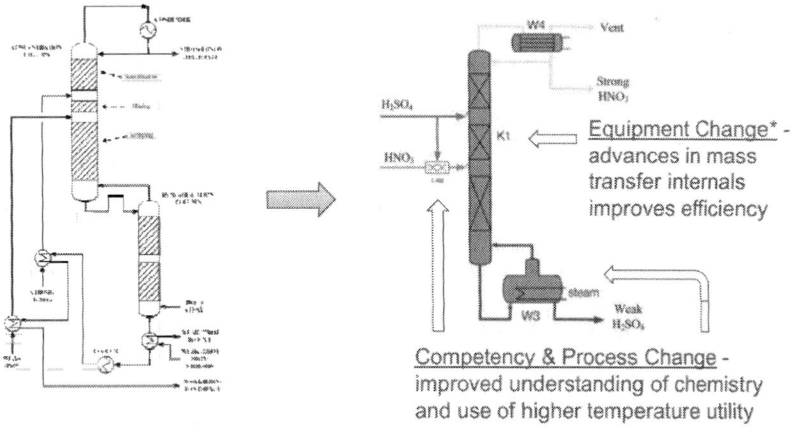

Leverage external expertise with proven commercialization experience to obtain readily available off the shelf improvement

* Schematic courtesy of De Dietrich Process Systems, Inc.

No. 9

LEVEL 2 (MODERNIZATION) EXAMPLE #2

Solvent Distillation Intensification

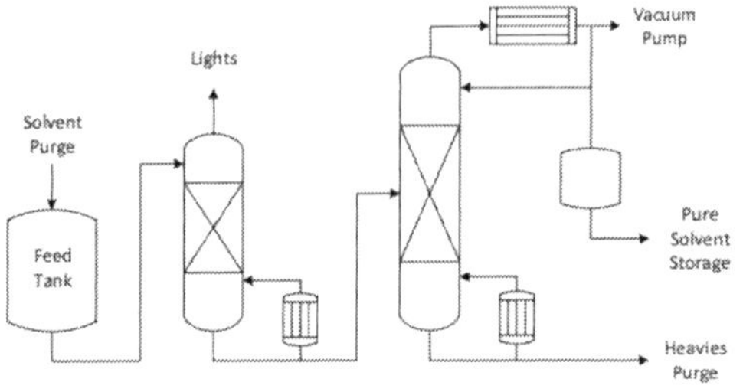

Two column system

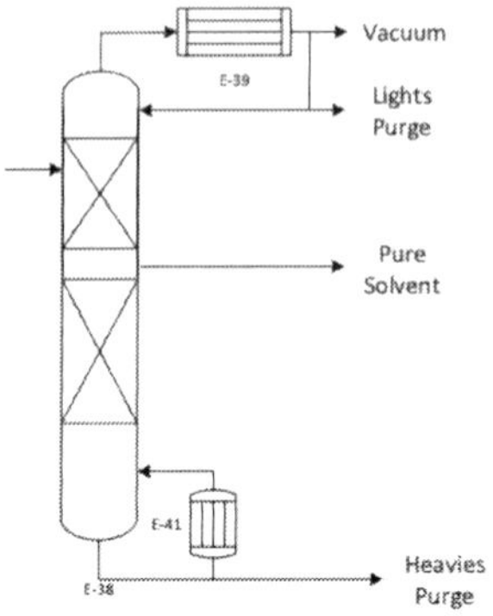

Single column with side draw

Solid Foundation + Expertise = Low Risk Intensification

LEVEL 3 (NEXT GEN) EXAMPLE #1

NOx Capture

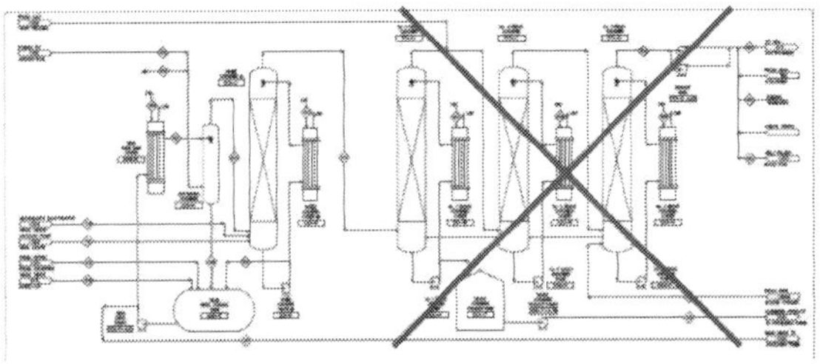

Gas absorption via multiple towers using two absorbents - early 80's BACT

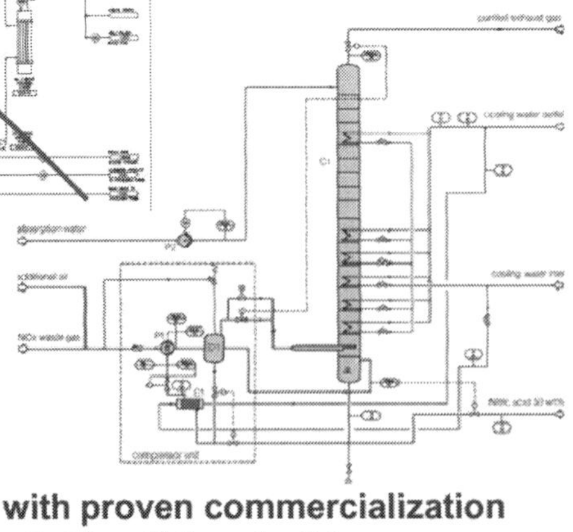

Improved understanding of chemistry

Operation at alternative conditions

Development of specialized internals

Leverage external expertise with proven commercialization experience to obtain readily available off the shelf improvement

* Schematic courtesy of De Dietrich Process Systems, Inc.

LEVEL 3 (NEXT GEN) EXAMPLES #2 & #3

COMBINING MULTIPLE OPERATIONS INTO ONE INTEGRATED UNIT

DIVIDED WALL DISTILLATION COLUMN

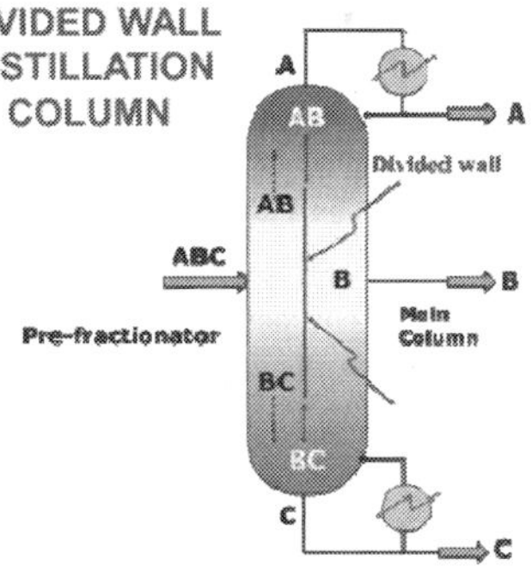

MULTIPLE EVAPORATORS AND STRIPPERS IN ONE UNIT*

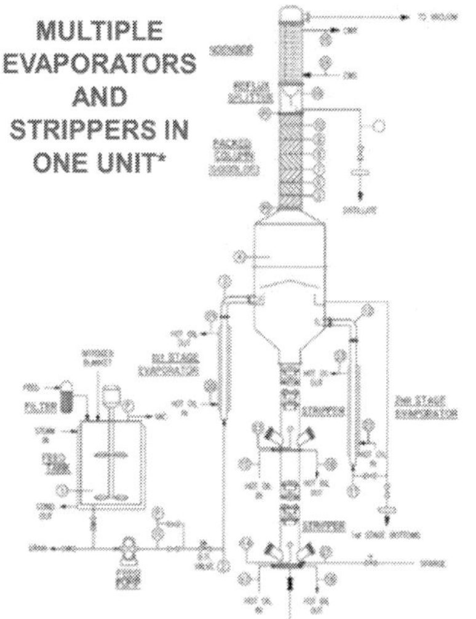

Advanced intensification will require firm foundational information, use of external expertise, a thorough pilot, and a multiple day "semiworks style" validation.

* Schematic courtesy of Artisan Industries Inc.

No. 12

LEVEL 3 (NEXT GEN) ADD'L EXAMPLES

- BATCH TO CONTINUOUS
- IN-LINE OR ON-LINE ANALYTICAL
- ADVANCED REACTOR DESIGNS
- MULTI-STAGE CENTRIFUGAL CONTACTORS
- MULTI-FUNCTIONAL EXTRUDERS

No. 13

BENEFITS

- Capital cost reduction of 20+% per kilogram for assets
- Operating cost reduction of 5 to 10% per kilogram
- Establish firm foundation for low/no-cost investment to expand existing assets

No. 14

THANK YOU FOR YOUR ATTENTION!!!

ANY QUESTIONS?

No. 15

415716 Integrated Design and Control of Reactive and Non-Reactive Distillation Processes

Abstract:

Process design and process control have been considered as independent problems for many years. In this context, a sequential approach is used where the process is designed first, followed by the control design. However, this sequential approach has its limitations related to dynamic constraint violations, for example, infeasible operating points, process overdesign or under-performance. Therefore, by using this approach, a robust performance is not always guaranteed. Furthermore, process design decisions can influence process control and operation (Huusom, 2015). To overcome these limitations, an alternative approach is to tackle process design and controllability issues simultaneously, in the early stages of process design. This simultaneous synthesis approach provides optimal/near optimal operation and more efficient control of conventional (non-reactive binary distillation columns) (Hamid et al., 2010) as well as complex chemical processes; for example, intensified processes such as reactive distillation (Mansouri et al., 2015). Most importantly, it identifies and eliminates potentially promising design alternatives that may have controllability problems later. To date, a number of methodologies have been proposed and applied on various problems to address the interactions between process design and control, and they range from optimization-based approaches to model-based methods (Sharifzadeh, 2013).

In this work, integrated design and control of non-reactive distillation, ternary compound reactive distillation and multi-component reactive distillation processes is considered systematically through a computer-aided framework. To assure that design decisions give the optimum operational and economic performance, operability and controllability issues are considered simultaneously with the process design issues. Operability issues are addressed to ensure a stable and reliable process design at pre-defined operational conditions whereas controllability is considered to maintain desired operating points of the process at any kind of imposed disturbance under normal operating conditions. First, to design non-reactive binary distillation columns, a set of conventional and simple design methods such as McCabe-Thiele and driving force approach (Bek-Pedersen and Gani, 2004) are selected. Next, these design methods are extended using element concept to also include ternary as well as multicomponent reactive distillation processes. The element concept (Pérez Cisneros et al., 1997) is used to translate a ternary system of compounds ($A + B \leftrightarrow C$) to a binary system of element (W_A and W_B). In the case of multicomponent reactive distillation processes the equivalent element concept is used to translate a multicomponent (multi-element) system (Jantharasuk et al., 2011) of compounds ($A + B \leftrightarrow C + D$(inert)) to a binary system of key elements (elements W_{HK} and W_{LK}). For an energy-efficient design, non-reactive driving force (for binary non-reactive distillation), reactive driving force (for ternary compound reactive distillation) and binary-equivalent driving force (for multicomponent reactive distillation) were employed. For both the McCabe-Thiele and driving force method, vapor-liquid equilibrium data are based on elements (except for binary non-reactive distillation column). ICAS-PDS is used to compute the reactive vapor-liquid equilibrium data set by consecutive calculation of reactive bubble points. It has been shown previously that designing a reactive distillation column at the maximum driving force will result in the minimum energy consumption (Bek-Pedersen and Gani, 2004). Note, that the same principles that apply to a binary non-reactive compound system are valid also for a binary-element or a binary-key-element system. Therefore, it is advantageous to employ the element based method for multicomponent reaction-separation systems.

The operation of the non-reactive distillation column, ternary reactive distillation column (binary-element) and multicomponent reactive distillation column (binary-key-element) is investigated at the highest driving force and other candidate points. It is shown analytically and through rigorous dynamic process simulation (using ICAS process simulation software and Aspen Plus) for all three cases that the sensitivity of the system to the disturbances in the feed at the highest driving force is less than any other candidate point. By application of this approach, it is shown that designing the non-reactive and reactive distillation processes at the maximum driving force results in an optimal design in terms of controllability and operability as well as an optimal/near optimal design from an energy point of view. It is verified that the reactive distillation design option is less sensitive to the disturbances in the feed at the highest driving force and has the inherent ability to reject disturbances.

References:

Bek-Pedersen, E., Gani, R., 2004. Design and synthesis of distillation systems using a driving-force-based approach. Chem. Eng. Process. Process Intensif. 43, 251–262. doi:10.1016/S0255-2701(03)00120-X

Hamid, M.K.A., Sin, G., Gani, R., 2010. Integration of process design and controller design for chemical processes using model-based methodology. Comput. Chem. Eng. 34, 683–699. doi:10.1016/j.compchemeng.2010.01.016

Huusom, J.K., 2015. Challenges and opportunities in integration of design and control. Comput. Chem. Eng. doi:10.1016/j.compchemeng.2015.03.019

Jantharasuk, A., Gani, R., Górak, A., Assabumrungrat, S., 2011. Methodology for design and analysis of reactive distillation involving multielement systems. Chem. Eng. Res. Des. 89, 1295–1307. doi:10.1016/j.cherd.2011.04.016

Mansouri, S.S., Sales-Cruz, M., Huusom, J.K., Woodley, J.M., Gani, R., 2015. Integrated Process Design and Control of Reactive Distillation Processes, in: 9th International Symposium on Advanced Control of Chemical Processes.

Pérez Cisneros, E.S., Gani, R., Michelsen, M.L., 1997. Reactive separation systems—I. Computation of physical and chemical equilibrium. Chem. Eng. Sci. 52, 527–543. doi:10.1016/S0009-2509(96)00424-1

Sharifzadeh, M., 2013. Integration of process design and control: A review. Chem. Eng. Res. Des. doi:10.1016/j.cherd.2013.05.007

(314e) Study of Fusel Oil Valorization Alternatives in Bioethanol Production

Rodriguez, G., Universidad Nacional de Colombia
Vargas, J. C., Universidad Nacional de Colombia
Gil, I. D., Universidad Nacional de Colombia
Originally presented on: 11/10/2015 10:10:00 - 10:35:00

Fusel oil is a byproduct of biomass fermentation in the industrial bioethanol production. It is a mixture of higher alcohols, mainly composed by isoamyl and other short-chain alcohols. Bioethanol industry is rapidly growing and fusel oil production too. Also, in countries where bioethanol is produced in large scale, byproduct utilization has become in an important issue in making the global process less polluting, more profitable and closer to the concept of biorefinery. Fusel oil has been considered as a low-value material that some ethanol producers burn to produce energy, however it can be used as a cheap and renewable source for the production of biosolvents, extractants, flavoring agents, medicinal and plasticizers. Colombian bioethanol industry produces currently around 1100 t per year of fusel oil and it is expected that this production continues growing because bioethanol is used as additive for gasoline in percentages established by local politics that stimulates the biofuels production.

Analysis and conceptual design of different alternatives to produce value-added products from fusel oil is presented. In the initial steps, experimental work in phase equilibrium and esterification kinetics of alcohols mixture found in fusel oil is made. From the experimental results, parameters for the calculation of phase equilibrium and reaction kinetics were obtained and used into the design step. A conceptual design was performed with the construction of residual curves maps (RCM), including and without including the chemical reaction effects, by means of process simulation in Aspen Plus platform. In this way, various technological alternatives are proposed for the separation of alcohols contained in fusel oil and for the production of isoamyl esters (acetate, propionate, benzoate, butyrate, among others). Finally, pilot plant facilities were constructed in order to study and validate the results of the fusel oil separation (one conventional distillation column) and reactive distillation (two reactive distillation columns) steps.

AIChE 2015

Annual Meeting
Salt Lake City, UT
November 08 - 13

Coupling of Ethanol Dehydration and Catalytic Combustion of Methane in a Multifunctional Reactor

FEI University
Chemical Engineering Department
Sao Paulo, Brazil
Luis Fernando Novazzi
Georges Jacques-Marie Formon

Outline

- Motivation and objectives
- Fundamentals / literature review
- Mathematical modeling
- Results and Discussion
- Conclusions

Motivation

- Process intensification / integration: analysis and design of equipment and processing techniques with better efficiency and sustainability
- Coupling of exothermic and endothermic reactions in a multifunctional reactor: gains on capital and operational costs
- Green ethylene, from ethanol dehydration

Objectives

- Main objective: coupling of ethanol dehydration (endothermic) and catalytic combustion of methane (exothermic), in a multifunctional reactor
- Specific objectives
  - Mathematical modeling and simulation of the reactor
  - Analysis of reaction conversion and yield
  - Flow patterns in the reactor

Fundamentals

- The integration of reactions in a multifunctional reactor with recuperative coupling is similar to shell and tube heat exchangers
- One set of reactions flows in the tubes region and the other in the shell side, typically in a fixed bed configuration
- The heat transfer area enables thermal coupling whereas the possible presence of a membrane leads to mass coupling, usually in equilibrium reactions

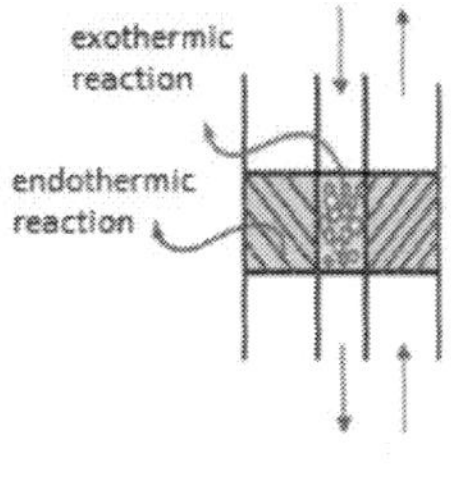

Fundamentals

- Ethylene is an important raw material in petrochemical plants and can be produced through ethanol dehydration, an alternative route when compared to steam cracking of hydrocarbons.
- Ethanol dehydration is an endothermic reaction, which comprises several reaction subsets and is carried out in a fixed bed of alumina, in a temperature range from 513 K to 723 K (Kagyrmanova *et al.*, 2011)

$$C_2H_5OH \rightarrow C_2H_4 + H_2O \quad (1)$$

$$2C_2H_5OH \rightarrow (C_2H_5)_2O + H_2O \quad (2)$$

$$(C_2H_5)_2O \rightarrow 2C_2H_4 + H_2O \quad (3)$$

$$C_2H_5OH \rightarrow C_2H_4O + H_2 \quad (4)$$

$$2C_2H_4 \rightarrow C_4H_8 \quad (5)$$

Literature review

- Although a shell and tube scheme for multifunctional reactors is more common, Zanfir and Gavriilidis (2001) studied the coupling of ethane dehydration and methane catalytic combustion, in a plate heat exchanger
- Important contributions in recuperative coupling without membrane were made by Kolios *et al.* (2002) and Annaland *et al.* (2002), who studied methane steam reforming and propylene production, respectively
- Rahimpour *et al.* (2012) presented an extensive literature review of reaction integration, concerning direct, regenerative and recuperative coupling
  - Hydrogenation of benzene to cyclohexane / dehydrogenation of ethylbenzene to styrene
  - Fischer-Tropsch synthesis / dehydrogenation of cyclohexane to benzene
  - Methanation reactions / decomposition of ammonia

Mathematical modeling

- Assumptions
  - One dimensional plug flow
  - Ideal gas behavior
  - Constant overall heat transfer coefficient U
  - Pseudo homogeneous reactions
  - Steady state

Mathematical modeling

- Mass balance and kinetics
  - Ethanol dehydration

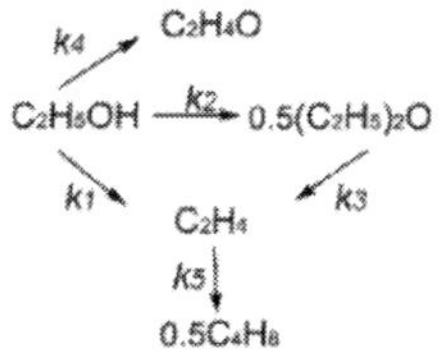

$$\frac{dF_j}{dW} = \sum_{i=1}^{q} r_{ij}$$

$$r_i = A_i\, e^{(-E_i/RT)} P_j^n$$

Reaction	Equation	A_i / (mol·atm^{-n}·kg^{-1}·s^{-1})	E_i / (kJ·mol^{-1})	n	ΔH_r^o / (kJ·mol^{-1})
1	$C_2H_5OH \rightarrow C_2H_4 + H_2O$	$5.41 \cdot 10^6$	147.7	1	45.646
2	$2C_2H_5OH \rightarrow (C_2H_5)_2O + H_2O$	$9.55 \cdot 10^7$	101.0	2	-12.007
3	$(C_2H_5)_2O \rightarrow 2C_2H_4 + H_2O$	$2.79 \cdot 10^{10}$	135.0	1	115.306
4	$C_2H_5OH \rightarrow C_2H_4O + H_2$	$2.78 \cdot 10^8$	138.4	1	68.550
5	$2C_2H_4 \rightarrow C_4H_8$	$1.45 \cdot 10^7$	113.7	2	55.660

Source: Kagyrmanova *et al.*, 2011

Mathematical modeling

- Mass balance and kinetics
  - Catalytic methane combustion

$$\frac{dF_{met}}{dW} = r_i$$

$$r_i = A_i\, e^{(-E_i/RT)} C_{met}^n$$

Reaction	Equation	A_i / s^{-1}	E_i / (kJ·mol^{-1})	n	ΔH_r^o / (kJ·mol^{-1})
6	$CH_4 + 2O_2 \rightarrow CO_2 + 2H_2O$	$1.79\cdot10^6$	81.0	1	-802.62

Source: Abbasi *et al.*, 2012

Mathematical modeling

- Energy balances

$$\frac{dT_{endo}}{dW} = \left(\frac{1}{\sum_{j=1}^{m} F_j C_{P,j}}\right)\left[\frac{4\,U}{\rho_{b,endo} D}\,(T_{exo} - T_{endo}) + \sum_{i=1}^{q} r_{ij}\Delta H_{r,i}\right]$$

$$\frac{dT_{exo}}{dW} = \left(\frac{1}{\sum_{j=1}^{m} F_j C_{P,j}}\right)\left[\frac{4\,U}{\rho_{b,exo} D}\,(T_{endo} - T_{exo}) + \sum_{i=1}^{q} r_{ij}\Delta H_{r,i}\right]$$

Mathematical modeling

- Pressure drop

$$\beta_0 = \frac{G(1-\Phi)}{\rho_0 D_p \Phi^3}\left[\frac{150(1-\Phi)\mu}{D_p} + 1{,}75G\right]$$

$$\frac{dP}{dW} = -\frac{\beta_0}{A_t \rho_b}\left(\frac{P0}{P}\right)\left(\frac{T}{T0}\right)\left(\frac{F_T}{F0_T}\right)$$

Physical properties

$$C_{p,j} = C1 + C2\left[\frac{C3/T}{\sinh(C3/T)}\right]^2 + C4\left[\frac{C5/T}{\cosh(C5/T)}\right]^2$$

$$\mu_j = \frac{K1\, T^{K2}}{1 + {K3}/{T} + {K4}/{T^2}}$$

Mathematical modeling

- Reaction conversion X and ethylene yield Y

$$X = \frac{FO_{ethanol} - F_{ethanol}}{FO_{ethanol}}$$

$$Y = \frac{F_{ethylene}}{FO_{ethanol} - F_{ethanol}}$$

Operational conditions

- Reactor feed
  - Exothermic side: methane / oxygen molar ratio of 8:1
  - Endothermic side: pure ethanol
- Overall heat transfer coefficient: 100 W/m^2·K
- Parallel flow

Reaction	P_0 / kPa	T_0 / K	τ_o / s	Catalyst shape factor	D_p / m
Endothermic	200	550 - 750	0.14	0.81	0.03
Exothermic	250	750 - 950	1.67	0.50	0.05

Results and discussion

Feed temperature effect

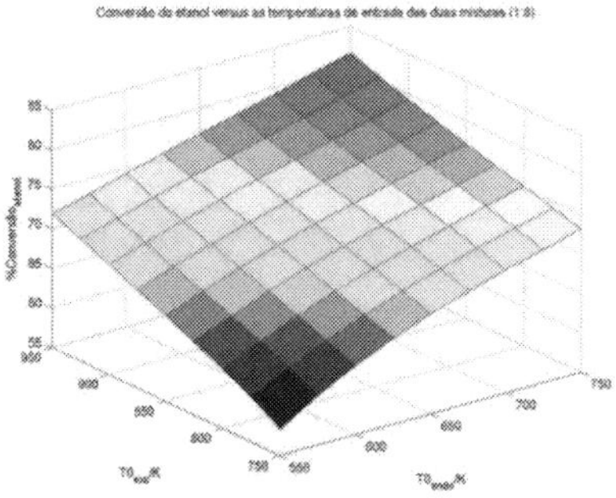

Reaction conversion

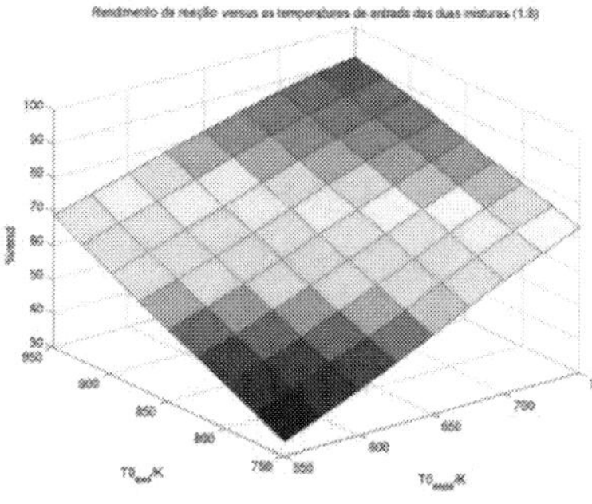

Ethylene yield

Results and discussion

- Spatial time effect ($T_{o,endo}$ = 550 K; $T_{o,exo}$ = 900 K)

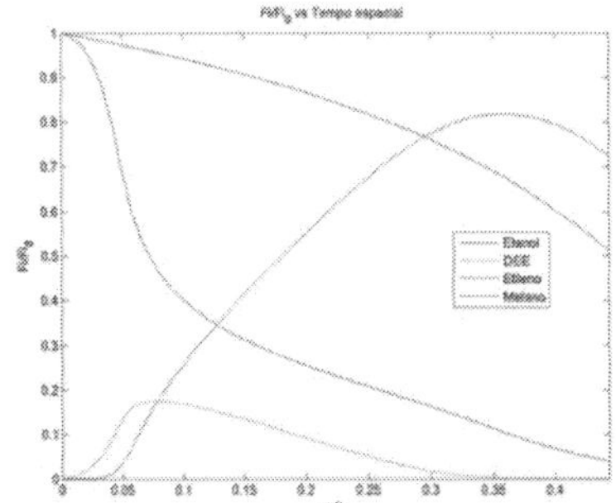

Stream flow

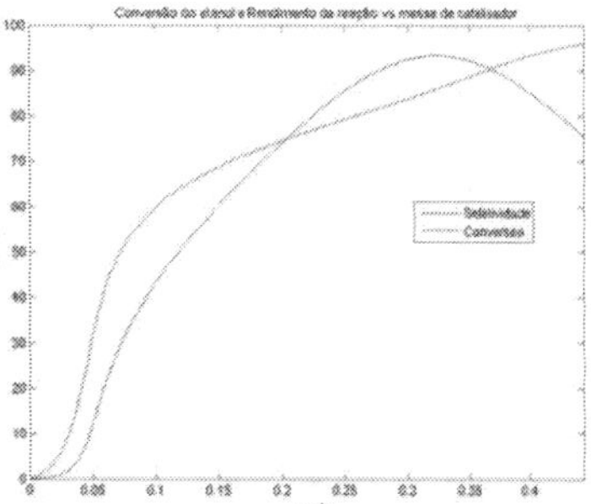

Reaction conversion (red) and ethylene yield (blue)

Results and discussion

- Temperature profile ($T_{o,endo}$ = 550 K; $T_{o,exo}$ = 900 K)

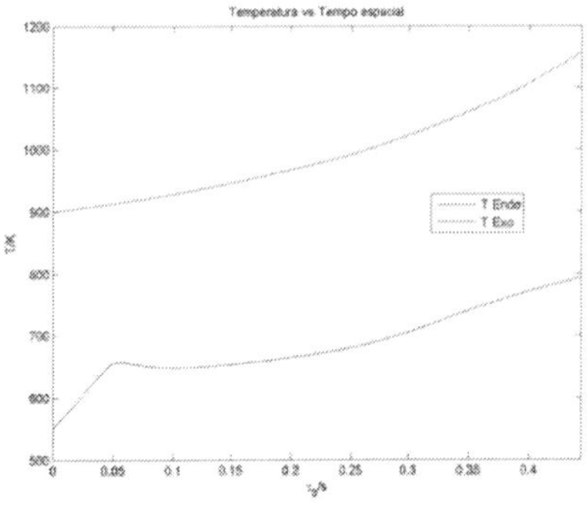

Results and discussion

- Spatial time effect ($T_{o,endo}$ = 650 K; $T_{o,exo}$ = 900 K)

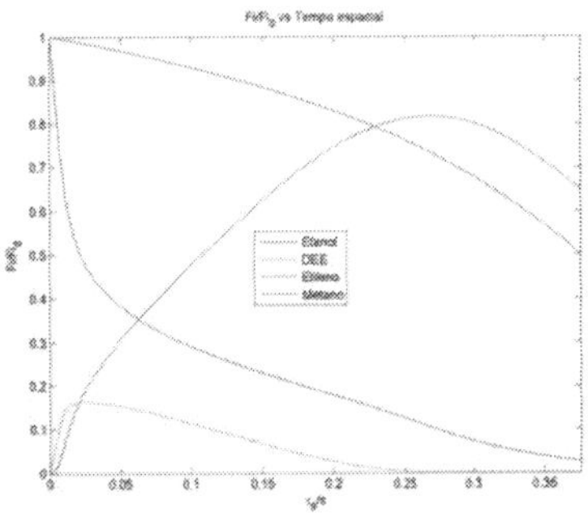

Stream flow

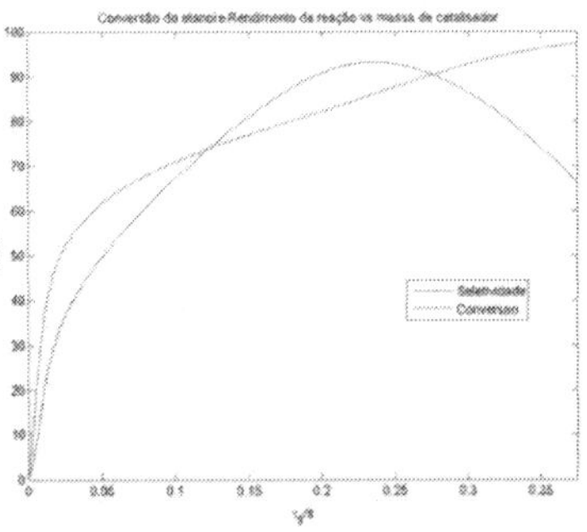

Reaction conversion (red) and ethylene yield (black)

Results and discussion

- Temperature profile ($T_{o,endo}$ = 650 K; $T_{o,exo}$ = 900 K)

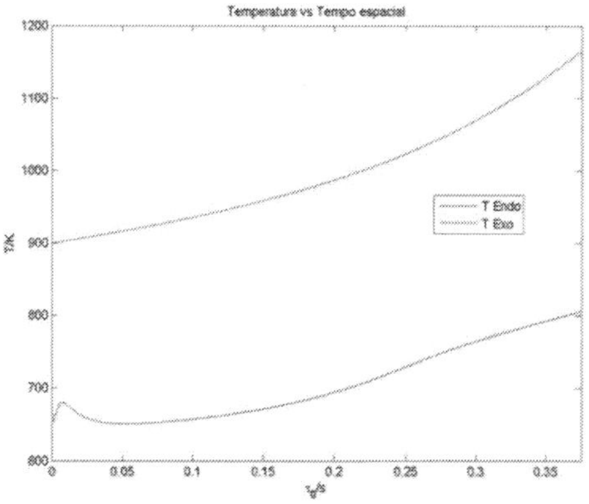

Results and discussion

- Spatial time effect ($T_{o,endo}$ = 750 K; $T_{o,exo}$ = 900 K)

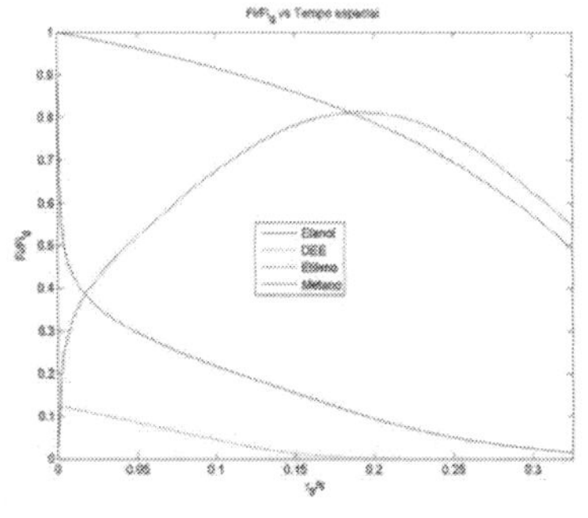

Stream flow

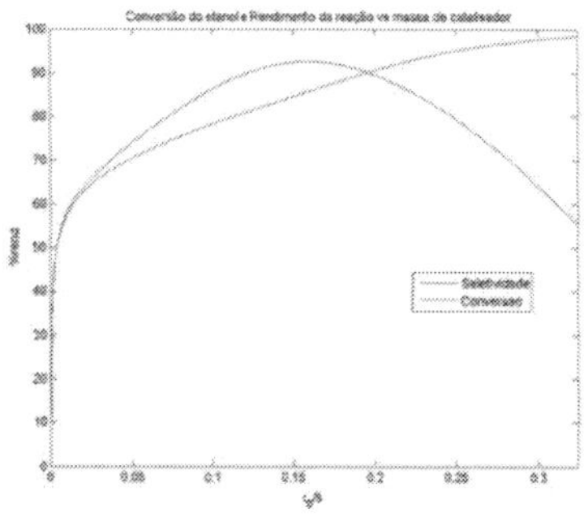

Reaction conversion (red) and ethylene yield (black)

Results and discussion

- Temperature profile ($T_{o,endo}$ = 750 K; $T_{o,exo}$ = 900 K)

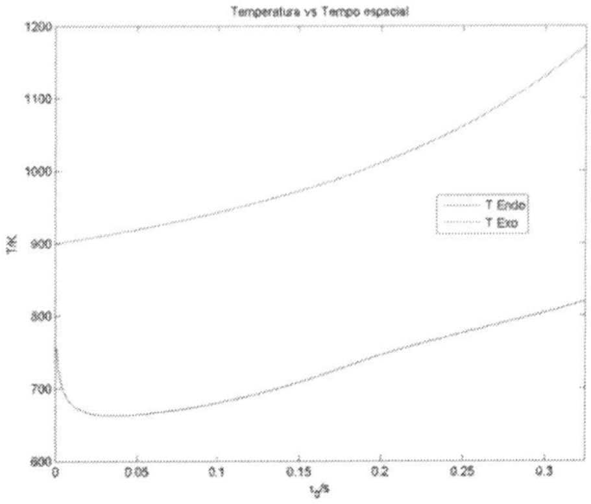

Results and discussion

- Reaction conversion X and ethylene yield Y ($T_{o,exo}$ = 900 K)

$T_{0,endo}$ / K	$\tau_{0,endo}$ / s	X	Y
550	0.323	0.860	0.933
650	0.235	0.859	0.932
750	0.159	0.857	0.926

Results and discussion

- Temperature profiles for different flow patterns ($T_{o,endo}$ = 650 K; $T_{o,exo}$ = 900 K)

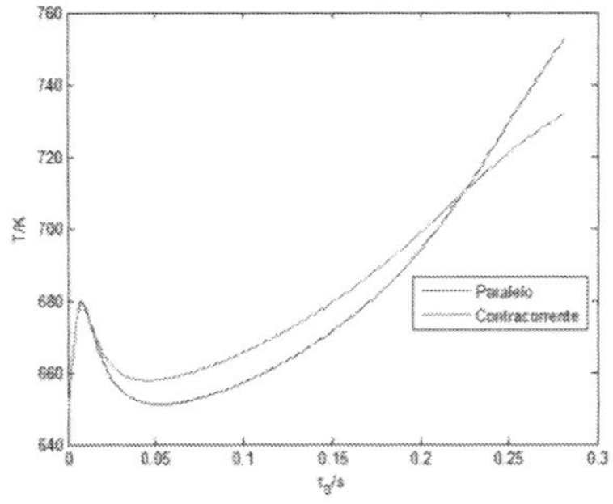

Endothermic reactions

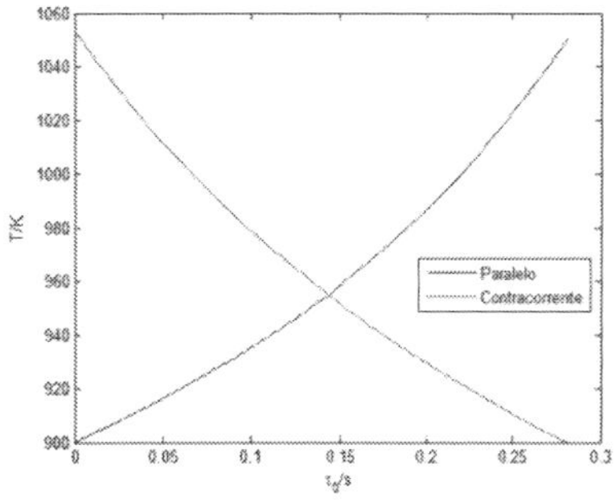

Exothermic reaction

Results and discussion

- Reaction yield *Y* for different flow patterns ($T_{o,endo}$ = 650 K; $T_{o,exo}$ = 900 K)

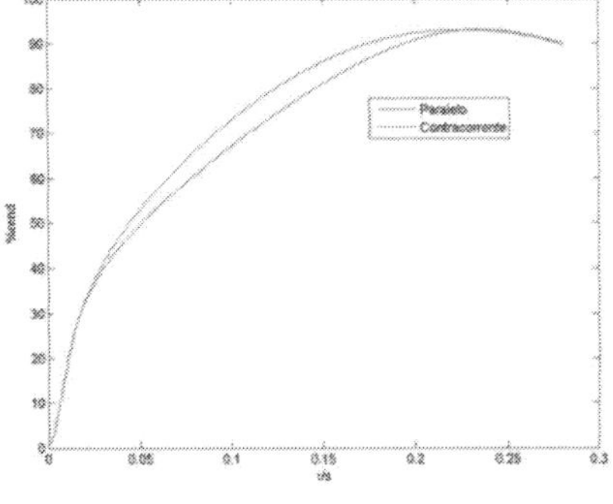

Conclusions

- Coupling of ethanol dehydration and catalytic methane combustion in a multifunctional reactor was evaluated under several operational conditions
- High temperatures favor ethanol conversion as well as ethylene yield
- Spatial times can be adjusted, resulting in an optimal conversion (86%) and ethylene yield (93%)
- When reaction conversion and yield are considered, countercurrent flow is slightly better than parallel flow
- Dynamic issues are also important !

References

ABBASI, R.; WU, L.; WANKE, S.E.; HAYES, R.E. Kinetics of methane combustion over Pt and Pt-Pd catalysts. *Chemical Engineering Research and Design*, v.90, p.1930-1942, 2012

ANNALAND, M.V.S.; NIJSSEN, R.C. A novel reverse flow reactor coupling endothermic and exothermic reactions: an experimental study. *Chemical Engineering Science*, v.57, p.4967-4985, 2002

KAGYRMANOVA, A.P.; CHUMACHENKO, V.A.; KOROTKIKH, V.N.; KASHKIN, V.N.; NOSKOV, A.S. Catalytic dehydration of bioethanol to ethylene: pilot-scale studies and process simulation. *Chemical Engineering Journal*, v.176, p.188-194, 2011

KOLIOS, G.; FRAUHAMMER, J.; EIGENBERGER, G. Efficient reactor concepts for coupling of endothermic and exothermic reactions. *Chemical Engineering Science*, v.57, p.1505-1510, 2002

RAHIMPOUR, M.R.; DEHNAVI, M.R.; ALLAHGHOLIPOUR, F.; IRANSHAHI, D.; JOKAR, S.M. Assessment and comparison of different catalytic coupling exothermic and endothermic reactions: A review. *Applied Energy*, v.99, p.496-512, 2012

ZANFIR, M.; GAVRIILIDIS, A. Modelling of a catalytic plate reactor for dehydrogenation-combustion coupling. *Chemical Engineering Science*, v.56, p.2671-2683, 2001

Acknowledgments

The authors wish to acknowledge the institutional support of FEI University

Thank you!

425115 Isocyanatoorganosilanes – Chemistry and Process Technology

Abstract:

Isocyanatoorganosilanes, commonly used in industrial settings as adhesion promoters between organic polymers and inorganic materials, are of great commercial interest in the coatings, adhesives and sealants markets. Isocyanatosilanes can be made through three different routes: 1) corresponding carbamatosilane thermal cracking with loss of methanol, 2) phosgenation of a corresponding aminosilane or 3) reaction of a chloropropylsilane with cyanate salts. Commercial isocyanatoorganosilanes are primarily made via thermal cracking of corresponding carbamatoorganosilanes. Other routes have not been commercialized due to either safety issues or economical inefficiency. The advantages and disadvantages of these routes, as well as a patent literature review, will be discussed.

419298 Optimization of Semi-Batch Reactive Distillation for the Synthesis of Isopropyl Acetate and Ethyl Acetate

Abstract:

Batch reactive distillation (BRD) is an integrated process which combines reaction and distillation in a batch process. Semi-batch reactive distillation (SBRD) is a alternative design from BRD in which reactant or entrainer is added as a side feed continuously. Besides the advantages of BRD, SBRD can circumvent limitations due to azeotropes.

In the synthesis of isopropyl acetate (IPAC) by BRD, the lowest-boiling stationary point is a ternary azeotrope located in a two-liquid region. After liquid-liquid separation the organic phase still contains 3% reactant and the aqueous phase contains only 95% water. Qi and Malone (2011) showed that the concentration of isopropyl alcohol in the distillate can be reduced efficiently by introducing acetic acid continuously as a side feed. A second non-reactive inverted batch distillation (IBD) can be employed to separate the product IPAC from water with high purity. Hence, SBRD has the potential to improve the process significantly.

The synthesis of ethyl acetate (ETAC) is similar, but the low-boiling ternary azeotrope is located outside of the two-liquid region. Therefore, conventional BRD is not feasible at all. However, if HAC is introduced as a side-feed, the composition of alcohol in the distillate can again be dramatically reduced so that the liquid-liquid split becomes feasable (Tang et al, 2003). Again a second non-reactive IBD comlumn is required to separate the product ETAC from water.

For both systems a tradeoff is encountered between the operation of the SBRD and the subsequent IBD. Specifically, if the constraint on the ammount of alcohol in the SBRD product is relaxed, the operation of the SBRD becomes easier but the operation of the subsequent IBD becomes more difficult. Therefore in order to determine the most economical operating policy, it is necessary to consider both steps. In this work we study the optimization of both processes (IPAC and ETAC) to find the operating policy that minimize the energy consumption (closely related to the operating cost). Because the constraint on the alcohol impurity is the most important variable affecting both columns, designs are optimized for different values of the constraint and then the best overall design is determined.

References:

Qi, W.; Malone, M. F., Semibatch Reactive Distillation for Isopropyl Acetate Synthesis. *Industrial & Engineering Chemistry Research* **2011,** 50, (3), 1272-1277.

Tang, Y. T.; Huang, H. P.; Chien, I. L., Design of a complete ethyl acetate reactive distillation system. *Journal of Chemical Engineering of Japan* **2003**, 36, (11), 1352-1363.

423194 Aging Effect of Bead-Shaped Activated Carbon Exposed to Volatile Organic Compounds

Abstract:

Aging Effect of Bead-shaped Activated Carbon

Maithri Venkat[1], Prof. Johannes Schwank[1], Dr. Hiroko Ohtani[2]

[1]University of Michigan Ann Arbor, MI; [2]Ford Motor Company, Dearborn, MI

Abstract

The removal of volatile organic compounds (VOC) produced during painting operations poses significant challenges in the automobile industry. Bead-shaped Activated Carbon (BAC) adsorption is one of the most efficient methods used for removing and concentrating VOCs from air. However, the lifetime of this adsorbent is considerably reduced following multiple adsorption-regeneration cycles, thereby affecting the economic feasibility of the process. The research primarily focuses on a fundamental surface chemistry based investigation to assess the molecular mechanism for the performance deterioration of activated carbon and to identify its root cause. As part of these efforts, several VOC molecules have been analyzed for their adsorption tendencies, temperature programmed desorption rate, isotherm measurement and microporous surface area distribution. Furthermore, a systematic study has been conducted for the evaluation of competitive adsorption, characterization of the surface morphology using scanning electron microscopy, optimization of adsorption capability for aged adsorbent and elimination of heel buildup in micropores.

With the aid of inferences gained from the aforementioned experimental analyses, this research facilitates the development of an improved VOC abatement operation.

Keywords: Bead-shaped Activated Carbon, heel buildup, competitive adsorption, aging effect

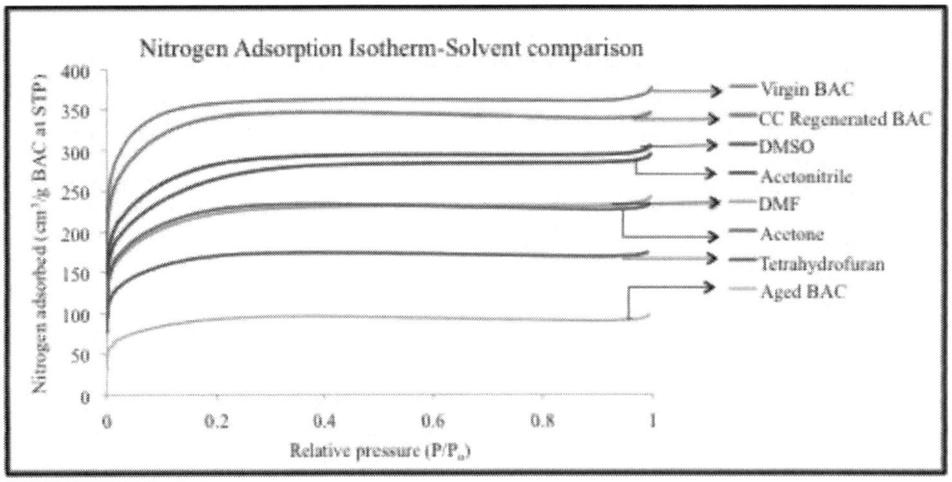

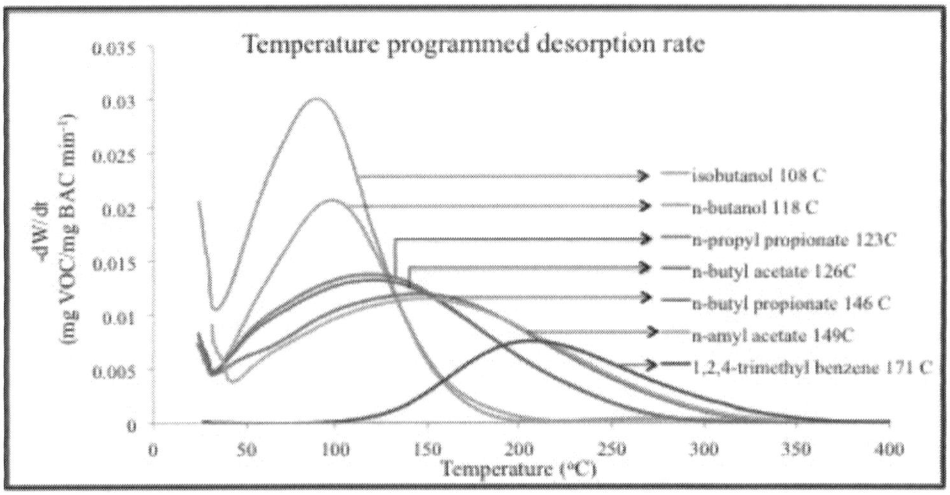

413125 Thermodynamically Favorable Synthesis of Alkali Sulfide Nanomaterials for Next Generation Battery Technology through Reaction with Hydrogen Sulfide

Abstract:

Li_2S has a specific capacity of 1166 mAh/g, and could be paired with a lithium-free anode, addressing safety concerns and low Coulomb efficiency of lithium metal in Li-S batteries. Likewise, Na_2S is a promising cathode material for Na-S rechargeable batteries, which are considered as an option for stationary energy storage in the future. Existing technologies for M_2S production are complex and energy inefficient. In this presentation we describe the efficient synthesis of anhydrous, phase pure M_2S nanopowders through reaction of alkali metals with H_2S, a hazardous chemical produced in large quantities through desulfurization of natural gas and petroleum reserves. The ideal reaction is described by the following equation:

$$2M + H_2S \rightarrow H_2 + M_2S$$

where M = Li, Na, K. As the Gibbs energy of this reaction is very negative ($\Delta G_{rxn} < -300$ kJ/mol), the reaction is expected to be spontaneous and irreversible. A practical challenge of course is developing an efficient route to effectively introduce the alkali metals, which have limited surface areas in solid state. Here we demonstrate this approach by scrubbing an H_2S-containing gas stream with a solution of alkali metal-naphthalene complexes (Na-NAP, Li-NAP). It is demonstrated that the reaction between these species and H_2S proceeds spontaneously to completion at ambient temperature and pressure, scavenging H_2S to below detection limit. X-ray diffraction and electron microscopy confirmed that the solid product is phase pure, nanocrystalline M_2S powders, whose size and shape may be manipulated by reaction conditions. Detailed analysis of both the gas- and liquid phase byproducts using mass spectrometry, FTIR, NMR, and gas chromatography was used to better understand the reaction mechanism. A small amount of H_2 was observed to evolve directly during the reaction, but the majority of hydrogen reacts with naphthalene to form 1,4-dialin, which itself is useful as organic solvent and jet fuel. Studies are underway to assess the electrochemical properties of the resulting nanopowders. The proposed route offers a green chemistry solution that generates valuable products while remediating a hazardous waste.

433373 Triethyl Citrate Production from a Citrate Salt

Abstract:

Triethyl citrate is a colorless, odorless, citric acid derivative used as a stabilizing food additive and pharmaceutical and polymer plasticizer. Citric acid esters are commonly obtained by direct esterification of citric acid and alcohols (e. g. ethanol, butanol), using acid catalysts. The process occurs in a series-parallel reaction scheme, where the monoalkyl citrate is initially obtained, followed by the formation of the diester, and finally Trialkyl citrate. In the case of reacting with ethanol; methyl, diethyl and triethyl citrate (TEC) are the reaction products.

The Citric acid that is the raw material for these esters is produced by fermentation. The downstream processing for its purification is complex and depending on the separation technology, it may require precipitation with calcium, extraction with amines or solvents, chromatographic separation, evaporation and several crystallization steps. All these processes are energy intensive and make citric acid an expensive raw material for the production of commodities such as the plasticizers.

Alternatively, this work evaluated the production of TEC from a calcium salt usually obtained during the precipitation process. Calcium citrate is a fine-white solid, insoluble in water and in ethanol.

TEC production was carried out by an acidification process of the calcium salt in an alcoholic medium using sulfuric acid. During the process, a first reaction step of ion exchange mechanism occurred (reaction between sulfuric acid and calcium citrate to produce calcium sulfate and citric acid in solution), and then the simultaneous esterification of the dissolved citric acid occurred. The latter reaction is catalyzed by the sulfuric acid remaining in the reaction mixture.

The experimental study was carried out using the Box-Benhken method where three variables were considered: Temperature, ethanol to calcium salt molar ratio, and sulfuric acid loading. Experiments were carried out in agitated test tubes under isothermal conditions. To reproduce a batch experiment, for each experimental condition, several tubes were loaded with identical initial compositions, but removed from the isothermal system at different times. This allowed to obtain a kinetic profile to follow reaction performance. After calcium sulfate removal, the reacting media was sampled and analyzed by using liquid chromatography.

Results indicated that TEC can be obtained directly from the calcium salt, avoiding the purification of citric acid for the traditional esterification route. A kinetic model of the process was established and it can be used for process up-scaling.

November 8-13, 2015 AIChE Annual Meeting Presentation
Paper 354f

PROCESS DESIGN PACKAGE DEVELOPMENT FOR THE FIRST-OF-A-KIND TECHNOLOGY

MINGHUA YE (Ph.D.)
And
SERGIO MOHEDAS

CHEMISTRY THAT MATTERS™

PROCESS DESIGN PACKAGE PURPOSE

PDP = Process Design Package

Purpose: Develop and document technology information in enough details to allow EPC company to prepare FEED package for plant construction

No. 1

SIMPLIFIED PROJECT SEQUENCE

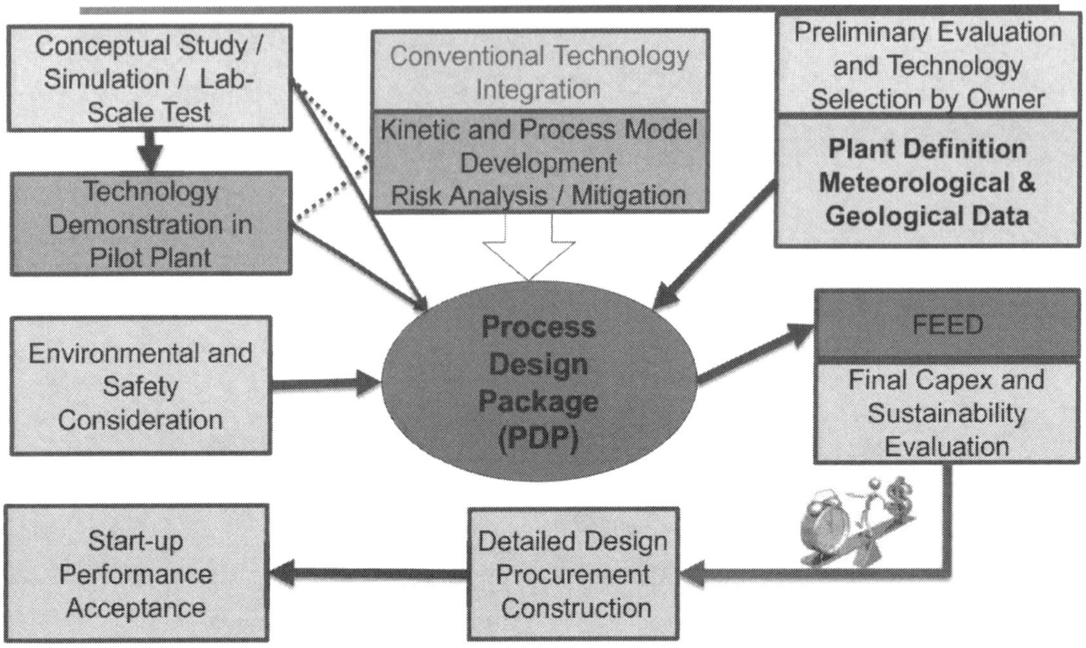

No. 2

PDP SCOPE

- **Focus on:** ISBL design of all priori technology and licensed process units required for plant construction and operation to produce specified product and to achieve expected performance
- **Content:** All necessary process & technology documents

No. 3

PDP CONTENT LIST

➢ PDP Content List:

- ✓ H&MB (controlling cases)
- ✓ Drawing (PFDs, P&IDs, Plot Plan, equipment sketches)
- ✓ Process-specific operating conditions
- ✓ Equipment Process Datasheets & Specifications
- ✓ Control, safety, and instrument specifications
- ✓ Supervision Operating Manual

No. 4

RISK ASSESSMENT AND MITIGATION

- ✓ Accuracy check of thermodynamic package vs reference or unit operation data
- ✓ Reliability and repeatable performance of reaction kinetics and catalyst
- ✓ Flexibility of the process design for feedstock fluctuation and potential process expansion
- ✓ Specifications on feedstock, utility, and special chemicals
- ✓ Material selection for pipeline and equipment
- ✓ HSE (healthy, safety, and environmental issues)

No. 5

REQUIRED INFORMATION FROM R&D OR TECHNOLOGY DEVELOPER

- **Reaction:**
  - ✓ Description of Process Chemistry / Reaction kinetics
  - ✓ Catalyst conditioning (startup) and regeneration procedures
  - ✓ Catalyst deactivation (normal and emergency shutdown)
  - ✓ Catalyst characteristics (regeneration period, normal life, temporary and permanent loss of activity due to poison, coking, etc.)
- **Thermodynamic Package:**
  - ✓ VLE, VLLE, LLE parameters for the process (Azeotropes, extractive distillations, non-idea characteristics)
- **Preliminary Process Simulation Model**
  - ✓ AspenPlus, Pro/II, Hysys, etc.
- **Proprietary Calculations or Properties**

No. 6

REQUIRED INFORMATION FROM LABORATORY AND PILOT PLANT

- **PFD, Process Description, and Procedures**
  - ✓ Pilot plant process flow diagram (PFD)
  - ✓ Pilot plant control and interlock philosophy
  - ✓ Analytical method/procedures and required equipment
  - ✓ Safety considerations: Catalyst / special chemical characteristics, MSDS, and emergency handling
  - ✓ Pilot plant startup, normal and emergency shutdown procedures
- **Test / Operation Records**
  - ✓ Findings / experiences (factors that impact unit performance)
  - ✓ Catalyst performance and life
  - ✓ Records of operating conditions, analyzed results, feed and product quality

No. 7

REQUIRED INFORMATION FROM COMMERCIAL PLANT

➢ Commercial Plant Definition

- ✓ Plant location, design capacity, on-stream time, site size
- ✓ Integration with existing facilities / Provision for future expansion

➢ Feedstock / Product / Battery Limits

- ✓ Feed source, rate and composition
- ✓ Availability of startup / shutdown materials
- ✓ Product / by-product specification
- ✓ Limits on emission or disposal of wastes

➢ Available Utilities:

- ✓ HP, MP, LP Steam, BFW, and other heating media
- ✓ Water (CW, SW,BFW, demineralized, potable)
- ✓ Plant air, Nitrogen, Electricity, Fuel Gas

No. 8

SITE INFORMATION

➢ Meteorological Information:

- ✓ Ambient dry bulb, and wet bulb temperatures
- ✓ Barometric pressure / Site elevation
- ✓ Relative humidity
- ✓ Wind direction and speed
- ✓ Rainfall and intensity / Sandstorm and dust

➢ Geological Information:

- ✓ Historical earthquake information
- ✓ Soil data / Special civil construction requirement

➢ Environmental and Safety Requirements:

- ✓ Waste water, closed drains, relief system
- ✓ Construction and shipping limits

No. 9

STANDARDS AND CODES

- **API** (American Petroleum Institute)
  - ✓ API 521 Guide for Pressure-Relieving and Depressuring System
- **ANSI/ISA** (American National Standard Institute / International Society of Automation)
  - ✓ S5.1-1984 (R 1992): Instrumentation Symbols and Identification Process Design Base and Criteria
- **ASME** (American Society of Mechanical Engineers)
  - ✓ B31.3 Process Piping Guide
- **ASTM**: American Society for Testing and Materials
  - ✓ Special standards of local government and individual company

No. 10

TOOLS (PROCESS AND DESIGN SOFTWARE)

- **AspenPlus**
- **Hysys**
- **CHEMCAD**
- **PRO/II, Provision**
- **In-House Proprietary Design Tools**
- **AutoCAD, SmartPlant, or other drafting software**
- **Other special simulation software (CFD, Chemikin, etc,)**

No. 11

DESIGN CONSIDERATIONS

- CONSIDER SEVERAL DESIGN CASES
  - Feed rate and composition variation
  - Environmental conditions
  - Turn-down, minimum, maximum capacity
- RELIABILITY / CONSERVATIVE DESIGN
- SAFETY, HEALTH, AND ENVIRONMENT
  - Identify failure modes (HAZOP Analysis)
  - Hazardous area isolation and evacuation
  - Unit protection for emergency shutdown
- EQUIPMENT FABRICATION / TRANSPORTATION
  - Equipment access / Working area for loading / unloading / Site assembly
  - Consider available transportation routes, and limitations (maximum size)

No. 12

REQUIRED PDP DELIVERABLES

- Plant elements (project scope)
- Process design basis
- Process description
- Heat & material balance
- Process equipment specifications
- ISBL HSE design considerations and MSDS
- Summary of utility and chemical requirement
- Summary of vapour, liquid, and solid waste emission
- PFD, MSD, P&IDs, and Plot Plan
- Process specifications for instruments and control valves
- Analyser specification
- Supervisory operating manual

No. 13

PROJECT EXECUTION AND MANAGEMENT

- **PDP Delivery Schedule**
- **Milestones**
  - KOM
  - H&MB /PFDs Review
  - P&IDs Review
  - HAZOP Review
  - PDP Review
  - SOM Review

- **PDP Close-out Report**

No. 14

SUMMARY

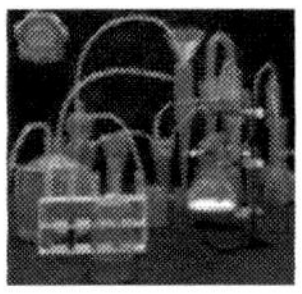

- **Overall Objective**: To develop and document technology information in enough details, based on which EPC company can prepare FEED (BDP) package for final cost estimation, project approval, plant detailed design and construction
- **Focus**: ISBL technical / licensed process units

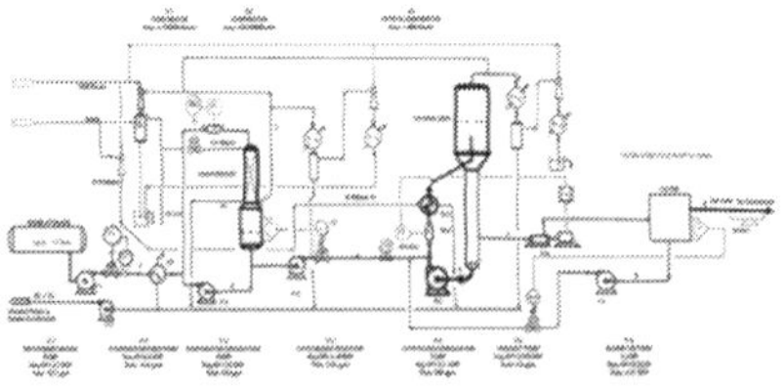

- **Task**: To prepare all necessary process & technology documents and drawings (PFDs, P&IDs etc.)
- **Requirement**: Technology + Plant Environmental + Regulation + Standards + Know-How
- **Resource**: Team work of technology development / R&D, Engineers (process, control, mechanical) with support of specialists
- **Schedule driven**: Must deliver work by dead-line

No. 15

DEVELOPING A PDP IS A TEAMWORK

No. 16

ADVENTURES IN COMMERCIALIZATION

FALL 2015 AICHE PRESENTATION

L. SERNA AND G. HEMMER

CHEMISTRY THAT MATTERS™

AUTHORS & CONTRIBUTORS

Greg Hemmer, Chief Scientist

SABIC's Innovative Plastics Business – Technology & Innovation, High Performance Polymer, Global Process Technology

Lourdes Serna, Senior Scientist

SABIC's Innovative Plastics Business – Technology & Innovation, Core Polymer, Global Process Technology

Viswanathan Kalyanaraman, Chief Scientist

SABIC's Innovative Plastics Business – Technology & Innovation, High Performance Polymer, Global Process Technology

Laura Schultz-Hume, Senior Manager

SABIC's Innovative Plastics Business – Technology & Innovation, High Performance Polymer, Global Process Technology

No. 1

IN THE BEGINNING – MARKET FACING SOLUTIONS

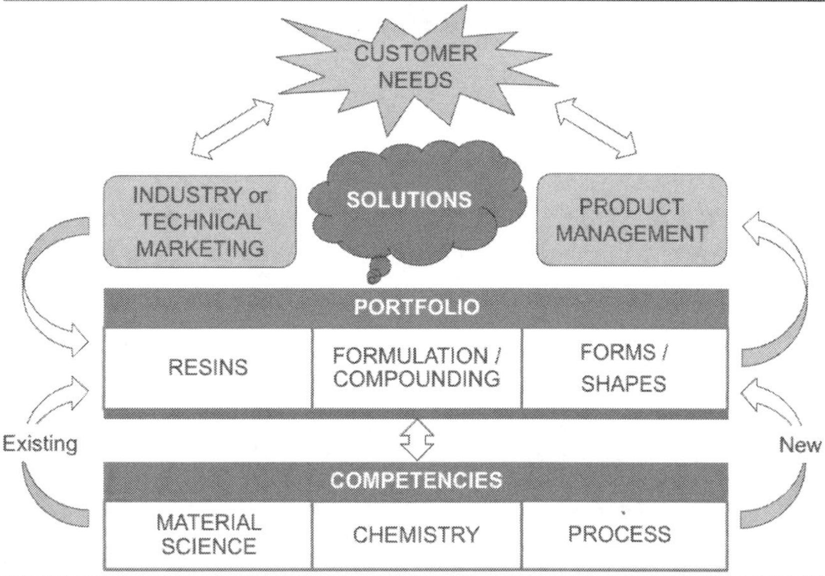

SOLUTIONS – MARKET BASED

ULTEM® Polyetherimide Powder (New Form)

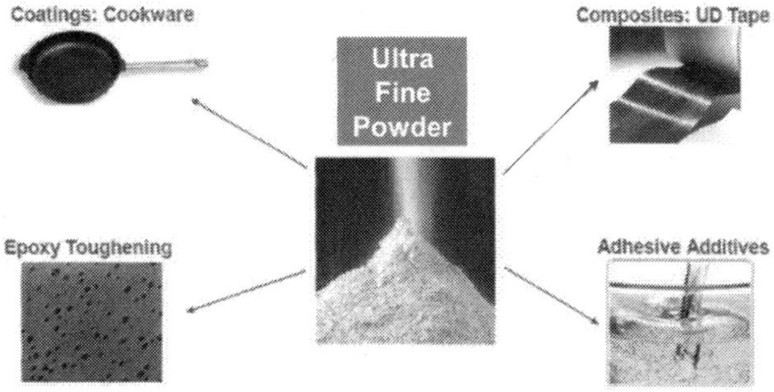

Specialty LEXAN® Polycarbonate (New Resin)

Bezels, lens covers, reflectors, housings, industrial lighting, electrical components, fuses

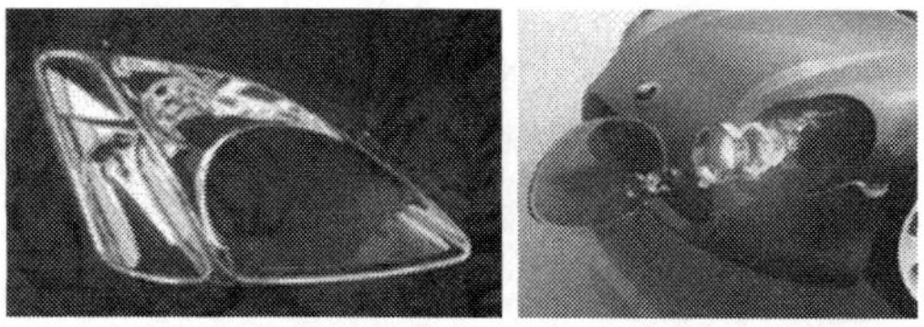

ACTIVITY COMMISSIONED

Stage-Gate™ Project Management Process

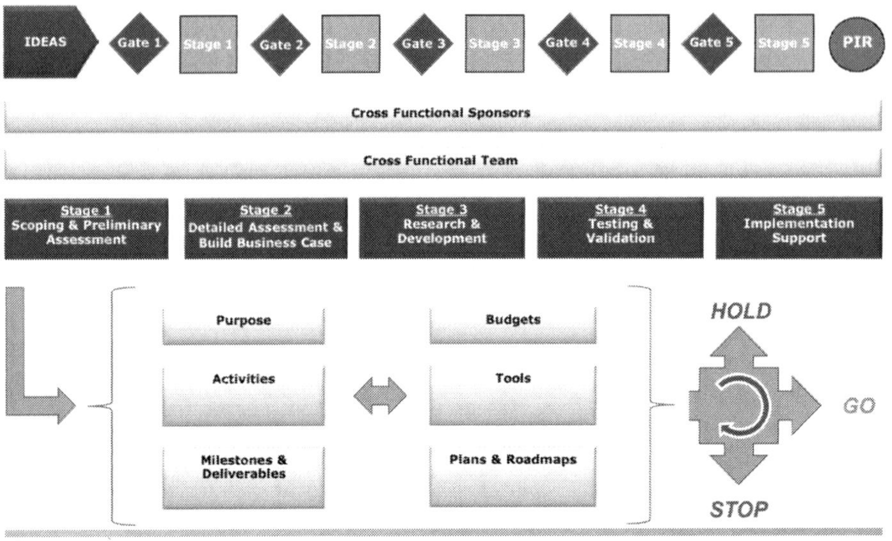

No. 4

DEVELOPMENT PLAN

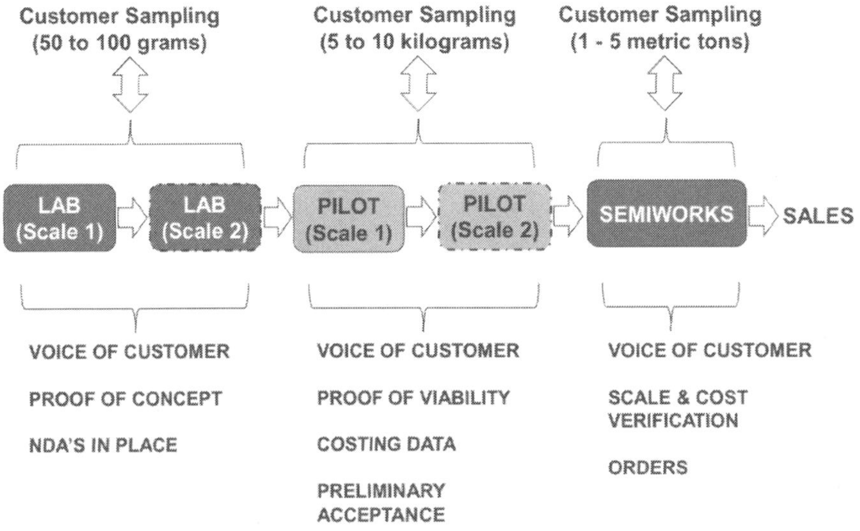

No. 5

DEVELOPMENT CYCLE

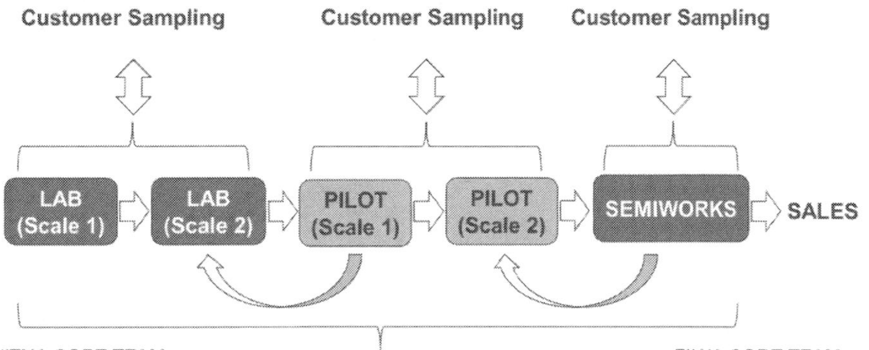

INITIAL CORE TEAM
- LEADER
- PRODUCT DEVELOPER
- PRODUCT MGR
- PROCESS TECH
- EHS&S / IP
- ANALYTICAL
- TECHNICIANS

COMPETENCIES		
MATERIAL SCIENCE	CHEMISTRY	PROCESS

FINAL CORE TEAM
- LEADER
- PRODUCT DEVELOPER
- PRODUCT MGR
- PROCESS TECH
- EHS&S / IP
- ANALYTICAL
- MANUFACTURING

No. 6

DEVELOPMENT ITERATIONS

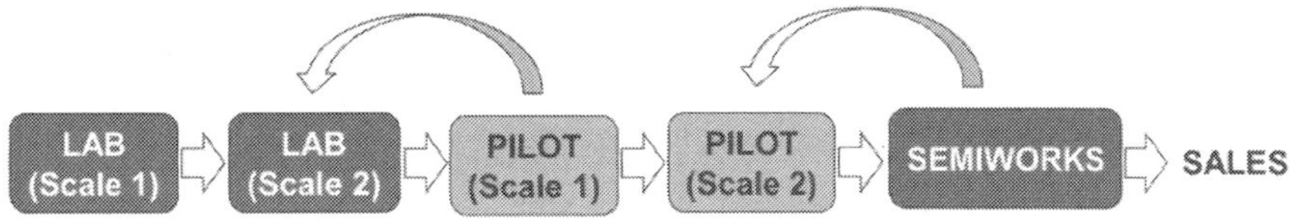

NEW RESIN DEVELOPMENT

Formulation – addition method found to influence product properties → set lab formulation to mimic pilot plant capability

Purification scaling – semiworks rate limitation encountered → designed pilot trial for specific response and identified as commercialization risk

Isolation scaling – commercial product had major property deltas → pilot trials executed to fully map operating space and strategize mitigation options

NEW FORM DEVELOPMENT

Formulation – quantity of ingredient "X" led to excessive foaming in pilot isolation → lab trials carried out to develop mitigation options

Isolation fundamentals – semiworks rate limitation encountered → design pilot trial for specific response; translate all details from pilot to semiworks

Understand scale factor/rate deltas between phases esp. w fixed equipment

Understand the fundamental principles governing the unit operation

No. 7

ADDITIONAL LESSONS LEARNED

ESTABLISH MULTI-FUNCTIONAL / MULTI-DISCIPLINE TEAM

GATED PROCESS FORCES USE OF INCLUSIVE TEAM

R&D

+ BUSINESS

+ PROCUREMENT

+ SUPPLY CHAIN

+ EHS&S

+ IP

+ ENGINEERING

+ MANUFACTURING

STRUCTURED DEVELOPMENT PLAN

GATED PROCESS FORCES RIGOR IN ACTIVITIES

VOICE OF CUSTOMER

+ ALIGNMENT

+ VALUE PROPOSITION

+ WELL DEFINED SCOPE

+ RISK MITIGATION (FMEA)

+ VALIDATION PLAN

+ ACCOUNTABILITY (MILESTONES)

COMMERCIALIZATION FACTORS

ALLOW SUFFICIENT TIME

LOCATION VS SUPPORT

EXPECT THE UNEXPECTED

BEWARE COST PROJECTIONS BASED ON PILOT DATA

AIR PERMITS / EHS VS SOLVENT SELECTION

EQUIPMENT BASE VS PROCESS NEEDS (ADAPTATIONS REQ'D?)

ANALYTICAL BASE VS SPECIFICATION

No. 8

THANK YOU FOR YOUR ATTENTION!!!

ANY QUESTIONS?

No. 9

Reactive Distillation with Pervaporation

An intensifying approach for carboxylic acid recovery from aqueous effluents

S. Lux, D. Painer, T. Winkler, M. Siebenhofer

Outline

- Motivation
- Process concept
- Experimental

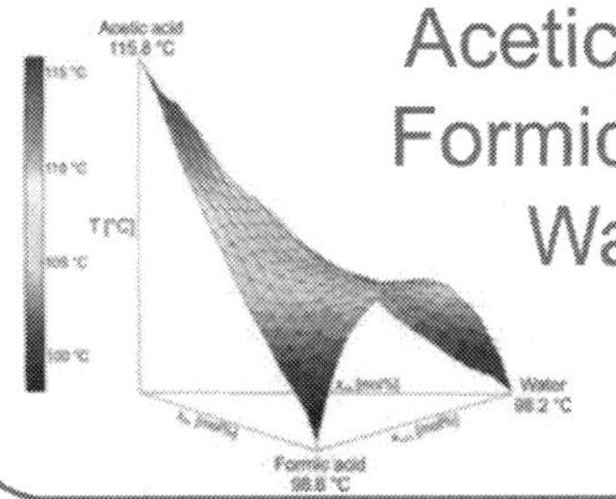

Acetic acid – Formic acid – Water

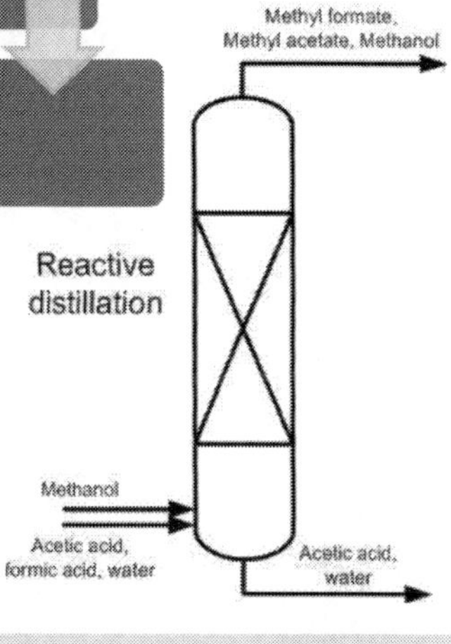

TU Graz

Motivation

Biomass as feedstock

- Complete utilization

Multicomponent mixtures

- Low concentration of valuable products
- Thermodynamic limitations

TU Graz

Motivation

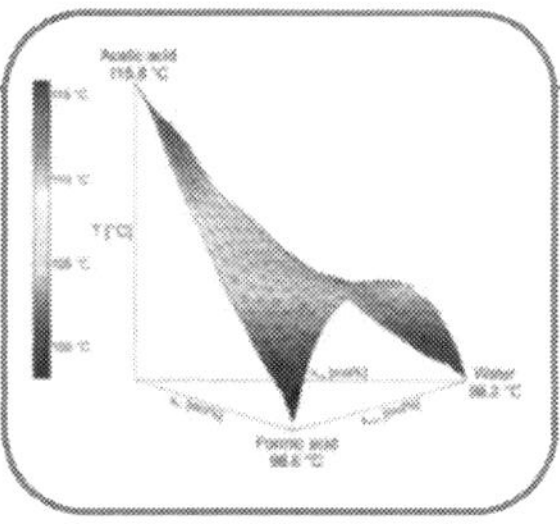

Separation of diluted azeotropic mixtures

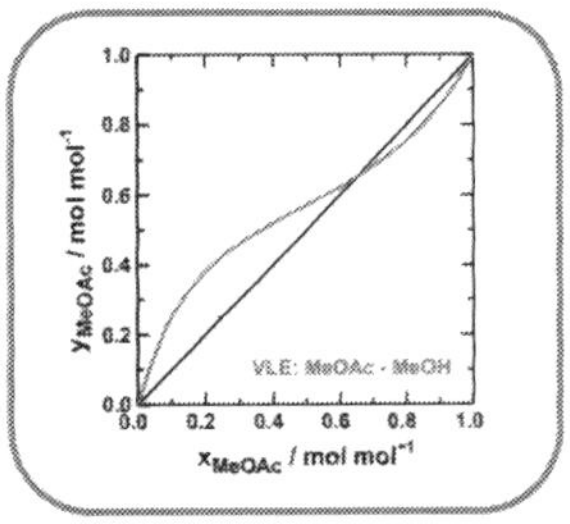

challenging

energy-intensive

costly

Process intensification

Extraction of carboxylic acids

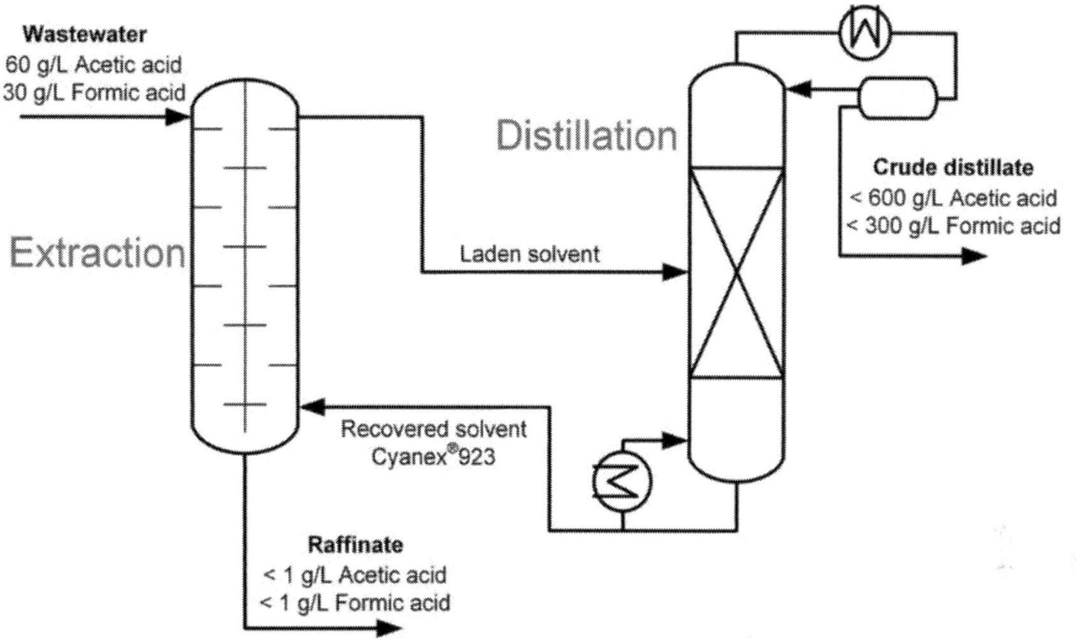

Extraction of carboxylic acids

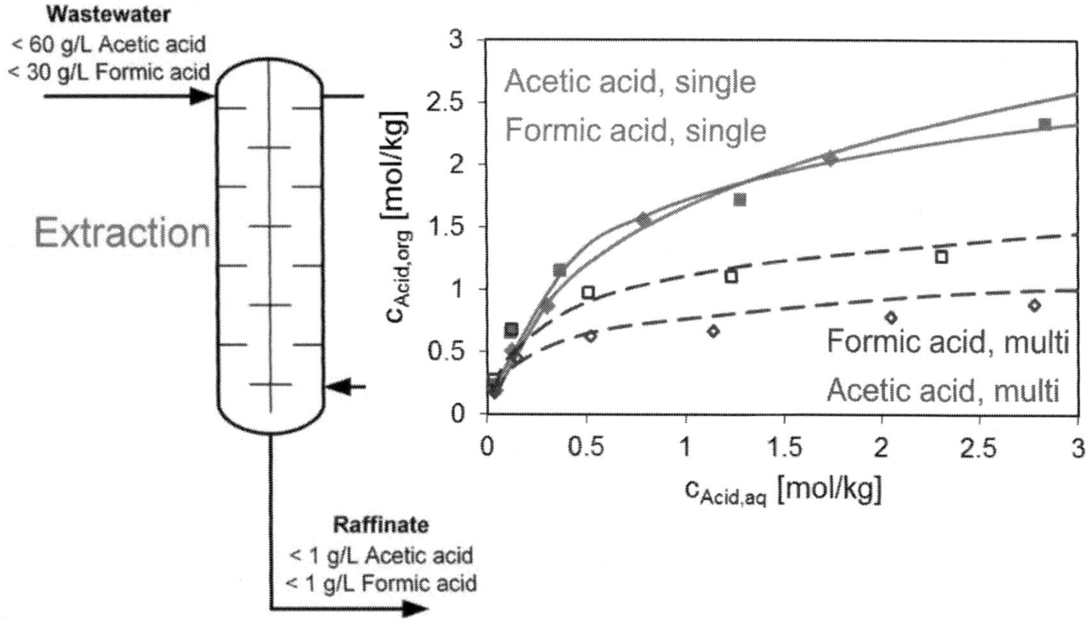

Extraction of carboxylic acids

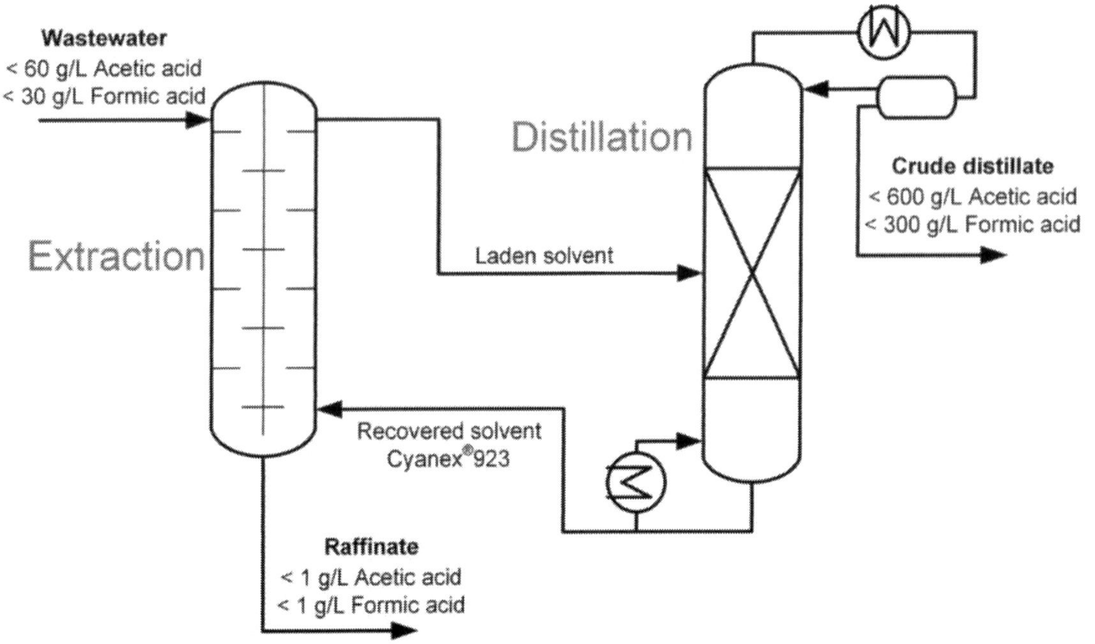

Extraction of carboxylic acids

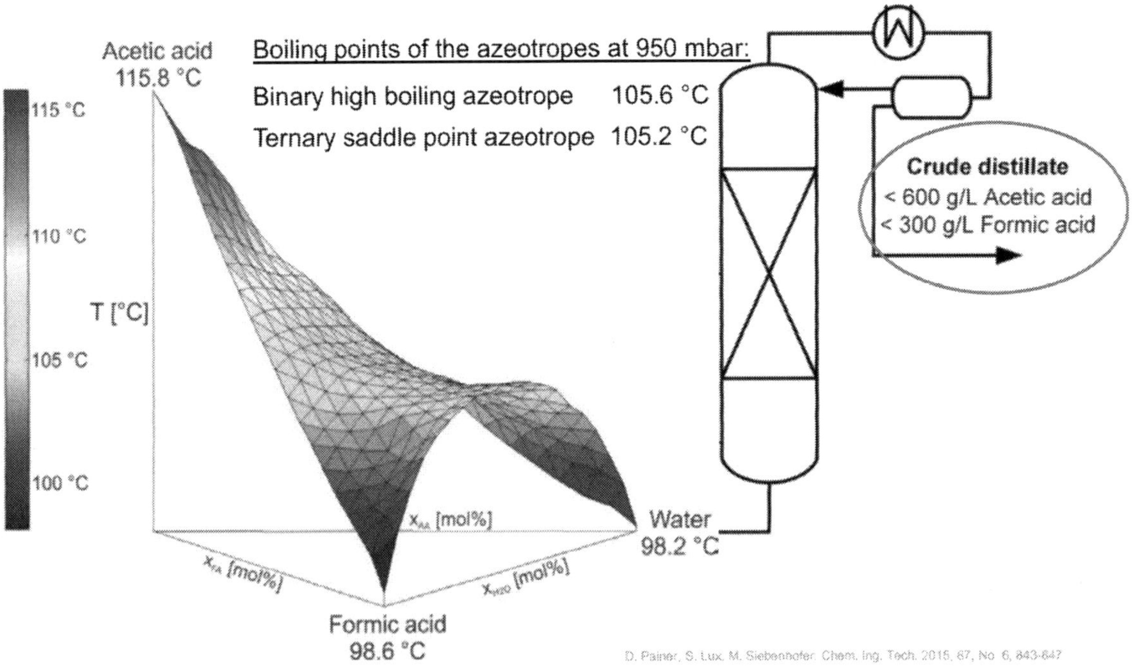

D. Painer, S. Lux, M. Siebenhofer: Chem. Ing. Tech. 2015, 67, No. 6, 843-847

Solver strategy

Change the properties!
Change the state!

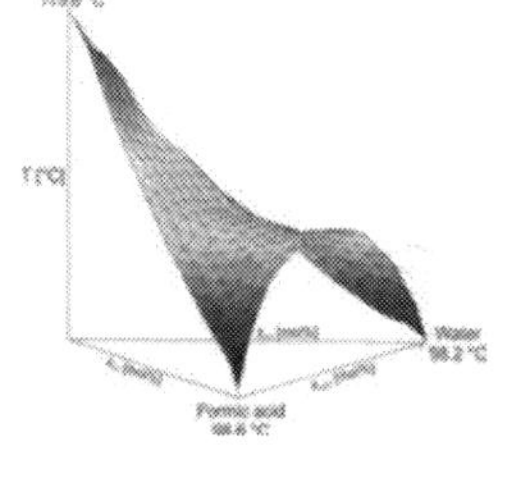

Building blocks of PI

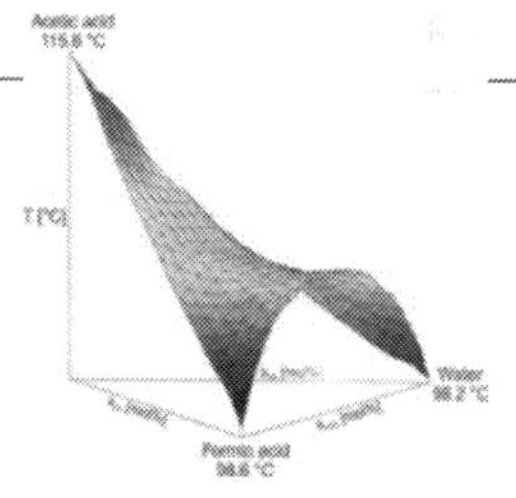

Chemical reaction
e.g. Esterification

Membrane separation
e.g. Pervaporation

Distillation

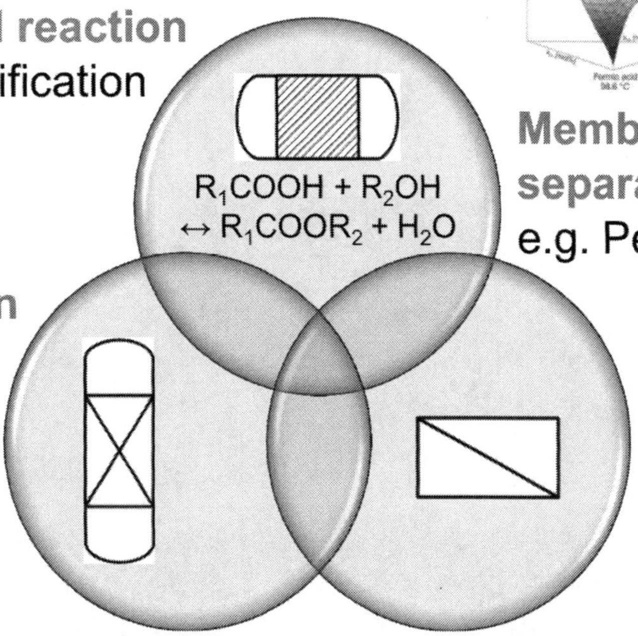

Recovery of carboxylic acids

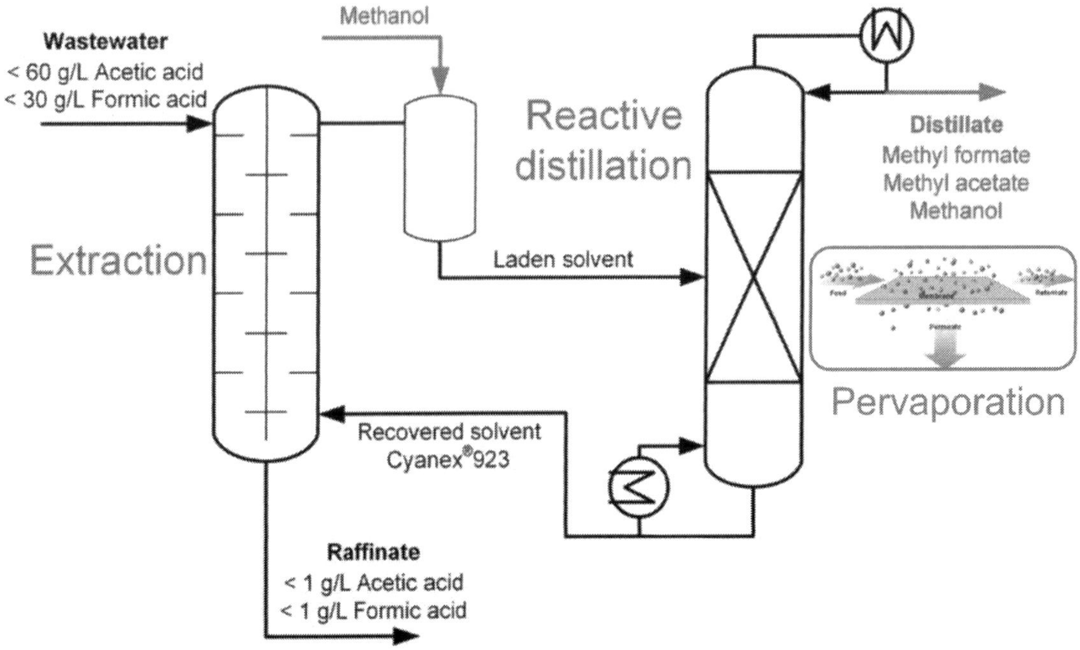

Change the properties

- Esterification with methanol

$$HCOOH + CH_3OH \overset{H^+}{\longleftrightarrow} HCOOCH_3 + H_2O$$

$$CH_3COOH + CH_3OH \overset{H^+}{\longleftrightarrow} CH_3COOCH_3 + H_2O$$

	Δhv [kJ/kg]	Boiling point [°C]
	Bp, 1.013 bar	1.013 bar
Water	2265	100
Formic acid	479	101
Acetic acid	398	117.9
Methanol	1097	64.6
Methyl formate	469	31.8
Methyl acetate	411	57

Esterification T = 40 °C, MeOH:Acid = 1

Formic acid esterification

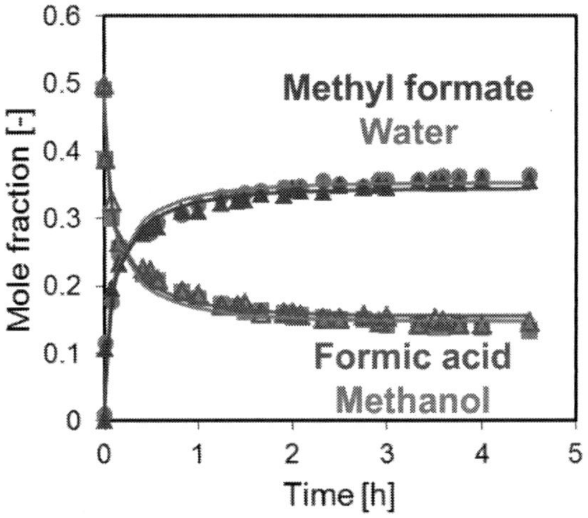

Acetic acid esterification

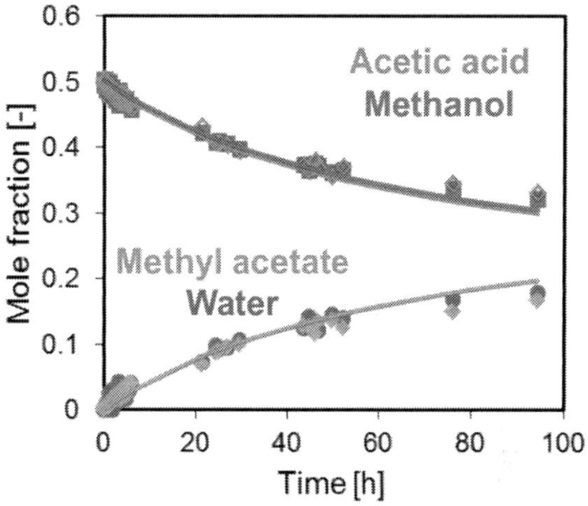

Simultaneous esterification T = 50°C

$$-r_A = \left(\frac{N_{ges}}{V}\right)^3 \cdot \sum (a_{Acid} \cdot K_{Diss}) \cdot (k_f \cdot a_{Acid} \cdot a_{MeOH} - k_b \cdot a_{Ester} \cdot a_{H_2O})$$

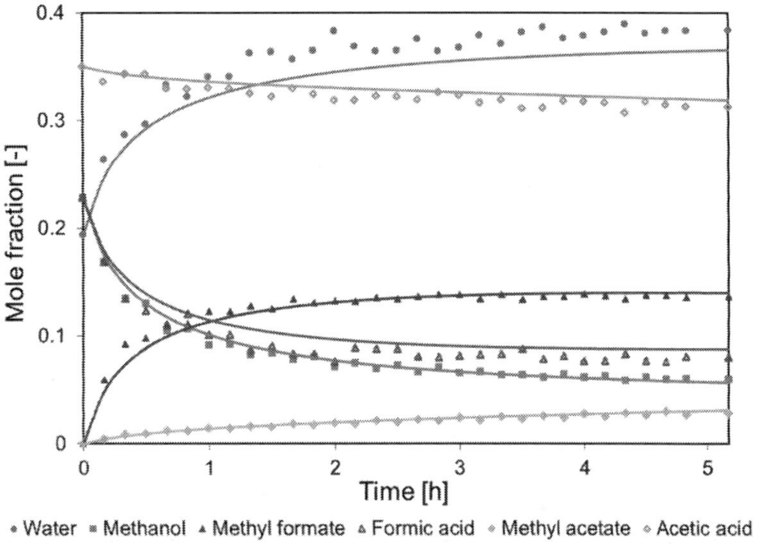

Reactive distillation

Solvent recovery via reactive distillation

PROOF OF CONCEPT

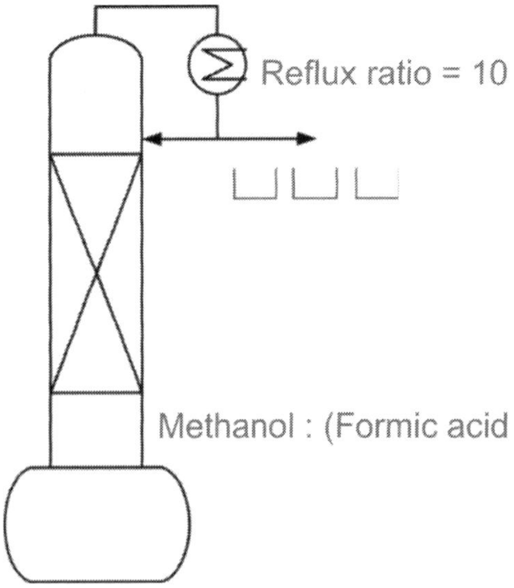

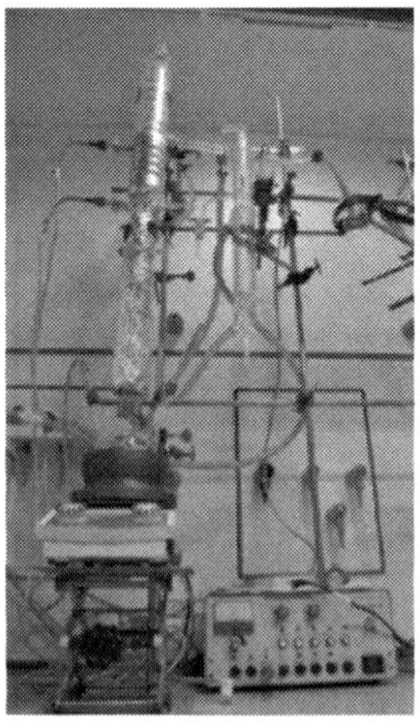

Methanol : (Formic acid + Acetic acid) = 1

Solvent:

70 wt.% Cyanex®923
30 wt.% Undecane

Solvent Load:
13 wt.% Formic acid
24 wt.% Acetic acid
3 wt.% Water

Esterification: Temperature effect

Formic acid esterification

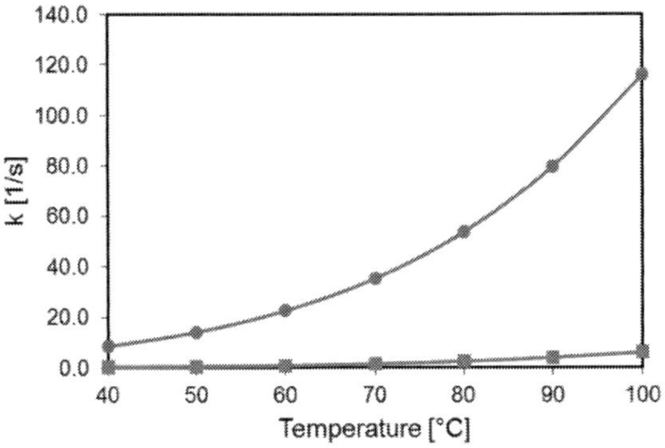

Acetic acid esterification

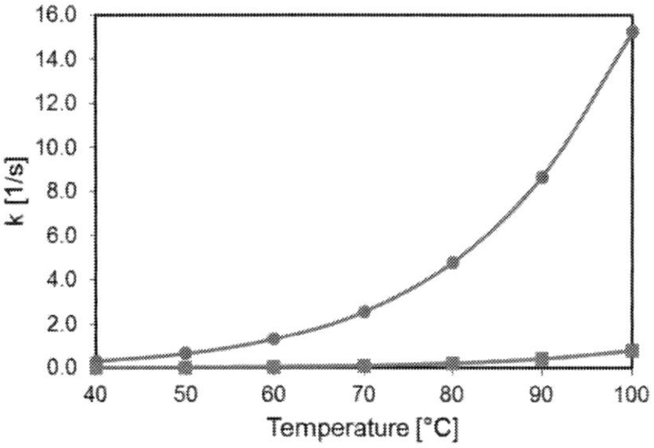

Forward reaction

Back reaction

Solvent recovery via reactive distillation

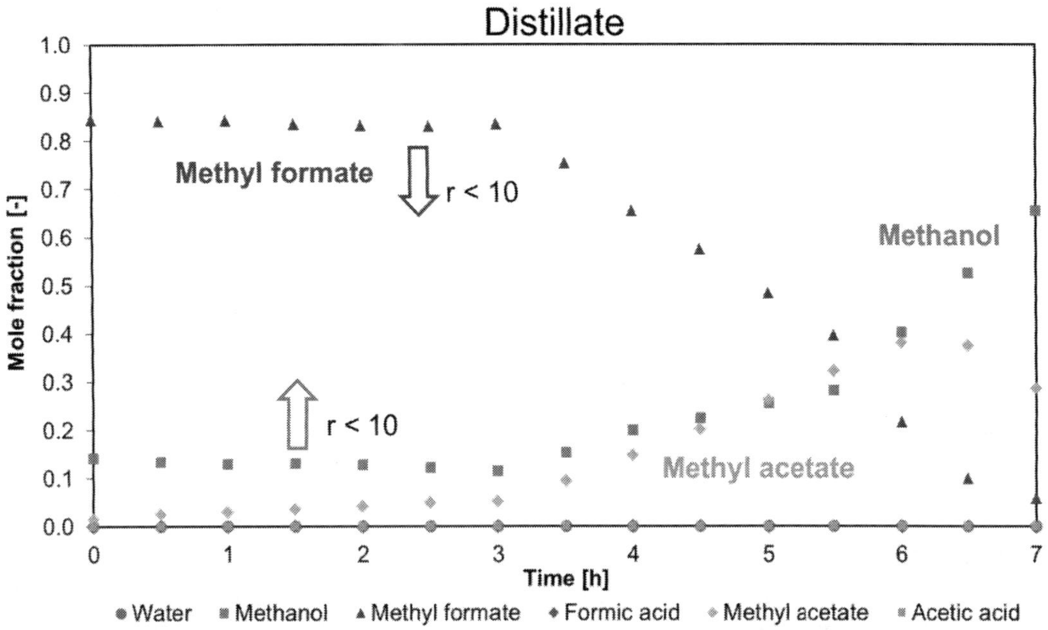

Solvent recovery via reactive distillation

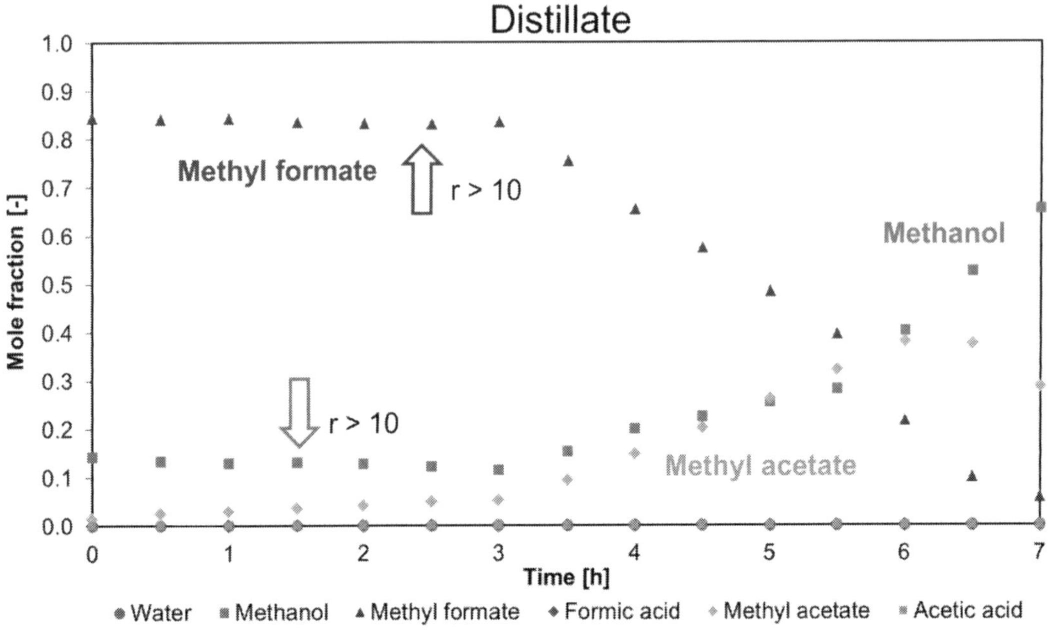

Pervaporation

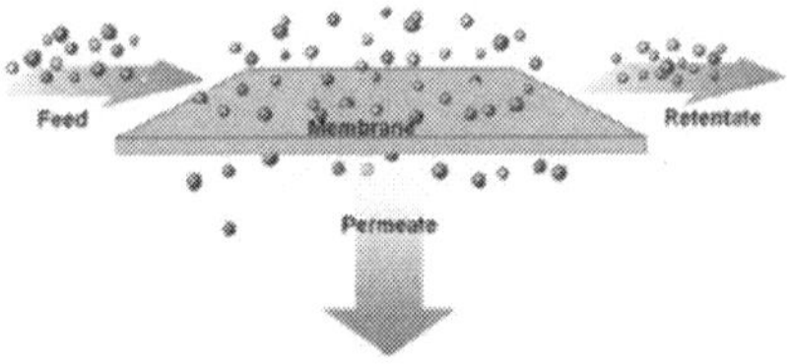

Pervaporation

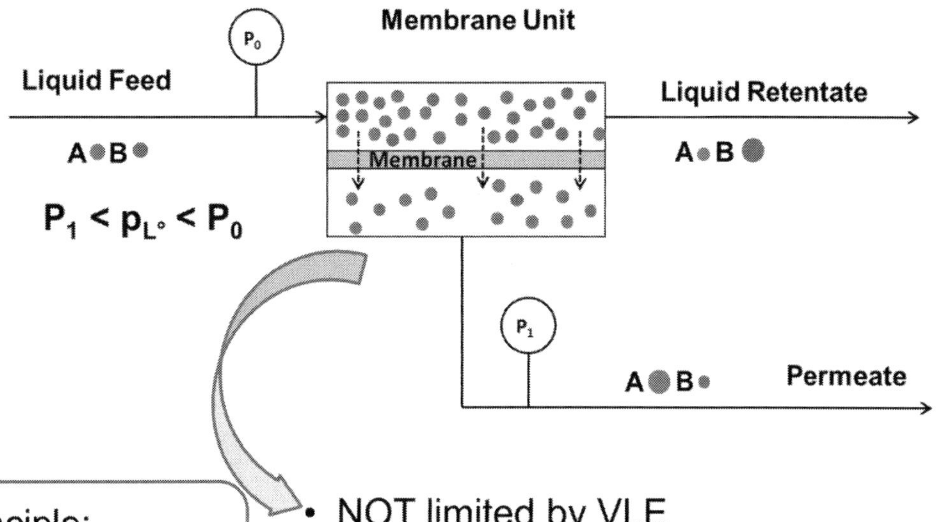

Principle:
Solution-diffusion properties

- NOT limited by VLE
- low operation temperature
- selective evaporation => energy savings

Membranes

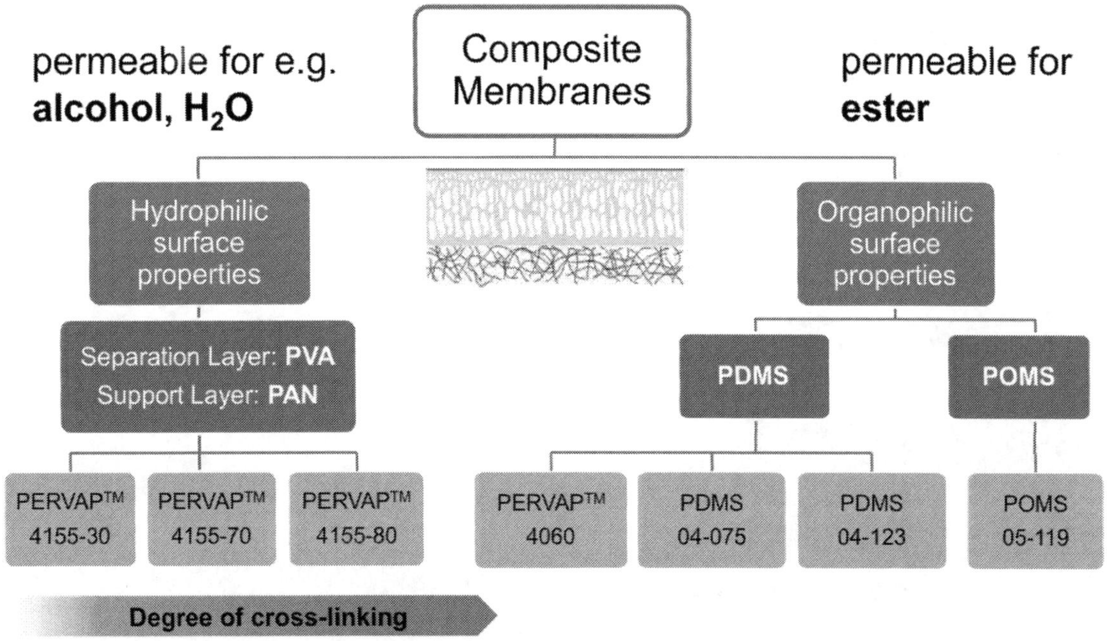

Membranes

permeable for e.g. **alcohol, H_2O**

Composite Membranes

permeable for ester

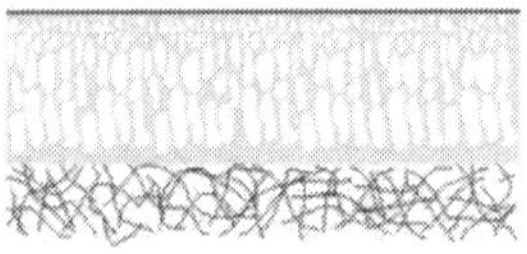

Hydrophilic surface properties

Separation Layer: **PVA**
Support Layer: **PAN**

PERVAP™ 4155-30 | PERVAP™ 4155-70 | PERVAP™ 4155-80

Organophilic surface properties

PDMS | POMS

PERVAP™ 4060 | PDMS 04-075 | PDMS 04-123 | POMS 05-119

Degree of cross-linking

Experimental Setup

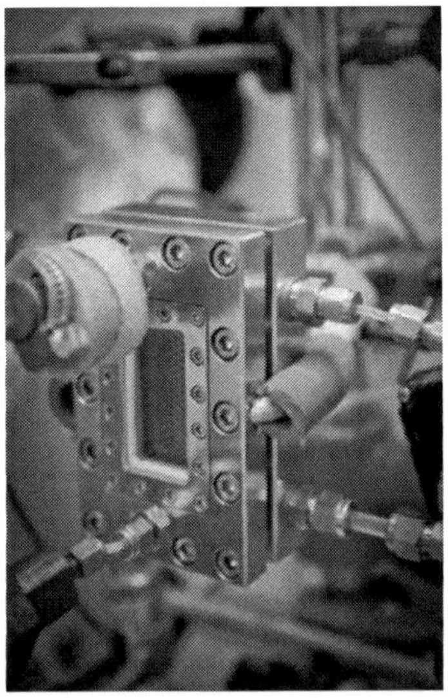

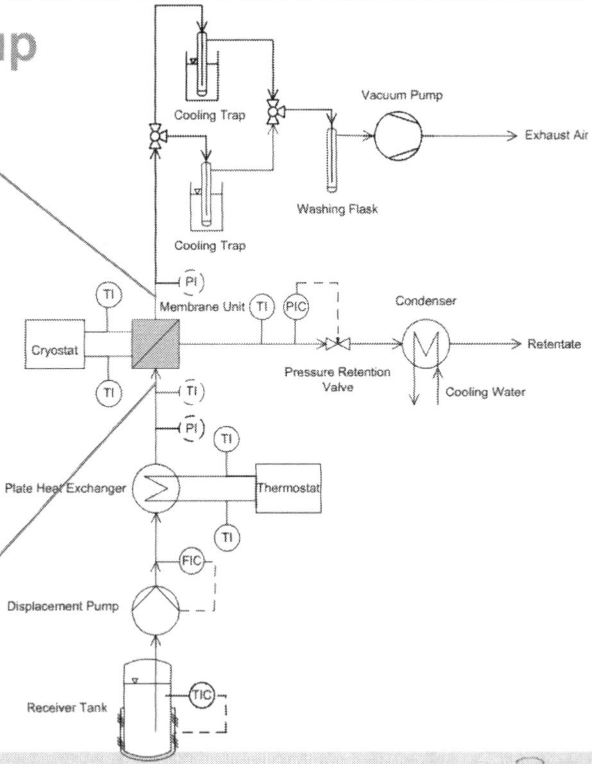

Membrane area: 27.5 cm^2

Experimental Setup

Recovery of carboxylic acids

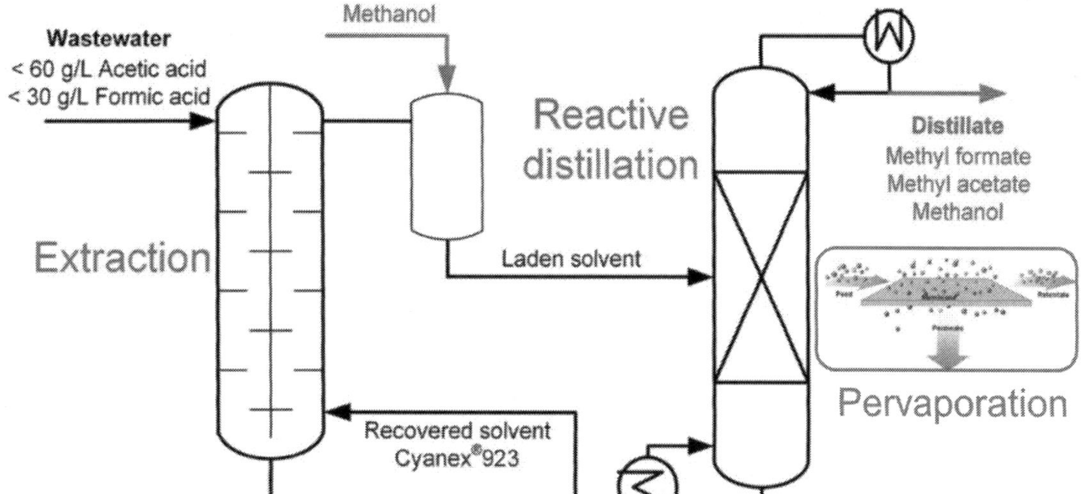

Recovery of carboxylic acids

Low boiling azeotrope:

Methanol – Methyl acetate

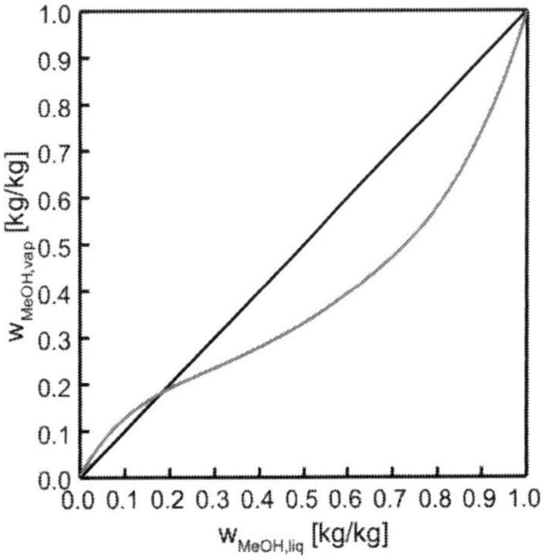

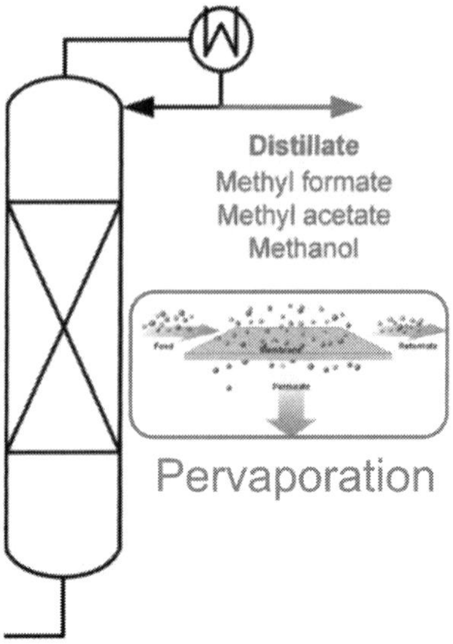

Hydrophilic Membranes: Methyl acetate – Methanol

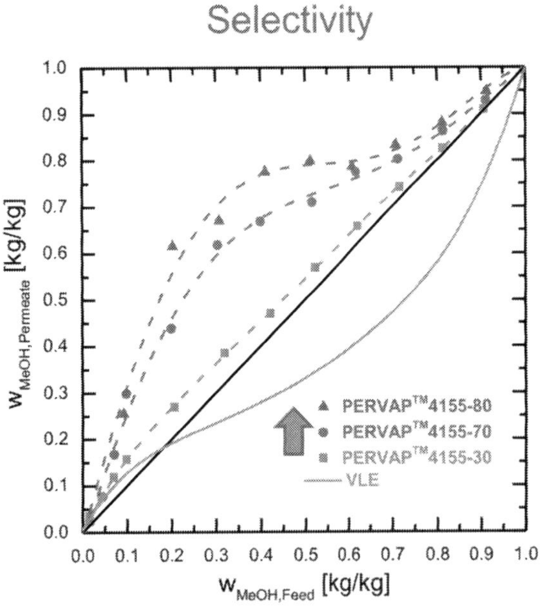

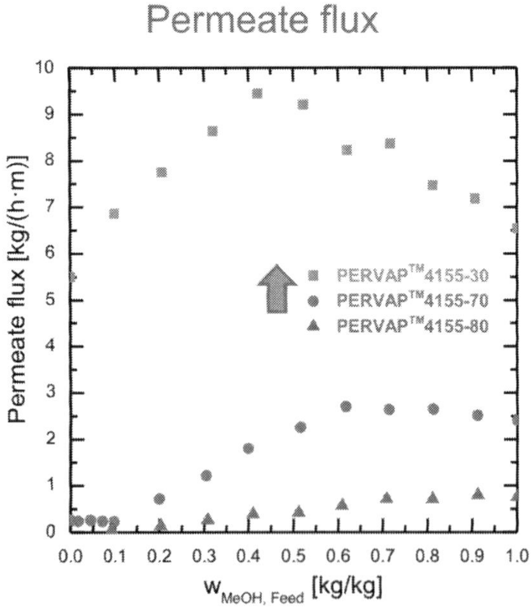

T = 50°C, p_{feed} = 1.5 bar, $p_{permeate}$ = 150 mbar, $\dot{V}_{feed}$ = 15 ml min^{-1}

Permeate Flux

MeOH + MeOAc
MeOH
MeOAc

PERVAP™4155-30

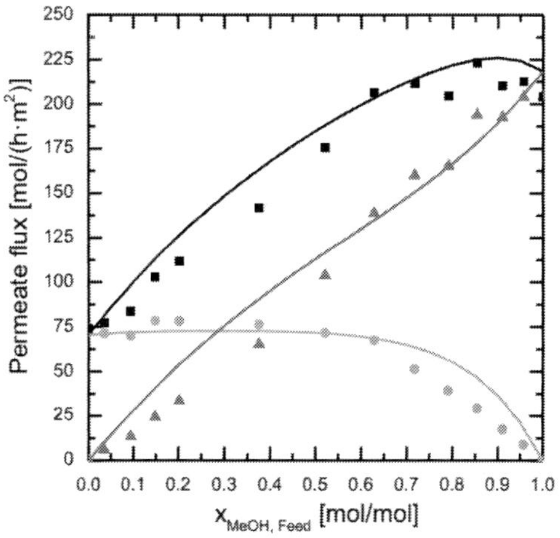

PERVAP™4155-80

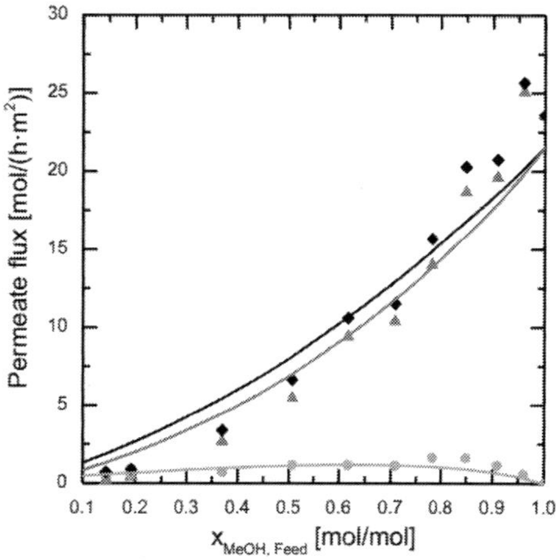

Solution-diffusion model:

$$J_i = \overline{Q_i} \cdot \Delta p_i \cdot S$$

swelling coefficient S

$$S = \frac{a + x_{SC,F}^{\ n}}{b - c \cdot x_{SC,F}^{\ n+1} + d \cdot x_{SC,F}^{\ n+2}}$$

Methyl formate–Methyl acetate–Methanol

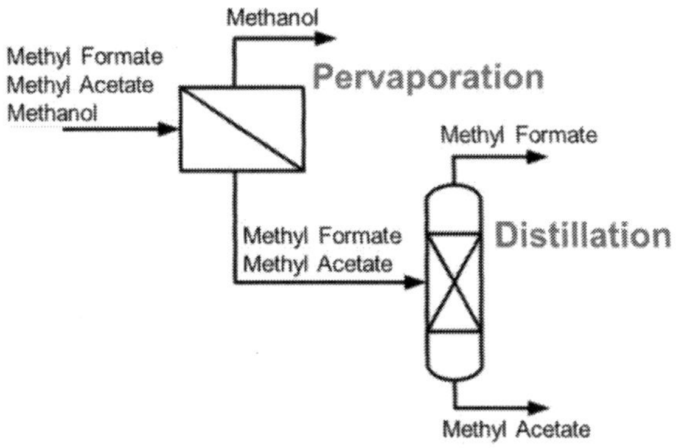

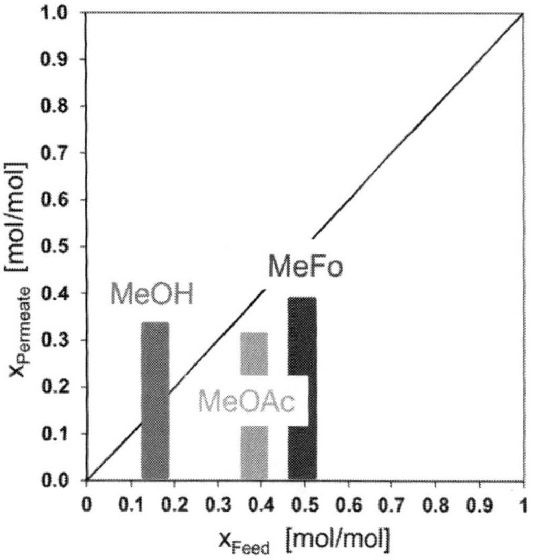

T = 40°C, p_{feed} = 1.5 bar,
$p_{permeate}$ = 50 mbar, $\dot{V}_{feed}$ = 15 ml/min

Pervaporation Curve Map

T = 40°C, p_{feed} = 2 bar,
$p_{permeate}$ = 50 mbar,
$\dot{V}_{feed}$ = 18 ml/min

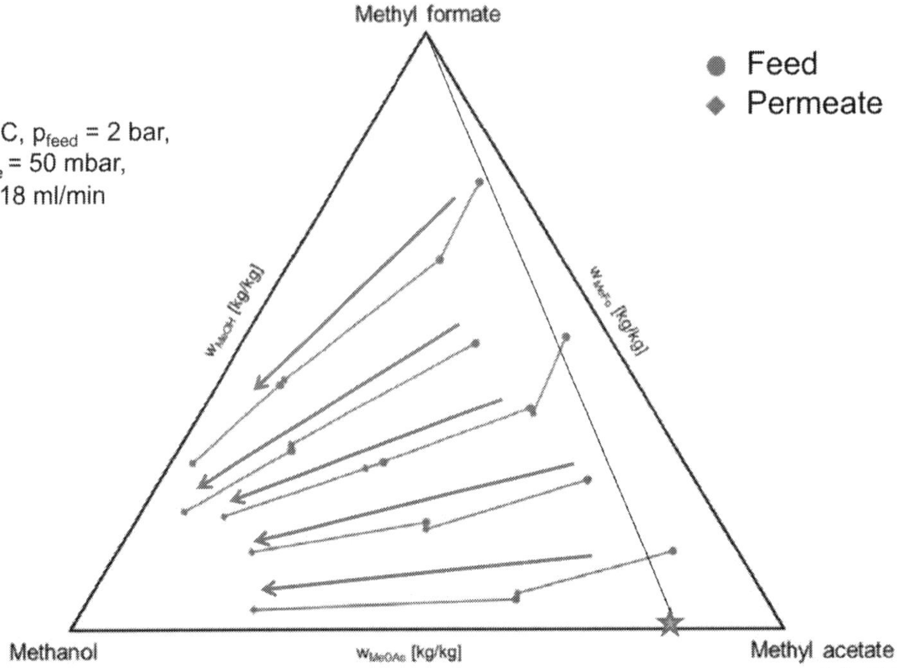

Summary

- Challenge: Separation of a diluted azeotropic mixture
- Process intensification solver strategy:
  - ✓ Change the properties
  - ✓ Change the state
- Recovery of carboxylic acids by reactive distillation
  - ✓ Valuable bulk products esters
- Distillate processing via pervaporation

Reactive Distillation with Pervaporation

Graz University of Technology
Institute of Chemical Engineering and Environmental Technology, NAWI Graz
Inffeldgasse 25c/I, A-8010 Graz
d.painer@tugraz.at

Acknowledgement: Sulzer Chemtech AG, Switzerland SULZER

421614 Performance of 3-D Printed Static Micro-Mixers for Polymerisation Reactions

Abstract:

Executive summary

The performance of two types of static mixers of 6 mm diameter, manufactured by 3D metal printing, have been investigated with respect to

1. Pressure drop
2. Mixing
3. Heat transfer
4. Residence time distribution.

Pressure drop is in line with bigger diameter static mixers. Mixing performance is not only depending on viscosity ratio, but also on flow rate and density difference of the liquids to be mixed. Heat transfer outperforms static mixers manufactured traditionally. The residence time distribution is characterised by a constant value of the dimensionless radial dispersion coefficient (Fourier number).

Introduction

Although some authors claim the first static mixers were already used at the end of the 19th century, most scientists and engineers agree that static mixers were used from the 70s of last century in process industries. The helical type design, which is known widespread under the name Kenics, was patented in 1965. Later other designs were introduced, leading to the multitude of static mixers currently available on the market[i].

Static mixers are nowadays used mainly for the effective mixing of liquids. In extruders they are used to equalise temperature differences in the polymer flow leaving the extruder screw.

The first patent describing the production of a polymer in a static mixer was filed in 1976 by DuPont. In this patent the polyamide production in a static mixer was claimed. The production of polystyrene in a static mixer was reported in 1985[ii].

Due to the favourable surface to volume ratio, tubular reactors have advantages when processing fast, highly exothermic reactions. Despite the advantageous thermal control in a tubular reactor, only few polymerisations are carried out in such a device on commercial scale. Partly, this is due to the inability of tubular reactors to process poly condensation satisfactorily as it requires the withdrawal of volatiles in course of the polymerisation, but also due to the fact that polymer chemistry is still developed in a glass flask mainly for specialty polymers. In the scale up procedure for such new polymers, change of manufacturing technology is generally not considered.

For the development of new polymer grades and formulations small scale equipment is attractive as the amount of material necessary for experimentation is limited and the equipment should fit into a fume hood. Recent developments in pump manufacturing, flow measurement and 3D printing of metals, makes it possible to manufacture such continuous flow equipment that suits the requirements of low volume, small size set ups for new polymer grade development. The behaviour of such equipment however, lies outside the scopes of any investigation published in literature.

Experimental setup

To investigate the mentioned performances, for each of the two static mixer designs (helical tape and X-bar), two experimental setups were built: one for measuring pressure drop and heat transfer and one for mixing performance and residence time distribution. All measurements were carried out with water and glycerol, or mixtures thereof on tubes of 6 mm diameter. Determining the heat transfer coefficient at low flow rates poses a challenge when doing so the "classical" way utilising a double pipe. Not only have two partial heat transfer coefficients to be considered, for low flow rates the temperature difference at the tube outlet diminishes. Both problems have been tackled by applying electrical heating and using the mixer tube itself as the heating element. The mixing performance and residence time distribution have been assessed by colorimetry using a cyan dye. The effluent of mixing tubes of different length was spread between two glass windows and the mixing pattern was acquired by digital photography. The mixing quality was determined by the pixel to pixel difference of extinction of red light relative to blue and green light. The residence time distribution was determined analogous by use of digital video. Both photographs and videos were evaluated using Matlab scripts.

Results

Pressure drops were measured using the heat transfer tubes without heating at variable flow. Both types of mixers were used of two different lengths: 10D and 35D. The results are in agreement with results obtained for larger diameter measurements, not manufactured by 3D printing.

Heat transfer turned out to be higher compared to published data in literature. The increase can be attributed to the different manufacturing technique. Traditionally, static mixers are inserts into tubes and there is no good contact to the tube wall. In the 3D printed tubes, the mixers and tubes are one piece of metal resulting in maximum possible thermal conductance from tube into the mixing element, enlarging the effective heat transfer area. Measuring for two different lengths, 10D and 35D, revealed no dependence of the Nusselt number on L/D of the tube.

In public literature, ***mixing performance*** has been published up to a viscosity ratio of 1:100. In practical use, mixing performance for viscosity ratios starting at 1:100 and up to 1:5000 are of interest. Therefore the mixing performance at viscosity ratios 1:100 and 1:1000 have been investigated. The result is a significant worse mixing performance at a viscosity ratio of 1:1000 as compared to a viscosity ratio of 1:100. Moreover, since the liquids that were used also showed a high density difference of over 200 kg/m^3, flow velocity effects were also observed. At low velocity, pooling was observed that gradually gets less when the flow velocity is increased. At last, the mixing arrangement, high viscous into low viscous or the other way round, that yields best results depends on both viscosity ratio and dosing ratio applied.

The residence time distribution was derived from the response curves achieved by frame to frame analysis of the digital video footage. Residence time distribution in a tubular system is generally quantified as an axial dispersion coefficient superimposed onto plug flow. To our opinion, this approach is not suited for laminar flow systems like polymerisation reactors since the extreme at indefinite high axial dispersion coefficient yields the solution for a CSTR and not for a laminar tube flow model. We have the RTD quantified using a radial dispersion coefficient superimposed onto a laminar tube flow model[iii]. This approach yields the correct solution at both the extreme values for the dispersion coefficient. When analysing the value of the radial dispersion coefficient (D), a linear relationship with velocity is observed, resulting in a constant value for its dimensionless form, the Fourier number ($Fo = D*t_{av}/R^2$).

---

[i] Ghanem, A., Lemenand, T., Della Valle, D., Perhossaini, H.: Static Mixers: Mechanisms, Applications, and Characterization Methods - A Review, Chemical Engineering Research and Design, 92 (2014), 205 - 228

[ii] Tien, N.K., Flaschel, E., Renken, A.: Bulk Polymerisation of Styrene in a Static Mixer, Chemical Engineering Communications, 36 (1985),

[iii] Naumann, E. Bruce: Chemical reactor design, optimization, and scaleup (MGH, 2002), Chapter 8.

426599 Ceramic Microchannel Modules for Heat Transfer and Process Intensification

Abstract:

Process intensification, which reduces the number of unit operations, offers sgnificant benefits in capital and operating costs for numerous industrial applications. For a variety of applications, ceramic materials offer benefits of corrosion resistance, high temperature durability (>>700C), and moderate thermal conductivites suited for high efficiency, compact, microchannel components. Microchannel components offer high efficiency due to the small transport distances associated with them. This talk will describe the benefits of using ceramic, microchannel components for high efficiency heat exchange. In addition, examples of process intensification, such as combining partial oxidation and steam methane reforming in a single,ceramic, microchannel component will also be described. The basic approach to designing and fabricating scaleable, modular, ceramic microchannel based devices for industrial scale processing will be described.

419224 Process Intensification of the Synthesis of Poly(vinyl butyral) Using a Micro-Structured Chemical System

Abstract:

In this work, we established a novel micro-structured chemical system to intensify the synthesis process of poly(vinyl butyral) (PVB) based on the mixing enhancement principle of poly(vinyl alcohol) (PVA) – n-butanal aqueous solution with hydrochloric acid. Because of the high mixing performance, higher temperature (30~60°C) with much higher reaction rate was applied. By studying the parameters of flow rate, temperature, feeding ratios of n-butanal and H^+ to the hydroxyl groups in PVA, the apparent butyral group content in PVB reached to more than 60% with a residence time within 1 min in microreactor, and reached to 75% with one more hour's aging time, much shorter than the conventional 8 ~ 10 hours in low temperature batch reactors. The size of primary PVB particle was smaller than 10 μm, which might reduce the time and water consumption in the post washing process. This work provides a good example for operating reactions between macromolecules and small molecules in microreactor system.

421122 Intensified Formation Rate of Methane Hydrate Under Enhanced Heat Transfer Resulted from Phase Change of n-Tetradecane

Abstract:

Heat transfer strongly affect the rate of hydrate growth. A method of direct heat removal by phase change of n-tetradecane was used to intensify the heat transfer during hydrate formation. The growth kinetics of methane hydrate in aqueous slurries containing 25–45 wt% of solid n-tetradecane were investigated at pressures between 4.70 and 6.46 MPa (gauge) and near the fusion temperature of solid n-tetradecane. The growths of methane hydrate were obviously divided into constant and variable rate stages. Two rate laws were established to describe the whole growth process and correlate the experimental data. The methane hydrate growth rates described in this paper were significantly enhanced compared to those obtained from indirect heat removal methods under similar hydrate formation conditions.

427267 Study on the Hydrodynamic Character of Quasi Plug-Flow Reactor for Enzymatic Hydrolysis of Cellulose

Abstract:

Enzymatic hydrolysis of cellulose is one of the major technical obstacles which constraint the large-scale industrialization of cellulosic ethanol. It is necessary to carry out works on the details of reactor design for enzymatic hydrolysis of cellulose. In present study, a method combining experimental and computational fluid dynamics is applied to study the hydrodynamics character of reactors for enzymatic hydrolysis of cellulose.

Rheological equations for different stages of enzymolysis liquid were established through experimental study. The results showed that, the initial rheological equation of enzymolysis liquid is: $\tau=2.12012\gamma^{0.03267}$, the rheological equation of enzymolysis liquid after 10 hours is: $\tau=1.3403\gamma^{0.0228}$, the rheological equation of enzymolysis liquid after 20 hours is: $\tau=0.7540\gamma^{0.2051}$.

The characters of fluid flow for different impeller to reactor diameter ratios at different stirring speeds were studied. The results showed that the reactor using the combined impellers exhibited better fluid flow characteristics. Uniform velocity and viscosity distribution were obtained and low velocity region and high viscosity region almost disappeared. The mixing effects and liquidity were improved.

This paper proposed a Quasi Plug-flow Reactor which was especially suitable for the cellulose enzymolysis liquid with high viscosity for the first time. Clapboards were set in the reactor in order to construct multi series reactor which can significantly improve the liquid velocity distribution and viscosity distribution of enzymatic reactor. The optimal combination of clapboards is type B, with optimal angle of 5 °.

The residence time distribution can be greatly improved if proper clapboard parameters were selected. Quasi plug-flow can be obtained in enzymatic hydrolysis reactor via optimizing the clapboard parameters.

The setting of baffles effectively reduces the fluid viscosity between mixing zone and the wall, which is beneficial to optimize the flow field. The optimal width, number, and inclined angle of baffles are $W/D_i = 0.1$, four and 0 °, respectively.

The simulation results of the effect of stirring speed and clapboard parameters on the fluid flow field distribution and power consumption provide a theoretical guidance for reactor design, geometry structure improvement and scale up.

433551 The Role of Forests in Enabling a Sustainable Chemical Facility - a National Analysis

Abstract:

Industrial facilities contribute to about 21% of GHG emissions and close to 50% of pollution in the entire US creating a need to identify novel ways to mitigate these emissions. Most efforts for reducing environmental impact focus on improving systems efficiency to reduce emissions and resource use. By far, the most common method to reduce emissions from industrial sources has been the use of end-of-pipe controls to reduce atmospheric pollutant concentration. Successful implementation of control technologies has helped in maintaining air quality standards to a bare minimum, just enough to meet regulatory limits. But despite such advances in technology, ecological degradation has shown few signs of slowing down with CO_2 concentration exceeding 400 ppm for the first time in a few million years.

This is mainly because most engineering and human activities have been ignoring the underlying support provided by nature. Ecological systems have the capacity to mitigate the impact of emissions created by engineering activities and supply resources to chemical processes adequate to sustain its functioning. For instance, trees and forests can capture air pollutants by interception of particles on their leaf surface while wetlands have the ability to treat wastewater streams as an alternative to conventional treatment systems. Industries are starting to acknowledge the economic value of ecosystem services and efforts are underway to incorporate the value of natural capital in business bottom line. For example, collaborative efforts between The Nature Conservancy group (TNC) and Dow Chemical have led to the realization that reforestation and land restoration is a novel and practical approach to improve and regulate air quality in an industrial setting. Besides, most industrial facilities have a large amount of land that is mostly left barren and with judicious restoration and protection methods these sites can provide ecosystem services that are beneficial to both the company and local society.

This talk explores the question, "Is there enough land available around industrial facilities that can be restored to mitigate emissions" To answer this question, we consider all point sources of emissions including CO, NO_2, O_3, PM_{10}, $PM_{2.5}$, and SO_2 emissions in contiguous US from the National Emissions Inventory database provided by the US EPA. Based on the location of each source, a rural or urban environment parameter was assigned to each point using the National Census data.

We use Gaussian plume models to predict the diffusivity of particles from point sources and to determine their maximum diffusivity from stacks at each location and determine optimal location for land restoration around facilities . Since particle diffusivity along downwind direction is a function of atmospheric stability, wind pattern and local weather conditions, locations with similar atmospheric characteristics were binned to different datasets to determine the maximum diffusivity of particles downwind. Based on the horizontal and vertical diffusivity of particles at multiple points downwind, a range of buffers around point sources were defined to estimate dispersion. For each point source, the longest buffer region was selected and based on the National Land Cover Dataset (NLCD 2011) barren, shrub and grass lands in the buffer regions were masked to determine the amount of land available for reforestation. To account for the amount of forest land that currently exists, forested land cover was also masked.

Nowak et.al [1] quantified a range of pollutant sequestration values in each county within contiguous US based on the native land cover in the region. Due to a large variability in data for pollutant sequestration between counties within each state, a minimum and a maximum sequestration range for each pollutant was used to identify the land footprint. Land footprint refers to the amount of land that has to be restored in-order to mitigate all the emissions generated by a facility in that particular region. To keep land restoration within size-able and practical limits, point sources within a minimal footprint less than 1 km^2 were observed further. Using a Euclidean Allocation tool in ArcGIS, the total available land for restoration within the buffer region for each point source was determined.

Preliminary results indicated that for more than 25% of these point sources, restoring the land around these facilities can provide enough services to offset the demand for emissions other than CO2 completely, while less than 2% of industrial facilities required only a slightly larger land area to be restored to offset its emissions. Transportation and warehousing facilities showed the highest potential where restoration of land provided enough ecosystem services to balance the demand followed by the manufacturing, mining and utility generation industries.

A region wise analysis of these facilities indicated that currently most of facilities in the south eastern part of the US have enough forest cover to provide ecosystem services to offset most of the emissions while facilities in the west and south western part of the US have a large amount of unused land that can be restored to provide ecosystem services to support facilities in that region.

These results indicate that land restoration and reforestation have the potential to be explored as a cost effective alternative to relying on conventional methods for improving air quality. Land restoration and preservation are becoming an integral part of land management practice to restore degraded ecosystems to its native habitat and are a vital way of increasing the availability of ecosystem services to humanity.

References:

[1] Nowak, D. J., Hirabayashi, S., Bodine, A., & Greenfield, E. (2014). Tree and forest effects on air quality and human health in the United States.Environmental Pollution, 193, 119-129.

422325 Techno-Economic Assessment of CO2 Conversion to Methanol, Urea and Formic Acid

Abstract:

Climate change will have world-wide implications. Therefore it is very important to investigate every possible route to reduce emissions of greenhouse gases, the root cause of the problem. Various courses of actions could be taken to reduce greenhouse gas emissions, which vary in the scale of their effect, their cost and potential distribution. These options are: improvements in energy efficiency in power generation or end-use, switching to fuels containing less carbon, substitution of electricity from carbon-free sources and last but not least the capture, storage and utilisation of carbon dioxide. The advantage of using the latter option is that the society would continue to utilise the existing low-cost infrastructure of the fossil fuels while minimising the environmental impact. After the capture of carbon dioxide there are options of storing it (ocean, aquifers, oil and gas fields) or utilising it in order to produce other valuable products.

Carbon Dioxide Utilisation (CCU) technologies will help with public support of CCS and will also help support technology development of CO_2, while simultaneously generating saleable products [1]. However, to date there has been very little systematic research involving process engineering into the techno-economic analysis of the processes. In this paper we will present different options of CO_2 utilisation routes as disposal of CO_2 capture emissions. A techno-economic assessment of three CO_2 utilisation (CCU) technologies for the production of methanol, urea and formic acid will be presented.

The three processes described were selected based on the estimated Technology Readiness Level (TRL) [2]. Based on this analysis the current best-practice technologies as the most-likely-to be commercialised are the catalytic hydrogenation (with renewable H_2) followed by distillation for the production of methanol, the CO_2 stripping process (Stamicarbon) for the production of urea and the CO_2 hydrogenation with renewable H_2 for the production of formic acid. The three processes are evaluated in terms of key performance indicators (KPIs): process efficiency, energy requirements, avoided CO_2 emissions, amount of CO_2 utilised, capital and operating costs and average levelised production costs (LPC).

The methanol process was simulated at a scale of 400,000 t/yr and the NRTL model was used to calculate the thermodynamic properties of the stream, while the kinetics scheme for the reactor model and the amount of catalyst were calculated based on the work of Van-Dal and Bouallou [3]. The urea process was simulated at a scale of 600,000 t/yr and the SR-POLAR model was used to describe the thermodynamic properties of the system. The formic acid process was simulated at a scale of 180,000 t/yr and the NRTL fluid package was used for the calculation of the thermodynamic properties. The three processes were modelled in the Aspen Plus and Hysys platform.

In addition to the results of the process modelling work we also present the results of an economic analysis related to the CCU product demand, market size and growth, production volume and pricing forecasts, and the sensitivities of the techno-economic process parameters to the price of CO_2 will also be discussed.

The key conclusion of this study is that, in the absence of significantly negative price associated with CO_2 utilisation, the three processes evaluated here do not as yet appear economically viable.

References

1. P. Styring, D. Jansen, H. de Coninck, H. Reith, K. Armstrong,Carbon Capture and Utilisation in the Green Economy: Using CO_2 to Manufacture Fuel, Chemicals and Materials,CO_2 Chem Media & Publishing Limited, Sheffield, UK (2011) ISBN: 978-0-9572588-1-5.

2. EC (European Commission), Horizon 2020, Work Programme 2014-2015, Technology Readiness Levels (TRL),General Annexes, 2014. URL: http://ec.europa.eu/research/participants/data/ref/h2020/wp/2014_2015/annexes/h2020-wp1415-annex-g-trl_en.pdf (Accessed on 21/02/2014).

3. E. S., Van-Dal and C. Bouallou, Design and simulation of a methanol production plant from CO_2 hydrogenation, Journal of Cleaner Production, 57 (2013), 38-45.

426171 Comparison of Different Optimal Designs of Single Mixed Refrigerant Processes

Abstract:

Improving energy efficiency of natural gas liquefaction processes is the major issue in LNG plant research as the processes consume a large amount of energy. A single mixed refrigerant process, one of the natural gas liquefaction processes, has been modeled and optimized in various literatures for improving process efficiency in terms of specific work (energy consumption per unit mass of LNG production). The direct comparison of the results obtained by the literature, however, is not easy because of the differences of process configurations, natural gas compositions, design bases, and optimization models. This study optimizes a variety of single mixed refrigerant processes for minimizing specific work with an equation-oriented modeling environment, and compares the process efficiencies using the specific works minimized by the optimizations. In addition to these, we analyzes the effects of i) process configurations of compression and multistream heat exchanger stages, ii) natural gas compositions, and iii) design bases such as pressure drops in the equipment, cooling water temperature, and equipment efficiencies on the single mixed refrigeration processes. As a result, we find some variables, which are highly sensitive or have a significant impact on the process efficiency, and discover active constraints preventing further improvement.

This research was supported by a grant from the LNG Plant R&D Center funded by the Ministry of Land, Infrastructure, and Transport (MOLIT) of the Korean government.

408147 A Novel Method of Continuous Crystallization to Avoid Filtration and Slurry Handling

Abstract:

Recently, development and elaboration of the unit operations necessary to implement continuous manufacturing of APIs and Drug Products have been being widely practiced successfully. However, fouling and clogging of lines in both continuous crystallization and flow chemistry still present a challenge in designing continuous processes. These persistent challenges are significant in handling slurries, traditional powder operations, and solid processing.

This work presents a novel unit operation designed to eliminate the need for slurry handling in the API manufacturing by utilization of a dynamic (falling) film based solution layer crystallization technique. Crystallization on an immobilized layer allows purification, solvent swapping and concentration of intermediate and final API streams to be conducted without the need for isolation or conveying of slurry or powder. In addition, in coupled with a powder free formulation operation this operation could facilitate the elimination of solids handling in drug manufacturing process altogether.

The results for different APIs and crystallization systems prove the performance of the technique for different process parameters, with high yield and level of purification, at high throughput and short processing (residence) time. In addition to the experimental results, a mathematical modeling study demonstrates the performance of the technique for different process parameters (temperature, flow-rate, supersaturation, growth-rate, and distribution coefficient) and the flexibility of the techniques for various process combinations, such as parallel, series, and scaled up/down.

409639 Indirect Ultrasonication in Continuous Slug-Flow Crystallization

Abstract:

In the pharmaceutical industry, the continuous generation of crystals of target size distribution has the potential to simplify and/or reduce post-crystallization operations [1]. The joining of two streams in dual impinging jet, coaxial, or radial micromixers has been demonstrated to produce seed crystals of uniform crystal size distribution (CSD) [2][3], but with a nucleation rate that depends on the liquid flow rate [2][3][4]. Ultrasonication is a technology for facilitating primary nucleation that has been useful in various industrial and academic studies [5]. The nucleation rate is a function of ultrasonication parameters rather than just liquid flow rate, which provides extra degrees of freedom for controlling the CSD. Ultrasonication-assisted crystallization seldom has clogging problems, and actually has the opposite effect and is commonly used to remove particles from containers for cleaning purposes [5]. Ultrasonication facilitates primary nucleation by inducing acoustic cavitation in the liquid solution using high-frequency mechanical vibrations from converters (connected sonication generator) [5].

In most studies of sonocrystallization, an ultrasonication probe (or horn) is in direct contact with the liquid solution, which may induce secondary nucleation from metal contact or contaminate the solution with metal after a long time of use [5]. This talk uses an ultrasonication horn and so has spatial localization of ultrasonication energy, but does not directly contact the horn with the liquid solution. This talk also describes the implementation of this indirect ultrasonication-assisted primary nucleation into a slug-flow cooling crystallizer to demonstrate rapid generation of a large amount of crystals of larger size than the nuclei. The main differences compared to high-quality past studies (e.g., [6]) is that this talk (i) uses an ultrasonic probe focused on a small spatial location instead of placing a long tube in a sonication bath, (ii) has order of magnitude lower time in which the solution is in contact with ultrasonication, and (iii) uses a simpler experimental system [7].

References

[1] Z. Sun, N. Ya, R. C. Adams, and F. S. Fang, “Particle size specifications for solid oral dosage forms: A regulatory perspective,” *Am. Pharm. Rev.*, vol. 13, no. 4, pp. 68–73, 2010.

[2] M. Jiang, Z. Zhu, E. Jimenez, C. D. Papageorgiou, J. Waetzig, A. Hardy, M. Langston, and R. D. Braatz, “Continuous-flow tubular crystallization in slugs spontaneously induced by hydrodynamics,” *Cryst. Growth Des.*, vol. 14, no. 2, pp. 851–860, 2014.

[3] M. Jiang, M. H. Wong, Z. Zhu, J. Zhang, L. Zhou, K. Wang, A. N. Ford Versypt, T. Si, L. M. Hasenberg, Y. E. Li, and R. D. Braatz, “Towards achieving a flattop crystal size distribution by continuous seeding and controlled growth,” *Chem. Eng. Sci.*, vol. 77, pp. 2–9, 2012.

[4] A. J. Mahajan and D. J. Kirwan, “Micromixing effects in a two-impinging-jets precipitator,” *AIChE J.*, vol. 42, no. 7, pp. 1801–1814, 1996.

[5] J. R. G. Sander, B. W. Zeiger, and K. S. Suslick, “Sonocrystallization and sonofragmentation,” *Ultrason. Sonochem.*, vol. 21, no. 6, pp. 1908–1915, 2014.

[6] R. J. P. Eder, S. Schrank, M. O. Besenhard, E. Roblegg, H. Gruber-Woelfler, and J. G. Khinast, “Continuous sonocrystallization of acetylsalicylic acid (ASA): Control of crystal size,” *Cryst. Growth Des.*, vol. 12, no. 10, pp. 4733–4738, 2012.

[7] M. Jiang, C. D. Papageorgiou, J. Waetzig, A. Hardy, M. Langston, and R. D. Braatz, “Indirect ultrasonication in continuous slug-flow crystallization,” *Cryst. Growth Des.*, vol. 15, no. 5, pp. 2486–2492, 2015.

411833 Kinetic Modeling and Scaleup Assessment of Seeded Solvent Mediated Form Conversion

Abstract:

This presentation focusses on the challenges of developing a scalable form conversion process. The slurry conversion of solid forms involves unique challenges compared to typical cooling and anti-solvent crystallizations. Form conversion is often coupled with thermodynamics constraints involving polymorphs or other solid forms. This presentation will discuss the importance of phase diagrams, procedures for monitoring the kinetics of form conversion using Raman spectroscopy, developing a mathematical model of form conversion, examining the impact of process parameters and developing a scalable process. In particular, it was found that solubility is not the most significant driver for the kinetics of the process.

418206 Improving the Separation of Organic Small Molecules Purified Via Sublimation

Abstract:

Conjugated small molecule materials are currently being investigated as organic semiconductors for use in next generation display and solid-state lighting applications. Widespread commercial adoption of these technologies is limited by significant materials costs, partially due to an expensive purification step. This work investigates the current industrial purification technique, thermal gradient sublimation, in order to better understand the underlying mechanisms and guide process development. The purification of two common charge transport layer materials, N,N'-bis(naphthalen-1-yl)-N,N'-bis(phenyl)-benzidine (NPD) and 4,4',4"-tris(carbazol-9-yl) triphenylamine (TCTA), is found to be limited by vapor phase transport and deposition. Interestingly, mechanisms in the solid phase such as diffusion within the solid feed, reaction at the feed particle surface, and mass transfer within the bed of feed particles are not rate limiting. The vapor phase transport and deposition process is consistent with diffusion. These results provide a basis for improving existing thermal gradient sublimation processes and suggest guidelines for the design and operation of future large scale purifications of high value, low volume specialty chemicals.

Chemtec

Improvement of Design Assessment Methodologies for Melt Static Crystallization Process

2015 AIChE Annual Meeting, Salt Lake City, UT

L. Guillou

November 10th 2015
Prepared by: Loïc Guillou

POE-TMP-007-E

MSC: Design Assessment vs. Design & Modeling

Design Assessment / Shortcut

- **Pre-evaluate**
- **Explore at low cost**
- **Comparatively rate (approx.)**

Design and Modeling

- **Secure your data**
- **Select ONE way to go and THE way to go**
- **Issue guarantees (or margins)**

2 approaches corresponding to 2 steps of a Project

Agenda

1. Who we are
2. Principle of MSC
3. Shortcut modelling for MSC processes
4. Shortcut modelling through statistic approach

Chemtec

Company Overview

Fives, an international industrial engineering Group

—

fives ultimate machines ultimate factory

Fives designs and supplies **machines, process equipment** and **production lines** for the **world's biggest industrial players.**
Fives supplies **PROABD® MSC plants** since the 50's.

Fives, a pioneer which constantly innovates

—

Fives' **strength** is to develop

— **proprietary technologies**

— with **high added value for the customers**

— and **high energy and environmental performances**

Innovation data:

More than **33 million €** spent on R&D in 2014 (outside projects)

132 innovations patented in 3 years

532 patent families in force

1 859 patents in force

Stein Digit@l Furnace AT

FCB Horomill® grinder

Chemtec

Principle of MSC

Principle of Melt Static Crystallization operation

- The Melt Static Crystallization (MSC) is a batch process for **separation and purification**
- The MSC is ruled by **the solid-liquid equilibrium** curve of the compounds, crystallization **kinetics** and **rheology**
- A MSC stage is divided into different steps, each managed through **heat load control**

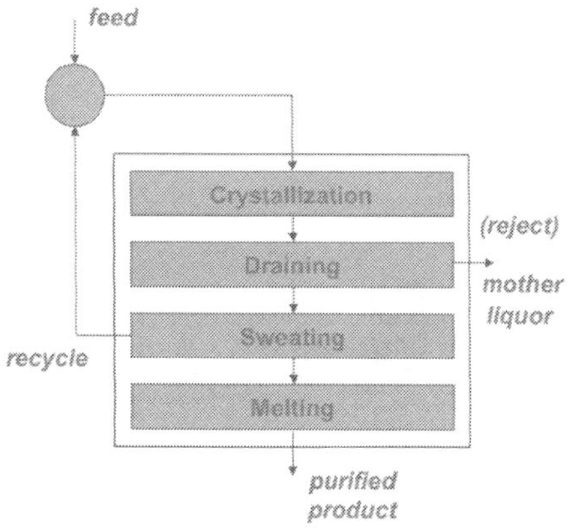

You want to know more? Visit our website and watch our MSC video

Multi-stages MSC = Fractional crystallization

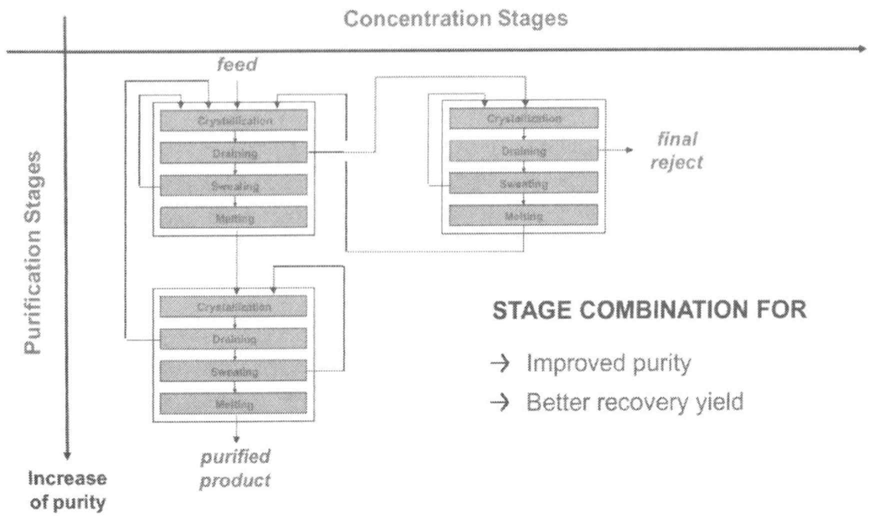

STAGE COMBINATION FOR

→ Improved purity

→ Better recovery yield

The PROABD® MSC unit

A PROABD® MSC unit is composed of three parts synchronized by dedicated **advanced automation** :

- Product handling
- Crystallization group with **proprietary equipment**
- Utilities

Such a unit :

- is **fully integrated** to a continuous production line
- ranges in size from a **few liters to hundreds of m³** of crystallizer volume

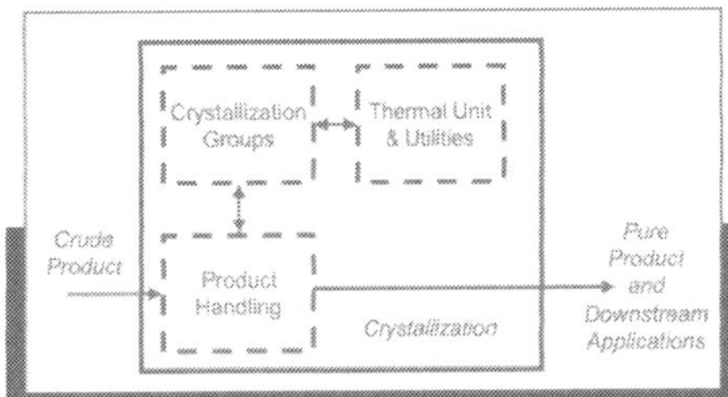

An unlimited list of compounds purified at lab & pilot scale

- Carbochemicals & Petrochemical
  - naphthalene,
  - anthracene,
  - carbazole,
  - fluorene,
  - cresol
  - xylenes, …
- Polymer Industry
  - vinylformamide,
  - dicumyl peroxide,
  - benzoic acid
  - styrene,
  - methylene diphenyl diisocyanate (MDI),
  - toluene diisocyanate (TDI),
  - NCB (nitrochlorobenzene)
  - acrylic acid
  - bisphenol A,
  - L-lactid …
- Cosmetic Industry
  - levulinic acid
  - t-butylhydroquinone, …
- Food Industry
  - erythritol
  - menthol, …
- Fine Chemicals & Specialties
  - cyclododecanol
  - o-toluene sulfonyl chloride
  - dicyclopentadiene (DCPD)
  - chlorotoluene, …
- Agrochemicals
  - 2-methylimidazole
  - dichlorobenzene
  - o-phenylphenol
  - phosphoric acid, …
- Intermediates to Active Pharmaceutical & Ingredients
  - N-hydroxy-4-piperidinol
  - 2-picoline, …

Chemtec

Shortcut modelling of MSC processes

Typical short cut modeling approach

"Engineering judgment" [2] + SLE → **Flow sheeting [1]**

Design for 1 stage

$k_{eff} = C_s/C_l$ effective impurity distribution coefficient → **Solving mass balance**

C_s = concentration of impurities in solid
C_l = concentration of impurities in liquid

Solving energy balance

Use K factor to extrapolate for multistage [3]

Performances and sizing

[1] Gilbert S.W., AIChE J., 37 (1991), 8, 1205-1218
[2] W. Marquardt, et al., Chin. J. Chem. Eng., 16 (2008), 333–342.
[3] Wellinghoff G. et al., Chem. Ind. Tech., 63 (1991) 881

Solving the mass balance

The big deal with k_{eff}

1. Supposed to be constant…
2. …But is in fact compound dependant,…
3. …And process parameters dependant!

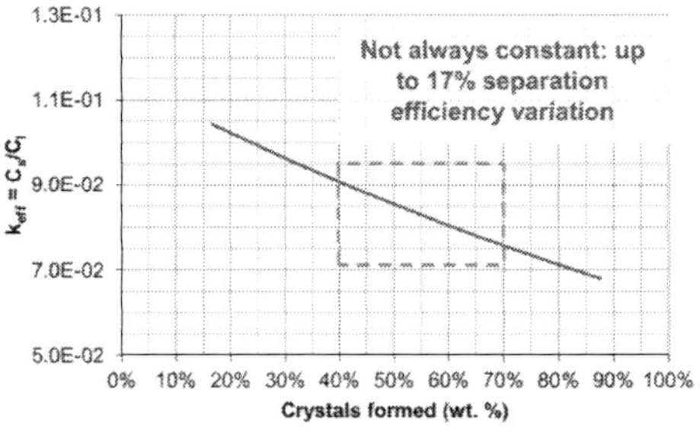

Lab rig for k_{eff} determination affects obtained values: ex. increased convection change growth behaviour vs. inclusions [5]

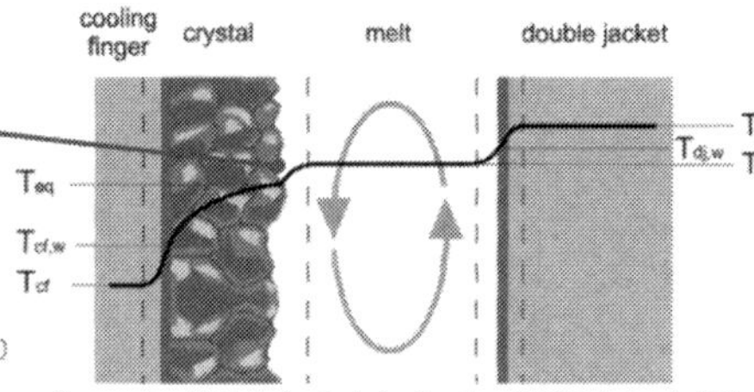

Example of triple jacket device for MSC lab tests [4]

[4] T. Beierling et al., Cryst. Growth Des. 2013, 13, 5229–5240
[5] A. Rich et al., Desal. & Water Treat., (2010) 13, 120–27

Sizing

fives

Sizing

1. How many stages required – K factor determination?
2. MSC is a batch process but nothing is time related in the shortcut method ?
3. Which trust into growth rates values obtained in unscalable lab rig?

➢ **Stage number and steps durations are missing but are a must have!**

➢ **Sweating duration is key and often neglected!**

➢ **Reported growth rate can be 10^{-6} to 10^{-4} m/s which is 10 to 100+ folds over estimated vs. real MSC growth rates!**

Data available is specific to few cases/compounds only...
... And amount of published data is not large enough...
... to allow supporting generalized MSC shortcut design!

Chemtec

fives

Shortcut modeling through statistic approach

Constituting a database – MSC methodology 101

fives

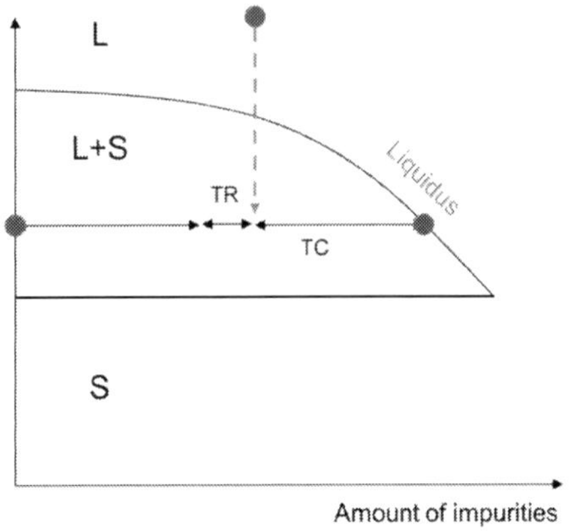

Instead of k_{eff}, use of mass ratio

1. TC mass of crystals formed
   TC is thermodynamic related
2. TR mass of mother liquor trapped
   TR is kinetics related

Constitution of a database on a specific lab rig with a fixed protocol

Constituting a database – MSC methodology 101

fives

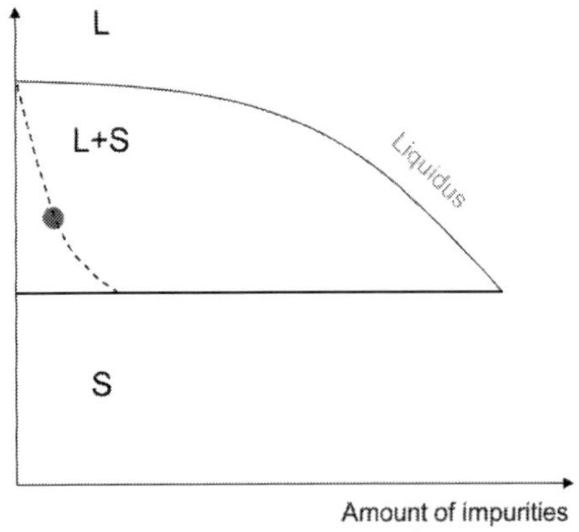

Instead of k_{eff}, use of mass ratio

1. TC mass of crystals formed
   TC is thermodynamic related
2. TR mass of mother liquor trapped
   TR is kinetics related

Final crystallizer content is calculated from TR and TC [6]

Method fully transposable to sweating

Constitution of a database on a specific lab rig with a fixed protocol

[6] Franke, M.B., et al., AIChE J. (2008), 54, 2925-2942

Constituting a database – turning k_{eff} pitfall

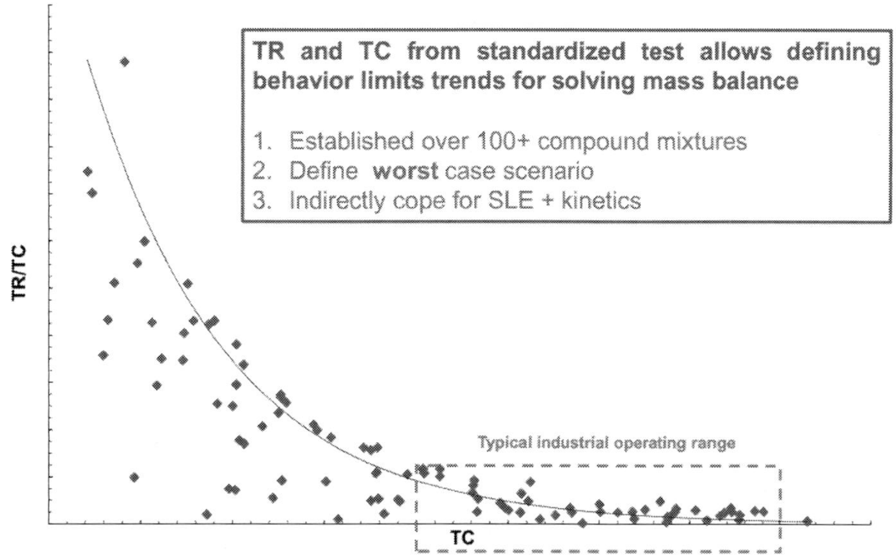

2015 AIChE Annual Meeting, Salt Lake City, UT

Chemtec 19

Constituting a database – turning K factor pitfall

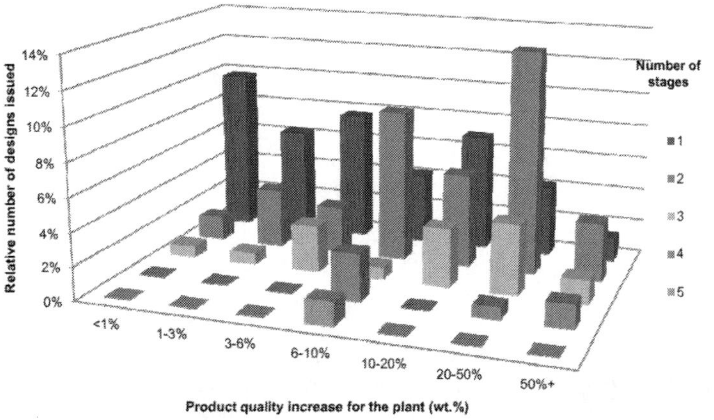

1. Large quality improvement ≠ large number of stages
2. Ultra purification ≠ large number of stages
3. **80% of designs = 1 or 2 stages only required**

K factor directly achievable through TR and TC

2015 AIChE Annual Meeting, Salt Lake City, UT

Chemtec 20

Constituting a database – turning step/stage duration pitfall

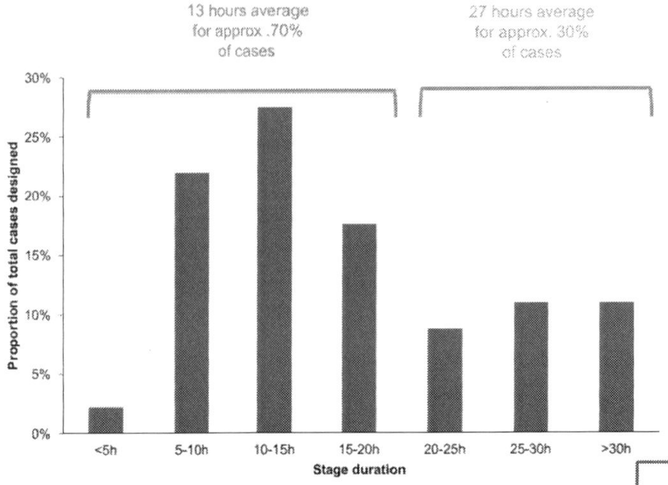

1. **13 hours duration for 70% of the cases**
2. Larger duration for lean feed or ultra-purifications

Time directly estimated from purification effort

MSC: Design Assessment vs. Design & Modelling

Design Assessment / Shortcut

- **Based on narrow number of flow schemes & using a few inputs …**
- **… Comparative design assessment can be sorted from trends & average metrics issued from large test database under normalized protocol**

This will be (latter) completed by adapted lab characterization but allows a rapid answer to the two most common questions:

- What may the process look like?
- How much may it cost?

fives ultimate machines ultimate factory

Loïc Guillou
Technology Manager
Chemtec

Loic.guillou@fivesgroup.com
T +33 3 89 36 32 00

Fives Proabd
CS 11437
20C rue de Chemnitz, Bât. B
Mulhouse Cedex , 68071 - **France**

435190 Continuous Separation of Enantiomers By Combined Crystallization-Distillation Systems: A Proof of Concept Study for Continuous Viedma Ripening

Abstract:

The separation of chiral molecules is of key importance in major industries, such as the pharmaceutical industry, where different enantiomers of a chiral compound can have markedly different biological effects. For instance, only one of the enantiomers of a drug might provide the therapeutic effect, while the other enantiomer shows little effect, no effect, or even harmful effects. Hence, the pharmaceutical industry is driven towards processes that allow separating enantiomers in a cost- and material-efficient way [1]. For systems of enantiomers that crystallize as conglomerates, the process of Viedma ripening has attracted considerable attention by the crystallization community [2,3]. In this process, a racemization reaction in solution (which may be catalyzed) and a complex interplay of crystal growth, dissolution, breakage and agglomeration lead to an enantiomerically pure product [4]. While the mechanism behind the process and how its performance can be improved were intensively investigated, a truly continuous process configuration of Viedma ripening has not been presented so far in the open literature.

In this work, we present such a continuous process configuration that relies on a combination of continuous crystallization, milling and distillation. Specifically, the process involves a mixed suspension mixed product removal crystallizer that is fed with a racemic solution of the substance. A suspension mill, operating in a loop around the crystallizer, serves as an attrition source. The outflow of the crystallizer is put forward to a continuous filtration unit where enantiopure crystals and the mother liquor are separated. The mother liquor is then subjected to a continuous distillation step in which part of the solvent is removed selectively, i.e., without the loss of chiral product or racemization catalyst. The remaining, concentrated mother liquor is recycled back to the crystallizer. Thus, the process allows obtaining enantiopure solids at a theoretical yield of 100%.

By using extensive process simulations based on population balance equation models [4], we show the start-up behavior (e.g., seeding requirements) and under which operating conditions (residence times in the crystallizer and the mills, feed concentration, milling intensity etc.) the process yields enantiopure solids in a continuous steady-state. We also present proof of concept experiments involving the amino acid asparagine in water as a model system including a racemization catalyst.

References

[1] H. Lorenz, A. Seidel-Morgenstern, *Processes to Separate Enantiomers*, Angew. Chem. Int. Ed, 53 (2014), 1218-1250.

[2] C. Viedma, *Chiral symmetry breaking during crystallization: Complete chiral purity induced by nonlinear autocatalysis and recycling*, Phys. Rev. Lett., 94 (2005), 065504.

[3] W.L. Noorduin, T. Izumi, A. Millemaggi, M. Leeman, H. Meekes, W.J.P. van Enckevort, R.M. Kellogg, B. Kaptein, E. Vlieg, D.G. Blackmond, *Emergence of a single solid chiral state from a nearly racemic amino acid derivative*, J. Am. Chem. Soc., 130 (2008), 1158-1159.

[4] M. Iggland, M. Mazzotti, *A Population Balance Model for Chiral Resolution via Viedma Ripening*, Cryst. Growth Des., 11 (2011), 4611-4622.

435733 Continuous Crystallization of L-Ascorbic Acid: Integration of Airlift Crystallizer with Membrane Distillation-Experiments and Results

Tuesday, November 10, 2015: 5:20 PM
155D (Salt Palace Convention Center)
Fatemeh Anisi, Process & Energy Department, Delft University of Technology, Delft, Netherlands and H.J.M. Kramer, P&E, Separation Technology, Delft University of Technology, Delft, Netherlands

Continuous Crystallization of L-ascorbic acid:

Integration of Airlift Crystallizer with Membrane Distillation-Experiments and Results

F. Anisi,[1] H.J.M Kramer[1] [1] *Department of Process and Energy, Faculty of Mechanical, Maritime and Materials Engineering, Delft University of Technology, Delft, the Netherlands*

Crystallization is one of the oldest important separation and purification processes in production of pharmaceutical compounds. The common problems in crystallisation processes are often include large variations in the product quality and inflexibility of the process. It is due to the entanglement of different phenomena such as primary nucleation, growth, secondary nucleation and so on. For a better control on the secondary nucleation and the growth of the crystals an airlift crystallizer has been developed, tested and analysed. Batch experiments with airlift crystallizer showed that rate of secondary nucleation is two orders of magnitude lower in such a crystallizer compared to conventional stirred crystallizer. Figure 1 shows how bimodal the CSD from stirred crystallizer is which confirms the higher rate of secondary nucleation in the stirred crystallizer. Therefore, it is a promising alternative for conventional ones.[1]

Membrane distillation technology can be used for supersaturation generation as an alternative to evaporation. Since it is not necessary to work with high temperatures, membranes are suitable options for thermo-labile compounds. During crystallization experiments supersaturation has been generated and controlled with continuous solvent removal and fed to the crystallizer. Seed crystals have grown by consuming the supersaturation.

Therefore, a continuous membrane-assisted crystallization using an airlift crystallizer is potentially believed to be more controllable, energy efficient and to produce a more uni-modal crystal size distribution.

In this contribution the airlift crystallizer and membrane-assisted unit are integrated in a skid. The set-up is capable of performing batch and continuous crystallization with airlift or stirred crystallizer. Continuous crystallization of L-ascorbic acid has been done in the airlift and stirred crystallizer while membrane controls the supersaturation during the process. The integrated system has been also modelled and various process conditions are tested.

Based on modelling results experiments are designed and show that generally use of membrane unit decreases the level of secondary nucleation but it is mainly suppressed in the airlift crystallizer rather than stirred one. The higher the residence time the larger the crystals grow. The results from the experiments are used to validate the model and furthermore to estimate the kinetic parameters and optimize the process via design of new experiments.

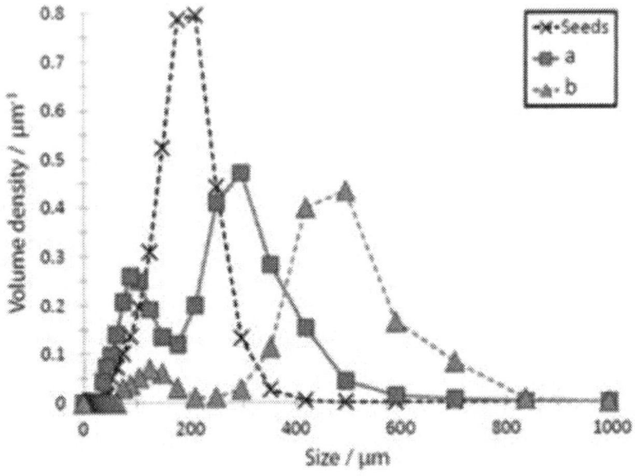

Figure 1:Crystal size distribution for different batch seeded cooling crystallization of L-ascorbic acid a) stirred crystallizer b) airlift crystallizer

USERPROPERTY * MERGEFORMAT [1] R. Lakerveld, J. Krochten, H. Kramer, An airlift crystallizer can suppress secondary nucleation at a higher supersaturation compared to a stirred crystallizer, Cryst. Growth Des. 14, 7 (2014)

437441 Reactor modeling and recipe optimization of polyether polyol processes: Polypropylene glycol

Abstract:

Polyether polyols are polymers that find application in a wide range of industries for the production of flexible foams, high resiliency molded foams, and viscoelastic foams, as well as in CASE (Coatings, Adhesives, Sealants, and Elastomers) applications. They are also known as polyglycols, whose uses include surfactants, dispersants, and lubricants, to name a few. A significant percentage of the world capacity in both propylene and ethylene oxide goes into these polymers, who find uses in products as advanced as self-healing fuel tanks for battlefield fuel transportation. With such widespread use, there is an increasing demand for these products using a manufacturing technology that is already mature and well established. The need for increased efficiency at low capital investment is significant.

This thesis study focuses on model-based optimization strategies to improve process performance of existing assets. Particularly, we develop dynamic optimization and optimal scheduling methods, and their integration, for industrial base-catalyzed alkoxylation processes. We develop rigorous dynamic reactor models for both homopolymerization and copolymerization operations. The reactor models are based on first-principles such as mass and heat balances, reaction kinetics and vapor-liquid equilibria. We derive reactor models with both the population balance method and method of moments. The obtained reactor models are validated using historical plant data. Polymerization recipes are optimized with the dynamic optimization algorithm, using the simultaneous collocation strategy and nonlinear programming techniques. Polymerization times are reduced by optimizing operating conditions such as the reactor temperature and monomer feed rates over time. Next, we study scheduling methods that involve multiple process units and products. The resource task network scheduling model is reformulated to the state space form that offers a good platform for incorporating dynamic models. Lastly for the integration study, we investigate a process with two parallel polymerization reactors and downstream storage and purification units. The dynamic behaviors of the two reactors are coupled through shared cooling resources. We formulate the integration problem by combining the state space resource task network model with the moment reactor model and solve it with the generalized Benders decomposition scheme.

This PhD project falls under the University Partnership Initiative between The Dow Chemical Company and Carnegie Mellon University. The technical presentation by Nie entitled "Reactor modeling and recipe optimization of polyether polyol processes: Polypropylene glycol" has received the Best Presentation award in the session "Modeling and Control of Polymer Processes" of the 2012 AICHE Annual Meeting. The implementation effort was recognized by Frost and Sullivan. The project team won the 2014 Manufacturing Leadership Award in the category of Engineering & Production Technology Leadership. The team has also received multiple awards internal to Dow. The recipe optimization effort continues to thrive in Dow, and the scheduling optimization technology implementation is under development.

437435 Photodirected Control of Topography using Thiol-ene "Click!" Chemistry

Tuesday, November 10, 2015: 4:10 PM
Salon I (Salt Lake Marriott Downtown at City Creek)

Stephen Ma, Chemical and Biomolecular Engineering, University of Delaware, Newark, DE, Norman J. Wagner, Chemical Engineering, University of Delaware, Newark, DE and Christopher J. Kloxin, Chemical and Biomolecular Engineering and Materials Science and Engineering, University of Delaware, Newark, DE

<u>Photodirected Control of Topography using Thiol-ene "Click!" Chemistry</u>

Stephen J. Ma1, Dr. Norman J. Wagner1,2, Dr. Christopher J. Kloxin1,2,3

1) University of Delaware, Department of Chemical and Biomolecular Engineering

2) Center for Molecular and Engineering Thermodynamics

3) University of Delaware, Department of Materials Science and Engineering

Wrinkling/buckling on elastomers represent a cost-effective approach to creating surface topography, leading to a broad range of applications in antifouling coatings,1,2 optical coatings and tunable lenses3,4, substrates for directed cell growth5,6, and enhancements in solar cell efficiency.7,8 One of the ongoing challenges of wrinkling is the limited ability to confine and orient the wrinkles toward the development of performance materials.

Using an acrylate-rich thiol-ene elastomer, embedded with photoinitiator and photoabsorber, we have developed a rapidly curing wrinkling system.9 Upon irradiating the strained elastomers with UV light, the photoinitiator will induce radical polymerization of the excess acrylates while the photoabsorber confines the light to a thin skin layer at the surface, creating the necessary conditions for

wrinkle formation on the order of seconds. In conjunction with photomasked UV light, these wrinkles can be easily confined and oriented to generate complex patterns of multiple distinct wrinkle wavelengths and wrinkle gradients. With the addition of a photoorthogonal photoinitiator, we can also 'lock' the shape of the elastomer through further bulk radical polymerization.

365 nm

Figure 1. Photopatterning is a very fast, inexpensive way to confine and orient wrinkles.

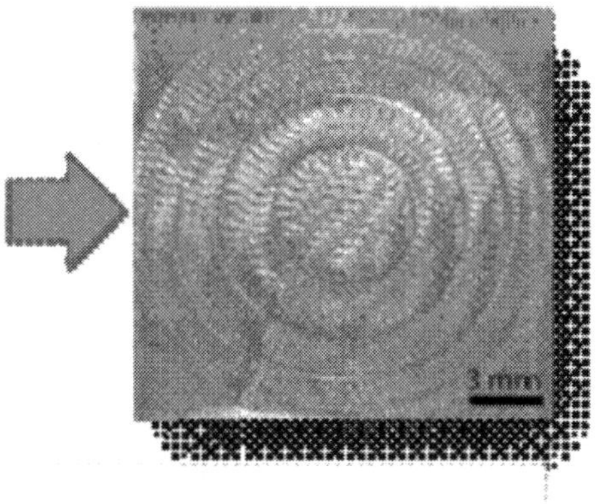

Beyond the ability to easily orient and align wrinkles through photopatterning, our wrinkle system is also highly versatile. The 'click' nature of thiol-ene chemistry ensures high tunability of elastic modulus, network architecture and surface functionality of the elastomer, through the types of monomers used and careful stoichiometric control. The presence of excess acrylates also enables post-functionalization of various moieties at the surface of the films.

Additionally, recent advances in particle synthesis through thiol-ene chemistry,10,11,12 has opened up new avenues in wrinkle formation. By covalently attaching these particles atop strained thiol-ene elastomers, it is possible to form wrinkles with an additional level of surface roughness on the size-scale of the particles themselves, thereby enhancing the surface properties of the materials. More interestingly, these particles offer new potential platforms for generating surface topography on 3D objects, enabling, for the first time, the experimental study of wrinkle formation on spheres.

References

(1) Efimenko, K., Finlay, J., Callow, M. E., Callow, J. A. and Genzer, J. *ACS Appl. Mater. Interfaces* **2009**, 1, 1031-

1040

(2) Efimenko, K., Rackaitis, M., Manias, E., Vaziri, A., Mahadevan, L. and Genzer, J. *Nature Mater.* **2005**, 4, 293-

297

(3) Chandra, D., Yang, S. and Lin, P.-C. *Appl. Phys. Lett.* **2007**, 91,

(4) van den Ende, D., Kamminga, J. D., Boersma, A., Andritsch, T. and Steeneken, P. G. *Adv Mater* **2013**, 25, 3438-

3442

(5) Guvendiren, M. and Burdick, J. A. *Biomaterials,* **2010**, 31, 6511-6518 (6) Guvendiren, M. and Burdick, J. A. *Material Views,* **2013**, 2, 155-164

(7) Kim, J. B., Kim, P., Pegard, N. C., Oh, S. J., Kagan, C. R., Fleischer, J. W., Stone, H. A. and Loo, Y. L. *Nat Photonics* **2012**, 6, 327-332

(8) Xie, K. Y. and Wei, B. Q. *Adv Mater* **2014**, 26, 3592-3617

(9) Ma, S. J., Mannino, S. J., Wagner, N. J. and Kloxin, C. J. *Macro Letters* **2013**, 2, 474-477

(10) Durham, O.Z., Krishnan, S., Shipp, D.A. *Macro Letters* **2012**, 1, 1134-1137

(11) Wang, C., Podgorski, M., Bowman, C.N. *Materials Horizons* **2014**, 1, 535-539

(12) Amato, D.V., Amato, D.N., Flynt, A.S., Patton, D.L. *Polymer Chemistry*, **2015**, Advance Article

437433 Polymerization of High-Molecular-Weight Polyvinylamine for CO_2-Selective Membranes with High Separation Performance

Abstract:

Polymer membranes for CO_2 separation from flue gas has been believed to be one of the most important solutions to greenhouse gas driven climate change. Facilitated transport membranes containing amines show a much higher CO_2/N_2 selectivity compared to solution-diffusion membranes. The polyamine membrane has to be thin enough to show a high CO_2 permeance. Polyvinylamine (PVAm) with high molecular weights (MWs) was successfully synthesized for the preparation of thin CO_2 separation membranes for CO_2 capture from flue gas. PVAm was synthesized through free radical polymerization using 2,2'-azobis(2-methylpropionamidine) dihydrochloride (AIBA) as the initiator. The synthesis process takes six steps: polymerization, acidic hydrolysis, precipitation, redissolution, hydroxide ion exchange, and vacuum filtration. PVAm polymers with different MWs were synthesized by adjusting the monomer concentration and initiator amount. The weight average MW of the synthesized PVAm was characterized by dynamic light scattering (DLS). The highest MW obtained was around 1,200,000. The effects of monomer concentration and initiator amount on the MW of the synthesized polymer will be discussed. The high MW PVAm synthesized in this work has shown a much higher viscosity than the commercially available PVAm (Lupamin® 9095, from BASF Corporation). The high MW PVAm solution with 3 wt.% polymer concentration showed a viscosity of 400 – 1400 cp, while Lupamin® 9095 solution with 3 wt.% polymer concentration showed a viscosity of only 50 cp. The high viscosity of the PVAm solution at a low concentration allowed the preparation of much thinner membranes. It could also help to reduce the penetration of polymer solution into the pores of the substrate, and further reduce the mass transfer resistance. Consequently, a much higher CO_2 permeance could be obtained from such a thin membrane with the thickness of 100 – 200 nm. Moreover, Lupamin® was obtained by basic hydrolysis using sodium hydroxide (NaOH), therefore the byproduct (sodium formate salt) was formed in the polymer product. The large amount of the non-reactive salt with CO_2 not only did not contribute to the facilitated transport of CO_2, but also could hamper the stability of the membrane due to the salting-out problem. In contrast, acidic hydrolysis was used in this work, and no byproduct was formed. Overall, the high MW PVAm synthesized in this work was advantageous to the membrane transport performance compared to the commercial available Lupamin® product. After the incorporation of mobile carriers (aminoacid salts) to the high MW polyvinylamine, the resulting membranes showed a high separation performance with a CO_2 permeance of about 1000 GPU and a CO_2/N_2 selectivity of more than 200 at 57 °C. The promising CO_2 separation performance from the amine-containing polymer membrane shows great potential for the application of CO_2 capture from flue gas.

437430 Use of a layer-by-layer (LbL) polymer coating technique that forms tunable, modular, nanoscale coatings

Abstract:

Harnessing the synergy between materials at the nanoscale can be a valuable tool in understanding and probing cellular phenomena and in driving specific processes that lead to repair and healing. One method to drive specific cellular process is to use polymers that are biodegradable and capable of serving as carriers of large macromolecules such as proteins. This work explores the use of a layer-by-layer (LbL) polymer coating technique that forms tunable, modular, nanoscale coatings. The process involves the formation of thin films through the alternating adsorption of positively and negatively charged multivalent species at room conditions. Each adsorbed layer in such films can range from 1 to 100 nm, and total therapeutic film thicknesses range from a few hundred nanometers to as much as 10 microns. The LbL technique was used to incorporate a broad range of functional polymers and biomacromolecules.

Polymers of the poly(β-amino ester) family were synthesized using Michael addition between a diacrylate and a primary amine. The polymers are hydrolytically degradable, positively charged and result in amino-acid like breakdown products. By incorporating these polymers in the LbL assembly, controlled release of growth factors was achieved to yield accelerated and highly integrated bone growth and regeneration including demonstrations in a surrogate implant model that illustrate the impact of delivery of these systems from metal implant surfaces for creating bone deposition and integration to achieve high strength bone-implant interfaces. The LbL approach allowed control of both the loading of biologic drug, and the timing of its release from the film. The work demonstrated that it is possible to tune simultaneous release of two drugs, each at different rates. Furthermore, the mechanical properties of the multilayer films, including wear resistance for orthopedic implants demonstrated that the strong cohesive nature of these films. LbL coatings are, by design electrostatically bonded to the surface, and are one to two orders of magnitude thinner while maintaining high loadings of biologic drug, which reduces many of these effective stresses. Very intricate and complex 3D surfaces were coated using the LbL approach, which involved the adsorption of polyions and biologics from very dilute aqueous solution. We have coated the interior pores within macroporous bone scaffolds and nanoporous polymer membranes and demonstrated a corresponding effect on tissue regeneration. This capability to coat porous substrates over a range of thicknesses from micron to centimeter scale allows the coating of a range of polymeric scaffolds with mechanical properties optimized for tissue repair. This work demonstrates the modular nature of LbL, in which the different components will be incorporated and evaluated. The mechanical and release components were isolated, thus enabling controlled release of sufficient factor from the implant surfaces to achieve true bone regeneration throughout the scaffold.

Collectively, this work provides insight into synthesizing polymeric nano- and maco- materials to probe and modify cellular interactions to control regenerative processes at varying length and time scales. Towards the development of next-generation biomedical therapies this polymerbased approach allows for engineering specific therapies with the ultimate goal to improve human health.

One Pot Synthesis of Biobased Fuel Additive Ethyl Levulinate by Direct Transformation of Bagasse in Ethanol Media

Shireen Quereshi[1], *Suman Dutta*[1*], *Tarun Kumar Naiya*[2]

1. *Department of Chemical Engineering, Indian School of Mines, Dhanbad, Jharkhand, India*
2. *Department of Petroleum Engineering, Indian School of Mines, Dhanbad, Jharkhand, India*

**Corresponding Author: ss.dutta@hotmail.com*

Abstract

Ethyl levulinate has emerged as a promising Biobased fuel additive which reduces dependence on non conventional fossil fuel sources. In this study we have demonstrated a one pot direct transformation of bagasse in ethanol media to produce ethyl levulinate at mild process conditions. A series of experimental studies have shown that mineral acid catalysts particularly sulfuric acid and HCl are more effective in increasing ethyl levulinate production. A detailed process insight study has also been done to demonstrate effects of different process parameters such as reaction time, catalyst loading & reaction temperature on ethyl levulinate yield. A 14.50 %(w /w) ethyl levulinate yield which represents a theoretical yield of 36.21% has been obtained from transformation of bagasse at a temperature of 125°C in a reaction time of 2 hours in the presence of sulfuric acid catalyst and ethanol media. H1-NMR & GC-FID analysis methods were used for qualitative & quantitative results analysis respectively. One pot transformation of bagasse into ethyl levulinate at such mild process conditions has been reported for the first time (to best of our knowledge).

Keywords: Ethyl Levulinate, Biomass Conversion, Esterification, Biodiesel, Biofuel Additives

Introduction

Increased energy demands and depletion of naturally occurring resources like coal, oil and natural gas has led to inclination towards development of fuel and chemicals from renewable sources. Recent study shows that fossil fuels are incredibly dense form of energy which is being utilized from several years but once they got extinct, it will take millions of years to be formed again[1]. Whereas, Biomass is an organic material and renewable resource having carbonaceous substance that can be used for the production of biofuel, fuel additives, chemicals, to generate electricity and other forms of power. Biomass is available in ample quantity and is considered as a good alternative to existing fuel sources[2]. In particular, lignocellulosic biomass which primarily consists of cellulose, hemicellulose and lignin is widely available and has no conflict with food as compared to first generation biofuel[3][4]. Cellulose are rigid, linear polymer consists of several subunits of glucose while hemicellulose is branched polymer consists of C_5 – C_6 carbon ,uronic acid and acetyl derivatives. Lignin stores maximum amount of energy in the biomass and it binds the cellulose as well as hemicellulose together[5]. Most of the time pretreatment is required for lignocellulosic biomass conversion due to the presence of lignin. In fact it acts like a upper binding surface in biomass and if lignin is not deploymerized then the carbonaceous cellulosic and hemicellulosic compound will not be free to take part in conversion process.Different approach has been applied and several routes for conversion of biomass into fuel and other useful chemicals are reported in litetaruter. However, they can be broadly classified into three major categories i.e. Thermochemical route, Acid hydrolysis and

Biological routes[1]. It is well known fact that conventional fossil fuel sources are made up of biomass that remains burried in the earth for several million years. Therefore, it can be anticipated that everything can be produced from lignocellulosic biomass which are currently produced from convetional sources. Tuck *et al.* has reported the systematic approach to produce a huge range of value added chemicals and fuels such as lubricants, plastic prescurosrs, surfactants, fibers and indusrial solvents etc[3]. Corma and co-workers has critically reviwed different processes including esterification, reduction, dehydration, decarbonylation, aldol condensation, decarboxylation and ketonization etc to convert biomass derived platform chemicals such as furans, polyols, fatty acids to produce fuel additives and chemicals[6].

Ethyl levulinate, mainly produced through esterification reaction, is considered among sustainable fuel additives due to its considerable cold flow properties. It is a potential Biobased fuel additive which is used for enhancing the cold flow properties of the fuel as well as it is also used in flavoring and fragrance industries[7]. Moreover, cloud point, pour point and cold filter plugging points of biofuel is improved by adding ethyl levulinate in different concentrations. Several studies shows that the cloud point , pour point and cloud filter plugging point is improved by reduction up to 4 – 5°C, 3 – 4°C and 3°C respectively [8], [9]. It also decreases smoke emissions, increases the efficiency of engine and also moderates the emission[10]. There are two ways in which the ethyl levulinate is produced (i) direct conversion of biomass to ethyl levulinate with acid catalyst in ethanol medium [10]–[12] and (ii) step wise conversion of biomass to levulinic acid with acid catalyst in water as solvent media then conversion of levulinic acid to ethyl levulinate with acid catalyst in ethanol medium[13]–[17]. The second process for ethyl levulinate which is mainly refered to as step wise procedure in which biomass is heated directly with acid catalyst in water as a medium then the levulinic acid is produced. Then the alcoholysis of levulinic acid with acid catalyst and ethanol as a medium produces ethyl levulinate. Whereas, C_5 content of lignocellulosic biomass follows a different path. Furthermore, lignin produces aromatics of different class (Refer to Figure-1). But major constraint of these processes are waste generation and low effieciency due to series of steps involved. First route i.e. direct conversion of lignocellulosic biomass into ethyl levulinate can be a better alternative beacause of one pot processing that causes less waste generation and imporved yields. In this work, we have reoprted a one pot process for for conversion of lignocellulsic bagasse in the presence of inexpensive mineral acid at mild temperatures that gives an advantage in terms of overall economics of the process.

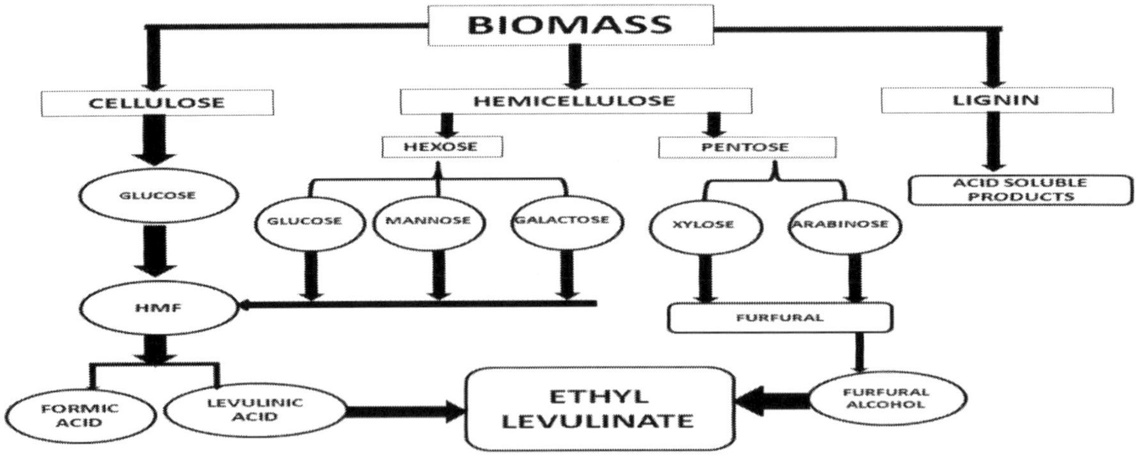

Figure 1 Biomass Conversion Route to Produce Ethyl Levulinate

Experimental Section

Materials and Methods:

Bagasse was procured from local vendor and dried in oven for 4 hours at 100°C temperature to remove moisture content from it. Thereafter, it was chopped into smaller pieces and grinded to powdered form. The powdered form of bagasse was weighed and again dried in the oven for one hour at 100°C temperature. No difference in initial and final weight was observed, therefore, it was assumed that bagasse is free from moisture and bone dry. Homogenous mineral acid catalysts sulfuric acid and HCl were purchased (Merck) and used without any further treatment. Ethyl levulinate, Levulinic acid, glucose and formic acid standard GC grade samples were purchased from Alfa Aesar for quantitative and qualitative analysis of obtained products. Solvent media Ethanol and Toluene of greater than 99% purity were bought from Merck.

Experimental Setup and Procedure:

Experimental setup consisted of borosilicate glass material assembled using commercially available round bottom flask, vertical condenser and other fittings. Oil bath heater was used to supply necessary energy and maintain reaction temperature in indirect mode. Provision for cooling water was made in order to condense vapors coming out during reaction and a continuous reflux was applied. In addition, oil bath heater was equipped with a magnetic stirrer for better agitation and uniform heat transfer. All experiments were performed using 4 g bone dried sample in 100 ml ethanol media at a temperature range of 100°C-125°C and in reaction time of 60-120 minutes. Sulfuric acid and HCl were used in 0.2M concentration of each for screening purpose. In addition, sulfuric acid in higher concentration of 0.4M, 06.M and 0.8M was also used for next set of experiments. Firstly bagasse was charged into round bottom flask followed by ethanol addition and mineral acid catalysts. Magnetic beads were also charged into round bottom flask for better heat transfer and to avoid bumping of liquid reaction mixture. The oil bath heater was started and set slightly above desired reaction temperature. Once, oil bath reached reaction temperature, round bottom flask containing reaction mixture was kept inside it that took approximately 10 minutes to reach and stabilize a desired reaction temperature. Simultaneously condenser connected with a cooling water source was installed above round bottom flask to condense and reflux lighter vapor components along with ethanol. After a fixed interval of reaction time, heat supply to round bottom flask was disconnected and it was immediately placed in a bucket of chilled water to prevent further reaction taking place.

Product Separation and Analysis:

Product mixture obtained after each experiment was filtered through fine filter paper and residue was collected in a separate crucible. Collected residue was squeezed to release trace amount of liquid contained in it and then was dried in an oven at 100°C for four hours. Bone dried residue was weighed again and individual weight of residue left after each experiment was noted. However, some soluble humin matter passes with the reactant mixture and color of filtered product was found to be very dark in nature. Therefore, a novel process for separation of soluble humin from product mixture and extraction of ethyl levulinate as reported by Tang *et al.* was applied for filtered product[19]. But before that filtered mixture was added with Calcium oxide to neutralize its high acid content. Then the product mixture was concentrated using Rota vapor (Buchi) and amount of ethanol vaporized along with product mixture in trace amount was collected in the flask attached down to the condenser. After some time when no more ethanol vapors were observed coming down to flask at low pressure then the concentrated sample was taken out from Rota vapor. The concentrated product obtained after initial separation was added with less polar solvent toluene in the ratio of 2:5 by volume in a separating funnel and the sample was then trembled and allowed to settle for at least 15 min till two layers were formed inside separating funnel namely light and heavy phase. The heavy phase at bottom consisted of undesired by-products and acid while the light phase at top consisted of toluene and ethyl levulinate. Thereafter, the light phase sample was concentrated again for the removal of toluene from product mixture to yield ethyl levulinate based on difference in their boiling points. The toluene vaporizes easily and separates out due to its highly volatile nature as compared to ethyl levulinate whereas; ethyl levulinate was left behind as a final product.

Isolated ethyl levulinate was qualititatively and quantitatively analyzed through different spectroscopic techniques including ^{1}H-NMR, GC-MS and GC-FID. NMR analysis was performed using 0.5 ml $CDCl_3$ solvent and 5 mg product of product mixture. Whereas, GC-MS analysis was done using Perkin Elmer instrument (Model: Clarus 600) at a maximum oven temperature of 250°C for total 28 minutes run. Initially oven temperature was kept at 60°C and a solvent delay of 4 minutes was provided. Thereafter, column was heated with a ramp rate of 10°C/minute to 180°C and a hold of 2 minutes was provided. Then again column was heated to 250°C with a ramp rate of 10°C/minute and kept on hold at final temperature for 2 minutes. Peaks at different retention times were noted and corresponding m/z ratio were noted. DB-5 column (Dimensions: 30mm X 0.250mm X 0.25μm, Agilent make) was used in GC-MS instrument where Helium carrier gas was used at a flow rate of 1ml/min and a split ratio of 1:10 (1 injected, 10 bypassed) was applied. Instrument was flushed with inert gas to remove traces of oxygen if any. Similar program was with slight modification was used for GC-FID analysis where column was used was DB-WAX (Agilent make). For GC-FID analysis, different concentration of standard ethyl levulinate was used to find retention time and to calculate response factor.

Results & Discussion

Catalyst Screening:

Initial experiments were performed to screen a suitable acid catalyst for direct transformation of bagasse into ethyl levulinate. Several catalysts such as Aluminum Chloride, Chromium Chloride, p-Toluenesulfonic acid, Hydrochloric acid and Sulfuric acid were used for conversion of bagasse in ethanol media at 100°C temperature for 1 hour in 0.2M acid concentration. Difference in initial and final weight of bagasse was measured in order to confirm if any reaction could take place. We did not found any significant change in initial and final weight of bagasse where Aluminum chloride, Chromium chloride, p-Toluenesulfonic acid and Hydrochloric acid were used as acid catalysts. However, there was a considerable change in initial and final weight of bagasse with sulfuric acid as catalyst. Although other catalysts than sulfuric acid may show significant effect at higher temperatures and longer reaction time, Sulfuric acid was found to be more suitable with high reactivity even at lower temperatures.

Overall Bagasse Conversion:

As discussed in previous section, sulfuric acid was found to be effective for converting bagasse into useful products even at lower temperatures and lower concentration. Therefore, we increased acid concentration to 0.4M, 0.6M and 0.8M in order to study higher acid concentration effects on overall conversion of bagasse irrespective of final product selectivity. Under optimum condition, we did not observed any significant difference in terms of overall bagasse conversion when reactions were performed at 125°C for two hours reaction time at 0.2M, 0.4M and 0.6M concentrations (Tabl-1).

Therefore, it can be concluded that increasing acid concentration is not helpful in terms of overall bagasse conversion. However, significant change in overall bagasse conversion was observed with 0.8M acid concentration. We are still exploring the possible reasons for contradiction in results with 0.8M sulfuric acid concentration. On the contrary, overall yield and theoretical yield of ethyl levulinate was found to be increasing with increase in acid concentration which shows that high acid concentration is favorable for higher product selectivity.

Table 1. Effect of acid concentration on overall bagasse conversion & EL Yield

Sulfuric acid concentration	Initial weight of bagasse (grams)	Final Weight of Bagasse (grams)	% Overall Conversion	EL Yield* (w/w)	Theoretical** EL Yield
0.2M	4.00	2.72	32.0%	1.0%	2.5%
0.4M	4.00	2.68	33.0%	4%	10.1%
0.6M	4.00	2.64	34.0%	10%	25.25%
0.8M	4.00	2.38	40.5%	14.50%	36.21%

*EL Yield (w/w) = (Weight of EL obtained/Initial Weight of Bagasse)*100
**Theoretical EL Yield = (Total EL obtained/Total Cellulose Content in Bagasse)*100
Note: Cellulose content was found to be 40%

Effect of Reaction Temperature and Time:

During all experiments we observed that like acid concentration, other two important parameters i.e. reaction temperature and reaction time are directly related with overall conversion rate of bagasse. An increasing trend in overall bagasse conversion was found with increase in temperature as well as reaction time (Table-2 & 3) which shows that either higher temperature or longer reaction time or a combination of both is required to obtain maximum bagasse conversion. Although reaction time effect was not very high as compared to temperature effects but it was found to be in increasing order. Similarly, Ethyl levulinate yield was also found to directly dependent on reaction temperature and time. We have found significant changes in EL yield even at slight changes in reaction temperature. Therefore, we anticipate that higher temperature will lead to higher yield of ethyl levulinate.

Table 2. Effect of reaction temperature, 4 g Bagasse, 100 ml Ethanol

Reaction Temperature °C	H_2SO_4 Concentration	Reaction Time, Minutes	% Overall Conversion	EL Yield* (w/w)	Theoretical** EL Yield
100	0.8M	120	24.0%	00%	00%
110	0.8M	120	29.0%	1.5%	3.75%
120	0.8M	120	36.0%	5.0%	12.05
125	0.8M	120	40.5%	14.50%	36.21%

Table 3. Effect of reaction time, 4g Bagasse, 100ml Ethanol

Reaction time in Minutes	H_2SO_4 Concentration	Reaction Temperature °C	% Overall Conversion	EL Yield* (w/w)	Theoretical** EL Yield
60	0.8M	125	35.7%	6.30%	15.75%
90	0.8M	125	38.2%	8.12%	20.30%
120	0.8M	125	40.5%	14.50%	36.21%

Effect of Reaction Temperature, Time and Acid Concentration on Ethyl Levulinate Yield:

Similar to overall bagasse conversion, an increase in ethyl levulinate yield was observed with increase in reaction temperature, acid concentration and time. Since overall bagasse conversion was higher, therefore, higher ethyl levulinate yield was according to our hypothesis. Although at lower temperature and shorter reaction time, ethyl levulinate yield was far less with respect to overall bagasse conversion but trend of increasing trend of ethyl levulinate yield was observed. But to our surprise, a higher ethyl levulinate yield was achieved with higher acid concentration which was found to be in contradiction with overall bagasse conversion trend with acid concentrations. It is evident from above discussions that acid concentration doesn't play major role in overall bagasse conversion but we found that acid concentration has critical role in terms of ethyl levulinate selectivity. Under optimum reaction conditions at 125°C temperature and a reaction time of two hours, we observed exponential trend in ethyl levulinate yield and maximum 14.5 wt% ethyl levulinate yield was achieved. Based on assumption that only cellulose was being converted into ethyl levulinate, this yield was found to be equivalent to 36.21% theoretical yield from cellulose. One possible reason may be availability of free acids to provide protonated ions in bulk which are critical for ethyl levulinate production. All experiments were repeated to verify accuracy of data.

Reaction Mechanism:

Based on results obtained from GC-MS data, it is not possible to define a definite reaction mechanism as wide ranges of products were shown in MS spectra (Figur-2) especially sulfurous compounds. However, based on major components and our desired end product, we anticipate that conversion of bagasse into ethyl levulinate followed a similar reaction mechanism which is followed in conversion of biomass into levulinic acid (refer to figure-2).

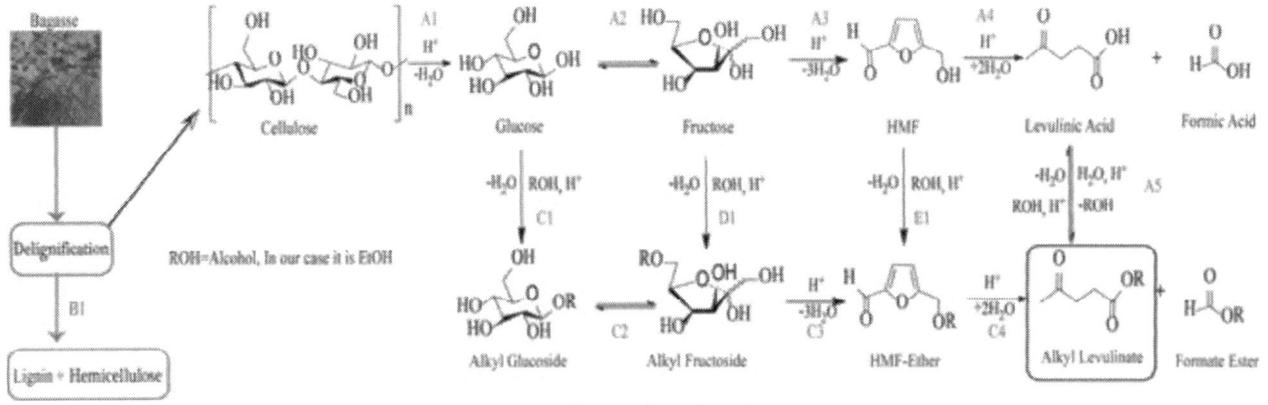

Figure 2 Reaction Mechanism for Ethyl Levulinate Production

This hypothesis was further confirmed by NIST library to confirm different products formed based on their m/z ratio and was found in order. We have observed that both formic acid and esters are present in end product along with ethyl levulinate and EMF. Therefore, it is believed that reaction has proceeded through both the pathways. However, more detailed analysis is still required as we expect that there may be some sulfurous compounds formed but not detected by GC-MS.

Conclusion:

- Conversion of Bagasse into ethyl levulinate is reported at mild process conditions.
- Sulfuric acid is more reactive in terms of overall bagasse conversion as compared to other acid catalysts.
- Higher reaction temperature and longer reaction time increases overall bagasse conversion as well as ethyl levulinate yield.
- Although acid concentration does not make significant effect in overall bagasse conversion under optimized conditions but higher acid concentration increases ethyl levulinate yield.
- Under optimum conditions, 14.5 wt% ethyl levulinate yield can be obtained at 125°C temperature and 2 hours reaction time in the presence of 0.8M sulfuric acid and ethanol media.

References

[1] J. C. Serrano-Ruiz and J. a. Dumesic, "Catalytic routes for the conversion of biomass into liquid hydrocarbon transportation fuels," *Energy Environ. Sci.*, vol. 4, no. 1, pp. 83–99, 2011.

[2] M. Imteyaz Alam, S. Gupta, E. Ahmad, and M. A. Haider, "Integrated Bio- and Chemocatalytic Processing for Biorenewable Chemicals and Fuels," in *Sustainable Catalytic Process*, B. Saha and M. Fan, Eds. Elsevier, 2015, pp. 157–177.

[3] C. O. Tuck, E. Pérez, I. T. Horváth, R. a Sheldon, and M. Poliakoff, "Valorization of biomass: deriving more value from waste," *Science*, vol. 337, no. 6095, pp. 695–699, Aug. 2012.

[4] D. M. Alonso, J. Q. Bond, and J. a. Dumesic, "Catalytic conversion of biomass to biofuels," *Green Chem.*, vol. 12, no. 9, pp. 1493–1513, 2010.

[5] S. Quereshi and S. Dutta, "Synthesis of Biomass Derived Platform Chemicals & Fuel Additives," in *5th Asia Oceana Conference on Green and Sustainable Chemistry, RSC*, 2015, p. 134.

[6] M. J. Climent, A. Corma, and S. Iborra, "Conversion of biomass platform molecules into fuel additives and liquid hydrocarbon fuels," *Green Chem.*, vol. 16, no. 2, pp. 516–547, 2014.

[7] L. Peng, L. Lin, J. Zhang, J. Shi, and S. Liu, "Applied Catalysis A : General Solid acid catalyzed glucose conversion to ethyl levulinate," *Appl. Catal. A Gen.*, vol. 397, pp. 259–265, 2011.

[8] H. Joshi, B. R. Moser, J. Toler, W. F. Smith, and T. Walker, "Ethyl levulinate : A potential bio-based diluent for biodiesel which improves cold flow properties," *Biomass and Bioenergy*, vol. 35, no. 7, pp. 3262–3266, 2011.

[9] L. Peng, X. Gao, and K. Chen, "Catalytic upgrading of renewable furfuryl alcohol to alkyl levulinates using AlCl3 as a facile, efficient, and reusable catalyst," *Fuel*, vol. 160, pp. 123–131, 2015.

[10] Z. W. Wang, T. Z. Lei, L. Liu, J. L. Zhu, X. F. He, and Z. F. Li, "Performance investigations of a diesel engine using ethyl levulinate-diesel blends," *BioResources*, vol. 7, no. 4, pp. 5972–5982, 2012.

[11] S. Saravanamurugan and A. Riisager, "Solid acid catalysed formation of ethyl levulinate and ethyl glucopyranoside from mono- and disaccharides," *Catal. Commun.*, vol. 17, pp. 71–75, 2012.

[12] C. Chang, G. Xu, and X. Jiang, "Production of ethyl levulinate by direct conversion of wheat straw in ethanol media," *Bioresour. Technol.*, vol. 121, pp. 93–99, 2012.

[13] C. Chang, G. Xu, W. Zhu, J. Bai, and S. Fang, "One-pot production of a liquid biofuel candidate — Ethyl levulinate from glucose and furfural residues using a combination of extremely low sulfuric acid and zeolite USY," *FUEL*, vol. 140, pp. 365–370, 2015.

[14] P. Neves, S. Lima, M. Pillinger, S. M. Rocha, J. Rocha, and A. a. Valente, "Conversion of furfuryl alcohol to ethyl levulinate using porous aluminosilicate acid catalysts," *Catal. Today*, vol. 218–219, pp. 76–84, 2013.

[15] K. Y. Nandiwale, S. K. Sonar, P. S. Niphadkar, P. N. Joshi, S. S. Deshpande, V. S. Patil, and V. V Bokade, "Catalytic upgrading of renewable levulinic acid to ethyl levulinate biodiesel using dodecatungstophosphoric acid supported on desilicated H-ZSM-5 as catalyst," *Appl. Catal. A Gen.*, vol. 460–461, pp. 90–98, 2013.

[16] G. Pasquale, P. Vázquez, G. Romanelli, and G. Baronetti, "Catalytic upgrading of levulinic acid to ethyl levulinate using reusable silica-included Wells-Dawson heteropolyacid as catalyst," *Catal. Commun.*, vol. 18, pp. 115–120, 2012.

[17] C. R. Patil, P. S. Niphadkar, V. V. Bokade, and P. N. Joshi, "Esterification of levulinic acid to ethyl levulinate over bimodal micro-mesoporous H/BEA zeolite derivatives," *Catal. Commun.*, vol. 43, pp. 188–191, 2014.

[18] F. D. Pileidis, M. Tabassum, S. Coutts, and M.-M. Ttitirici, "Esterification of levulinic acid into ethyl levulinate catalysed by sulfonated hydrothermal carbons," *Chinese J. Catal.*, vol. 35, no. 6, pp. 929–936, 2014.

[19] X. Tang, Y. Sun, X. Zeng, W. Hao, L. Lin, and S. Liu, "Novel Process for the Extraction of Ethyl Levulinate by Toluene with Less Humins from the Ethanolysis Products of Carbohydrates," *Energy & Fuels*, 2014.

425168 Direct Conversion of Hemicellulosic Materials to Biobutanol

Abstract:

Being the most abundant biomass, lignocellulosic materials exhibited great potential to be utilized for biochemical production. To shed light on the direct conversion of lignocellulosic biomass into biochemicals, here a newly discovered *Clostridium* sp. strain G117 was investigated on its capability to generate butanol with xylan as the sole carbon source. In this study, 0.86 ± 0.07 g/L butanol accompanied with hydrogen were directly produced from xylan (30 g/L) by strain G117 without prior enzymatic hydrolysis. After fermentation process optimization, 1.24 ± 0.37g/L butanol was obtained when fed with same amount of xylan within 5 days. Noteworthy, in contrast to the traditional ABE fermentation process, acetone production by strain G117 was significantly suppressed when grown on xylan, which could greatly simplify the downstream separation process. Moreover, much higher amount of butanol (2.94 g/L) was produced from xylo-oligosaccharides. In summary, *Clostridium* sp. stain G117 is a promising candidate in degradation of the sustainable feedstock - xylan and its oligosaccharides to produce considerable amount of butanol without formation of by-products such as ethanol and acetone. This study opens the door for direct conversion of lignocellulosic biomass to biochemicals in a single step with the elimination of fermentation by-products.

434151 Co-Liquefaction of Mixed Culture Microalgal Strains Under Subcritical Water Conditions

Abstract:

Hydrothermal liquefaction (HTL) is one of the productive thermochemical conversion processes capable of converting wet feedstock into renewable bio-oils. In this study, microalgae *Cyanidioschyzon merolae* (CM) and *Galdieria sulphuraria* (GS) were co-liquefied under hydrothermal conditions in a stainless-steel batch reactor. The temperatures and CM/GS mass ratios were varied from 150 to 300°C and 0 to 100% respectively. All the experiments were performed at 20% solid loading and at a reaction time of 30 min. The maximum bio-crude yield was obtained at 300°C and at a mass ratio of 4:1 (CM/GS). A positive synergetic effect is observed for the bio-crude yield, which is dependent on temperature and mass ratio. The biocrude oils were analyzed using Gas Chromatography Mass Spectrometry (GC-MS) and Fourier Transform Ion Cyclotron Resonance Mass Spectrometer (FT-ICR MS). FT-ICR MS analysis of the biocrude oils indicate that both the co-liquefied and pure individual algal oils are compositionally similar but their relative intensities changed with process parameters. Ultimate and proximate analysis was also performed for the biomass, biocrude oil and biochar.

NREL
NATIONAL RENEWABLE ENERGY LABORATORY

Improved Estimate of Air Pollutant Emissions from Biorefinery

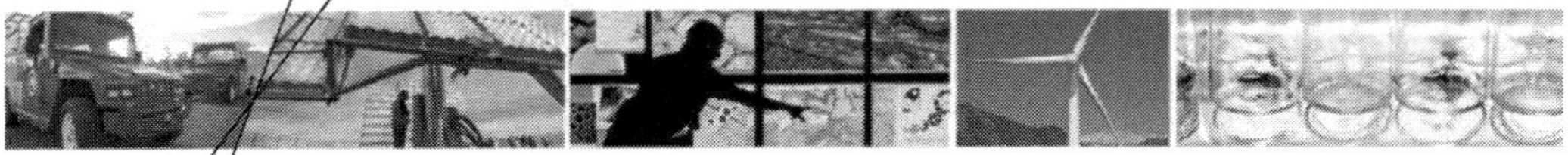

Eric C.D. Tan

National Renewable Energy Laboratory (NREL), Golden, CO

November 10, 2015
Session: Sustainable Fuel from Renewable Resources
2015 AIChE Annual Meeting, Salt Lake City, UT

NREL National Bioenergy Center Capabilities

Thermochemical Conversion
- Micro-reactors to pilot plants

Biochemical Conversion
- Bench scale to ton/day

Genomics Laboratory
- Tools for strain development

Biomass Characterization
- Wet chemical and NIR

Spectroscopy Facilities
- NMR, IR, TEM, MBMS

Biorefinery Analysis
- Advanced process design
- Techno-economic analysis
- Sustainability & Life-Cycle Assessment
- Strategic analysis

http://www.nrel.gov/biomass/capabilities.html

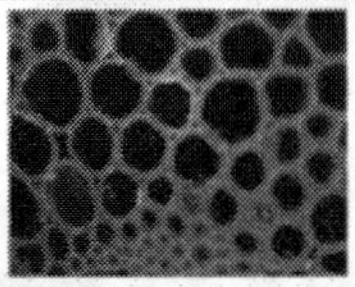

NREL has been developing biomass conversion technology

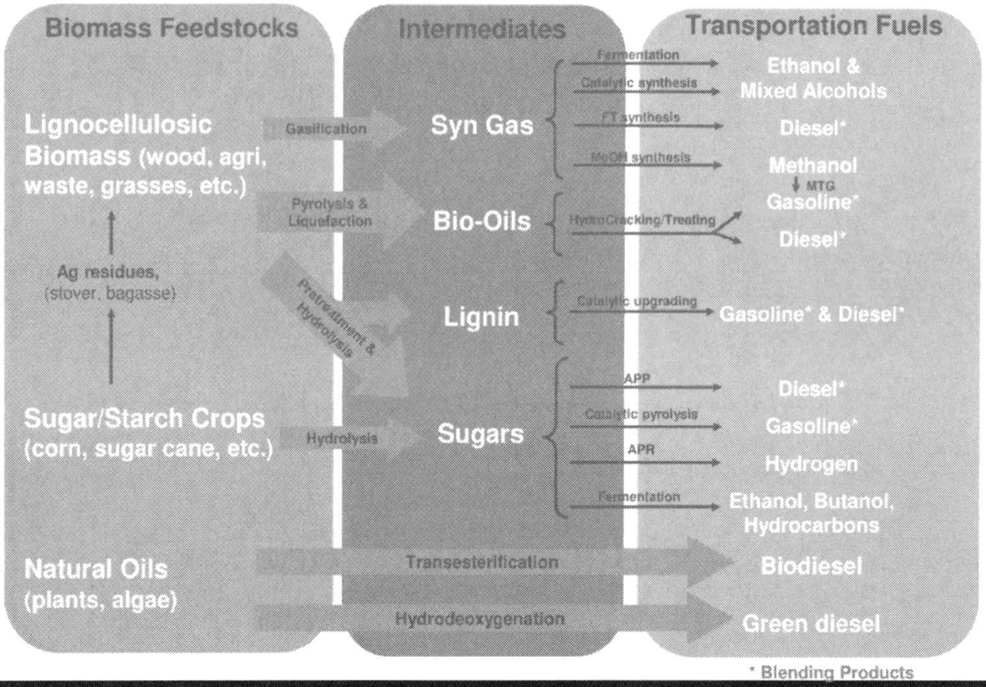

NREL Process Design Reports

http://www.nrel.gov/biomass/publications.html

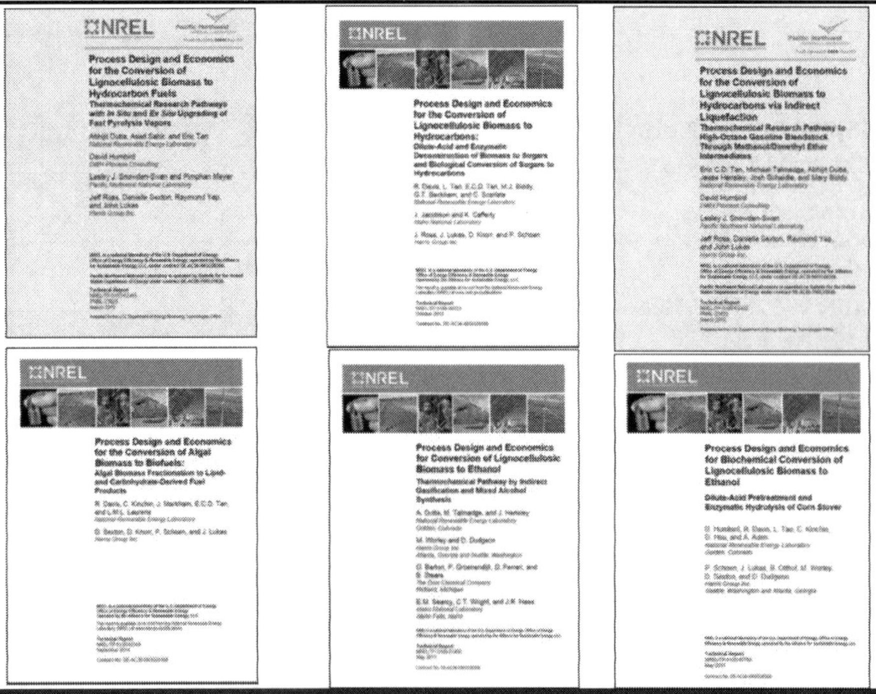

NREL as part of DOE's commitment to sustainability

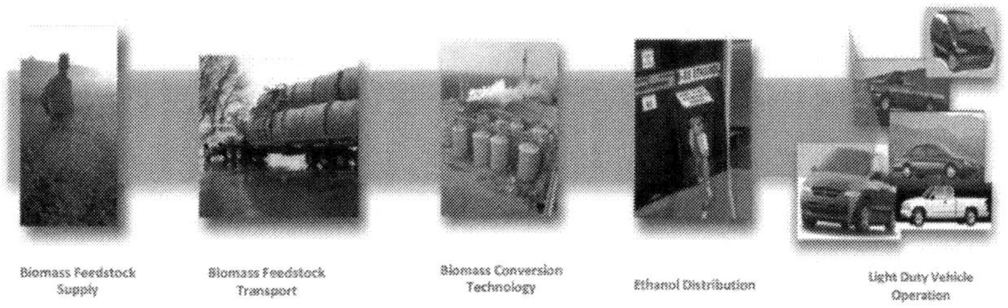

- Involving with quantifying and improving sustainability across the HC blendstocks supply-chain. (Biorefinery = main focus)
- Establishing sustainability baseline metrics for the biorefinery (BC & TC).
- Assessing lifecycle impact and environmental sustainability (e.g., tradeoffs between conversion technology options).
- Developing annual state-of-technology (SOT) sustainability assessments for the conversion platforms (based on conversion SOTs).

  >> Conversions SOTs – validated with bench- and/or pilot-scale test data.

Approach

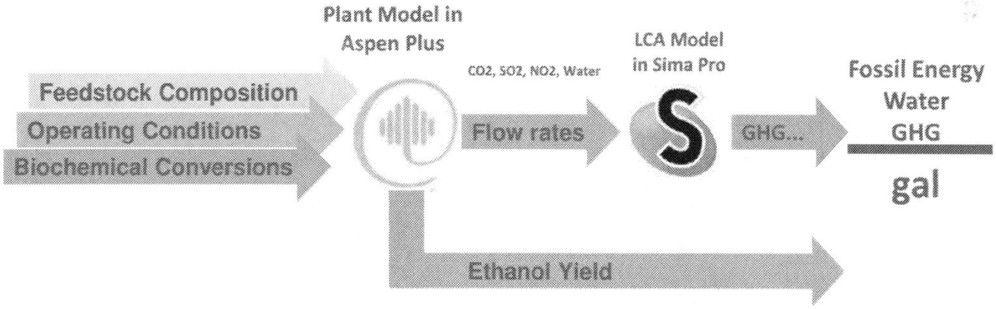

- **Aspen Plus was used to estimate:**

  >> Direct GHG emissions (CO_2),

  >> Direct water consumption

  >> Direct select criteria air pollutant (CAP) emissions (NO_2 and SO_2).
- **SimaPro was used to quantify both the direct and indirect (embedded) emissions of the TC and BC designs.**
  - Examples sustainability metrics evaluated:

    >> Net GHG emissions (kg CO_2 equivalent)

    >> Fossil energy demand

    >> Carbon conversion efficiency

    >> Consumptive water use

    >> Energy return on investment (EROI)

NREL's thermochemical process design (Mixed alcohols via Indirect Gasification of Woody Biomass)

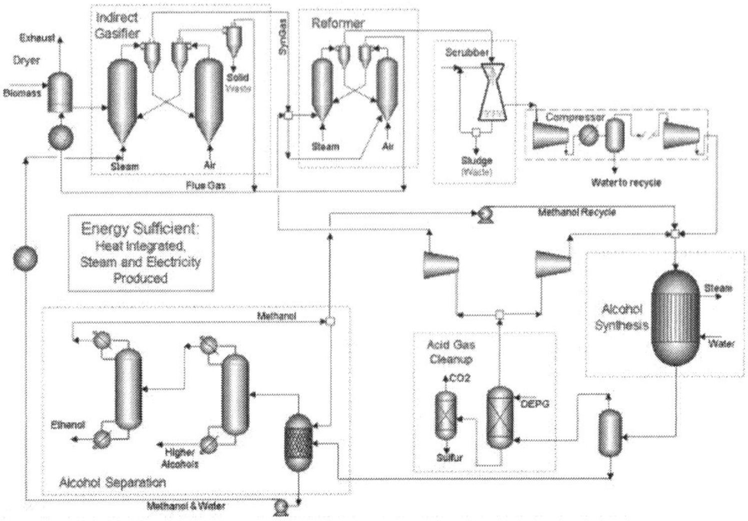

Illustration 1: Thermochemical conversion of woody biomass to mixed alcohols

LCA Output (tracking fossil GHG, excluding biogenic CO2)

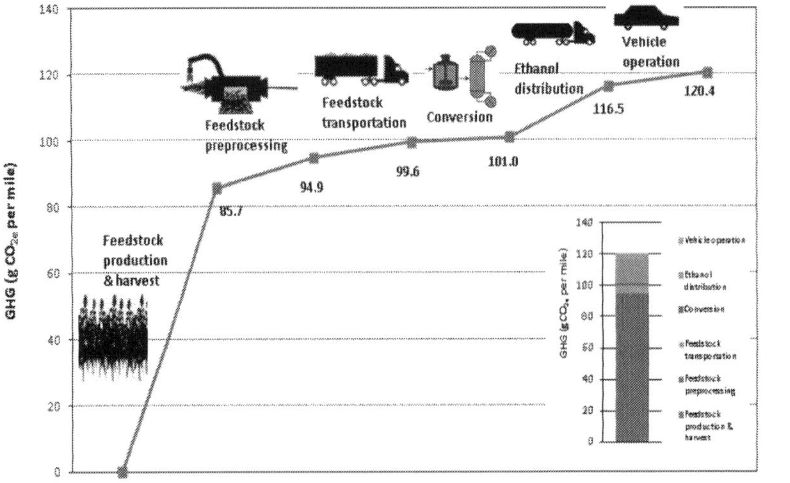

%GHG reduction = 74%

NREL's biochemical process design (dilute-acid, enzymatic hydrolysis and co-fermentation process)

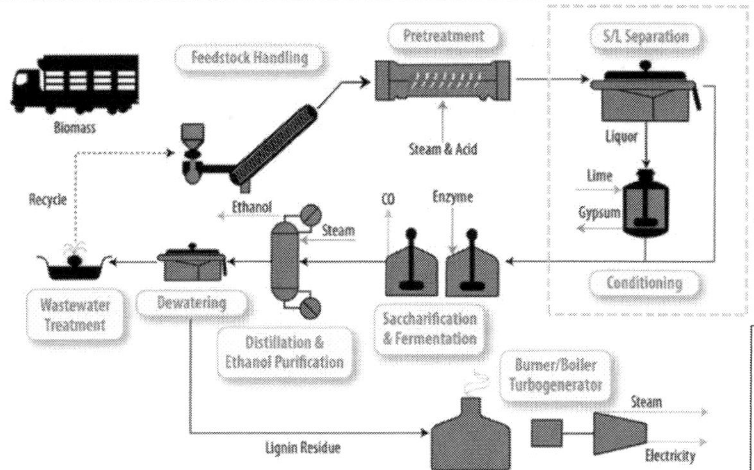

Hydrolysate is conditioned to neutralize acetic acid and other compounds liberated in pretreatment that are toxic to the fermenting organism.

Process Areas
- ✓Feedstock Handling
- ✓Pretreatment & Conditioning
- ✓Saccharification & Fermentation
- ✓Product Recovery & Purification
- ✓Wastewater Treatment
- ✓Combustor, boiler, and turbogenerator
- ✓Storage
- ✓Utilities

Illustration 2: Biochemical conversion of corn stover to ethanol

LCA Output (tracking fossil GHG, excluding biogenic CO2)

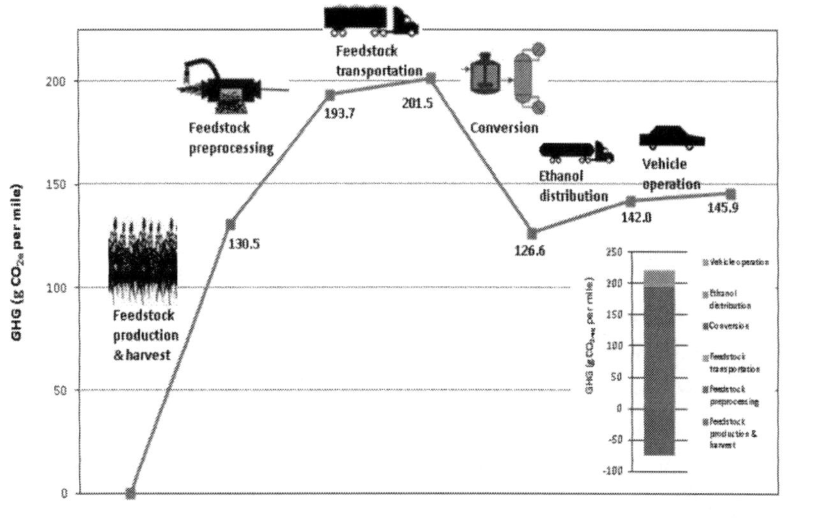

%GHG reduction = 69%

Direct Biorefinery Air Emission

- One of the challenges in quantifying air quality and human health impacts of advanced biofuel pathways is the lack of actual measurements of air emissions from biofuel conversion processes.
- The goal is to develop an air pollutant emissions inventory for the biorefineries.
- Six common criteria air pollutants
  - ✓ Ozone
  - ✓ Particulate Matter
  - ✓ Carbon Monoxide
  - ✓ Nitrogen Oxides
  - ✓ Sulfur Dioxide
  - ✓ Lead
- In NREL's conversion processes, the largest quantity of criteria air pollutant (CAP) emissions are from the combustors.

Direct Biorefinery Air Emission

- CO_2 and SO_2 emissions from combustion typically can be estimated with a high degree of certainty.
  - ✓ These emissions depend almost exclusively on the C and S content of the fuel, which is generally known with a high degree of precision.
  - ✓ Both C and S are almost completely oxidized during combustion to CO_2 and SO_2.
  - ✓ Emission rates can be estimated with a process simulation tool such as Aspen Plus.
- On the other hand, the combustion emissions of criteria air pollutants and non-CO_2 GHGs, such as N_2O depend on many factors, including
  - ✓ The precise technology applied when the fuel is combusted,
  - ✓ Incomplete combustion of a small fraction of the fuel, or
  - ✓ Complicated chemical and physical processes during the combustion.

Detailed kinetic modeling: Improved Estimates of Air Pollutant Emissions from Biorefinery

- To better characterize air pollutant emissions from biofuel conversion processes, we developed detailed gas-phase chemistry for biorefinery combustion.
- Detailed kinetic modeling can offer several capabilities. In addition to the more accurate prediction of air pollutant emissions, a detailed kinetic model also allows us to study the effect of the feed composition on these emissions.
- CHEMKIN is a valuable simulation/modeling tool for investigating complex systems involving homogenous and/or heterogeneous reactions such as combustion.

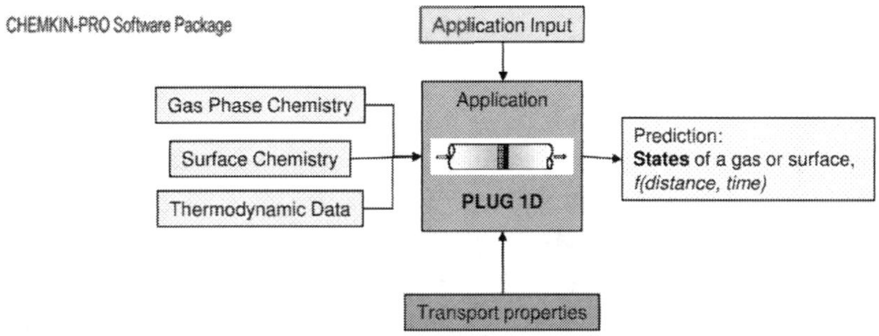

Illustration: Direct Biorefinery Air Emission from Lignin Combustor

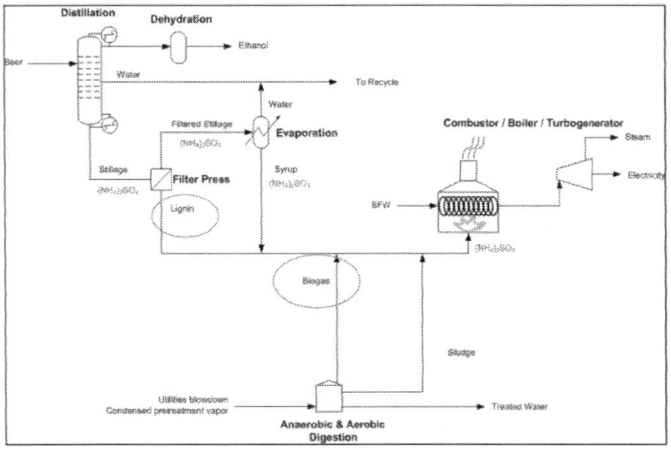

- The focus of this study is to develop a preliminary detailed reaction mechanism for lignin combustion.
- Lignin is the only biomass component included in the current mechanism and methane is used as the biogas surrogate.

Chemical Reaction Mechanisms

- **Developing a reaction mechanism from scratch essentially is not feasible.**
- **As a starting point, it is preferable to adopt an existing model. In this study, a chemical kinetic model is constructed by assembling various related reaction mechanisms provided in published literature.**
- **Lignin combustion consists of three sequential steps: devolatilization, secondary gas phase reactions of the released gas (and tar species), and (char) combustion.**
- **Due to its complex chemical structure, lignin can be represented by a mixture of model components. In this study, the reference species guibourtinidol (C15H14O4) is chosen as the surrogate for lignin.**

Lignin:
- Complex aromatic structure
- High energy content
- Resists biochemical conversion

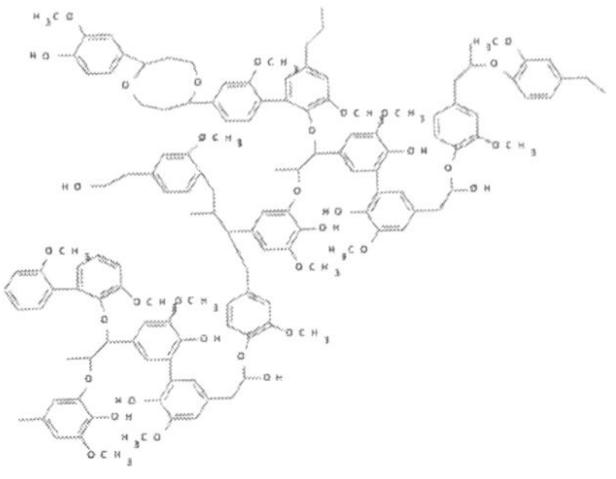

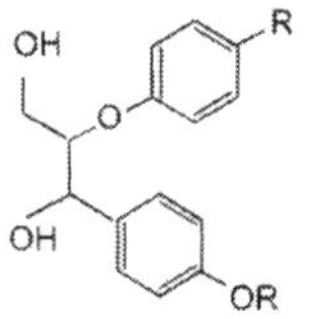

Lignin surrogate ($C_{15}H_{14}O_4$)

Algorithm for the development of a detailed reaction mechanism

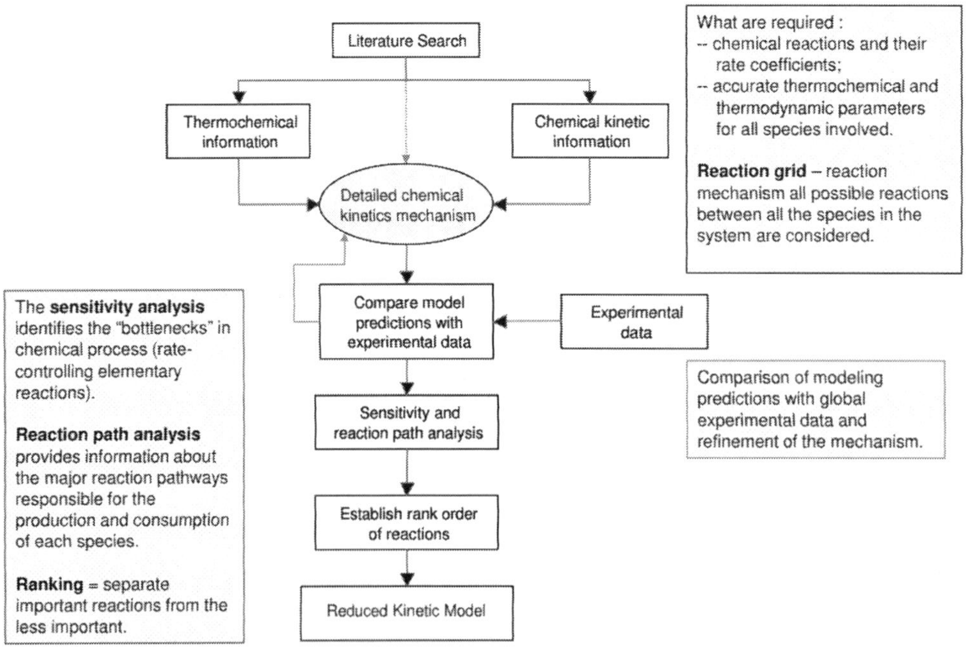

TowerPak Stirling Power Boiler system

- TowerPak Stirling Power Boiler system by Babcock & Wilcox
- This system features a live-bottom grated fuel bin to ensure drying and complete combustion of the wet solid fuel.
- The system was quoted to produce 525,000 lb/h (239,000 kg/h) of steam at 850 °F (454°C) and 900 psig.
- Boiler efficiency, defined as the percentage of the feed heating value that is converted to steam heat, is ~80%.

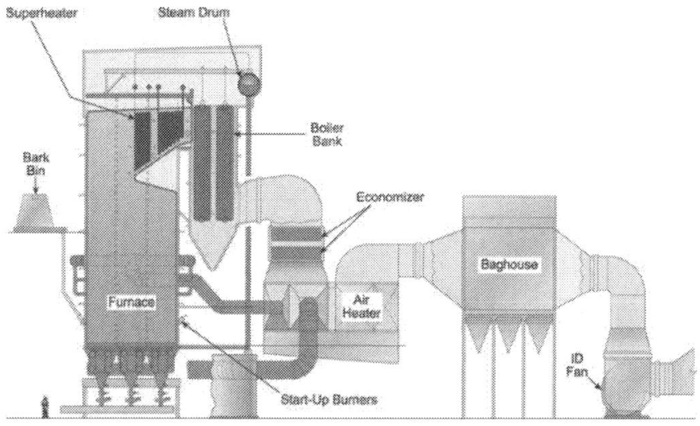

Source: http://www.babcock.com/products/Pages/Towerpak-boiler.aspx

Chemical Reaction Mechanisms

- **Lumped Kinetics of Lignin Devolatilization during Lignin Combustion** *(modified the kinetic developed by CRECK modeling group)*

Lignin Decomposition Reactions	Rate	Units
4Lignin => 2pCoumaryl + 3Phenol + 9Char	$k = 4 \times 10^{15} \exp(-48500/RT)$	$[s^{-1}]$
2Lignin => LIGcc + 3G{COH_2} + 3.5H_2 + 11Char	$k = 4 \times 10^{15} \exp(-48500/RT)$	$[s^{-1}]$
LIGcc => pCoumaryl + 0.5G{CO} + 1.5G{COH_2} + H_2O + 5Char	$k = 5 \times 10^{6} \exp(-31500/RT)$	$[s^{-1}]$
LIGcc => Phenol + G{COH_2} + 0.5H_2 +4Char	$k = 5 \times 10^{6} \exp(-31500/RT)$	$[s^{-1}]$
LIGcc => pCoumaryl + 3H_2O + 2.5H_2 + 13Char	$k = 5 \times 10^{6} \exp(-31500/RT)$	$[s^{-1}]$
LIGcc => 2.75CH_4 + C_2H_4 + CO + 4G{CO} + 6.25Char	$k = 5 \times 10^{6} \exp(-31500/RT)$	$[s^{-1}]$
G{CO} => CO	$k = 1 \times 10^{13} \exp(-50000/RT)$	$[s^{-1}]$
G{COH_2} => CO+H_2	$k = 5 \times 10^{11} \exp(-65000/RT)$	$[s^{-1}]$

G{CO} and G{COH_2} are species trapped in the metaplast and successively released to gas phase.
pCoumaryl and Phenol are represented with $C_9H_{10}O_2$ and $C_{11}H_{12}O_4$, respectively.
Units for Arrhenius preexponential factor and activation energy are presumed to be $mol\,l^{-1}s^{-1}$ and $cal\,mol^{-1}$, respectively.

- **The released volatile components from lignin will then undergo successive decomposition and/or combustion reactions. The detailed mechanism of the secondary gas-phase reaction and combustion of species released during lignin thermal degradation was obtained from literature (CRECK 2012).**

Chemical Reaction Mechanisms

- Reactions associated with nitrogen-content species (i.e., N2O, NO, NO2) were extracted from the GRI-MECH version 3.0 mechanism (GRI-Mech).
- The finalized mechanism contains 154 chemical species and 2,996 reactions. The reaction mechanism describes the combustion of lignin and methane (a component of biogas) as well as the formation of criteria air pollutants such as CO, NO, NO_2, volatile organic compounds (VOC), and unburned hydrocarbon (UHC).

$k = AT^B exp(-E/RT)$

	A	B	E
H+O2+N2<=>HO2+N2	2.600E+19	-1.240	.00
N+NO<=>N2+O	2.700E+13	.000	355.00
N+O2<=>NO+O	9.000E+09	1.000	6500.00
N+OH<=>NO+H	3.360E+13	.000	385.00
N2O+O<=>N2+O2	1.400E+12	.000	10810.00
N2O+O<=>2NO	2.900E+13	.000	23150.00
N2O+H<=>N2+OH	3.870E+14	.000	18880.00
N2O+OH<=>N2+HO2	2.000E+12	.000	21060.00
N2O(+M)<=>N2+O(+M)	7.910E+10	.000	56020.00
LOW / 6.370E+14 .000 56640.00/			
H2/2.00/ H2O/6.00/ CH4/2.00/ CO/1.50/ CO2/2.00/ C2H6/3.00/ AR/ .625/			
HO2+NO<=>NO2+OH	2.110E+12	.000	-480.00
NO+O+M<=>NO2+M	1.060E+20	-1.410	.00
H2/2.00/ H2O/6.00/ CH4/2.00/ CO/1.50/ CO2/2.00/ C2H6/3.00/ AR/ .70/			
NO2+O<=>NO+O2	3.900E+12	.000	-240.00
NO2+H<=>NO+OH	1.320E+14	.000	360.00
NH+O<=>NO+H	4.000E+13	.000	.00
NH+H<=>N+H2	3.200E+13	.000	330.00
NH+OH<=>HNO+H	2.000E+13	.000	.00
NH+OH<=>N+H2O	2.000E+09	1.200	.00
NH+O2<=>HNO+O	4.610E+05	2.000	6500.00
NH+O2<=>NO+OH	1.280E+06	1.500	100.00
NH+N<=>N2+H	1.500E+13	.000	.00
NH+H2O<=>HNO+H2	2.000E+13	.000	13850.00
NH+NO<=>N2+OH	2.160E+13	-.230	.00
NH+NO<=>N2O+H	3.650E+14	-.450	.00
NH2+O<=>OH+NH	3.000E+12	.000	.00
NH2+O<=>H+HNO	3.900E+13	.000	.00
NH2+H<=>NH+H2	4.000E+13	.000	3650.00
NH2+OH<=>NH+H2O	9.000E+07	1.500	-460.00
NNH<=>N2+H	3.300E+08	.000	.00
NNH+M<=>N2+H+M	1.300E+14	-.110	4980.00
H2/2.00/ H2O/6.00/ CH4/2.00/ CO/1.50/ CO2/2.00/ C2H6/3.00/ AR/ .70/			
NNH+O2<=>HO2+N2	5.000E+12	.000	.00
NNH+O<=>OH+N2	2.500E+13	.000	.00
NNH+O<=>NH+NO	7.000E+13	.000	.00
NNH+H<=>H2+N2	5.000E+13	.000	.00
NNH+OH<=>H2O+N2	2.000E+13	.000	.00
NNH+CH3<=>CH4+N2	2.500E+13	.000	.00
H+NO+M<=>HNO+M	4.480E+19	-1.320	740.00
H2/2.00/ H2O/6.00/ CH4/2.00/ CO/1.50/ CO2/2.00/ C2H6/3.00/ AR/ .70/			
HNO+O<=>NO+OH	2.500E+13	.000	.00
HNO+H<=>H2+NO	9.000E+11	.720	660.00
HNO+OH<=>NO+H2O	1.300E+07	1.900	-950.00
HNO+O2<=>HO2+NO	1.000E+13	.000	13000.00
CN+O<=>CO+N	7.700E+13	.000	.00
CN+OH<=>NCO+H	4.000E+13	.000	.00
CN+H2O<=>HCN+OH	8.000E+12	.000	7460.00
CN+O2<=>NCO+O	6.140E+12	.000	-440.00
CN+H2<=>HCN+H	2.950E+05	2.450	2240.00
NCO+O<=>NO+CO	2.350E+13	.000	.00
NCO+H<=>NH+CO	5.400E+13	.000	.00
NCO+OH<=>NO+H+CO	0.250E+13	.000	.00
NCO+N<=>N2+CO	2.000E+13	.000	.00
NCO+O2<=>NO+CO2	2.000E+12	.000	20000.00
NCO+M<=>N+CO+M	3.100E+14	.000	54050.00
H2/2.00/ H2O/6.00/ CH4/2.00/ CO/1.50/ CO2/2.00/ C2H6/3.00/ AR/ .70/			
NCO+NO<=>N2O+CO	1.900E+17	-1.520	740.00
NCO+NO<=>N2+CO2	3.800E+18	-2.000	800.00
HCN+M<=>H+CN+M	1.040E+29	-3.300	126600.00
H2/2.00/ H2O/6.00/ CH4/2.00/ CO/1.50/ CO2/2.00/ C2H6/3.00/ AR/ .70/			
HCN+O<=>NCO+H	2.030E+04	2.640	4980.00
HCN+O<=>NH+CO	5.070E+03	2.640	4980.00
HCN+O<=>CN+OH	3.910E+09	1.580	26600.00
HCN+OH<=>HOCN+H	1.100E+06	2.030	13370.00
HCN+OH<=>HNCO+H	4.400E+03	2.260	6400.00
HCN+OH<=>NH2+CO	1.600E+02	2.560	9000.00
H+HCN(+M)<=>H2CN(+M)	3.300E+13	.000	.00

Chemical Reaction Mechanisms

- **Reaction Rates and Kinetic Constants of the Char Combustion, as Estimated by the Two-Step Model**

	Global kinetics Arrehenius rate	A [s^{-1}]	E [kJ/mol]	n
Devolatilization	$R_1 = A_1 exp(-E_1/RT)Y_B$	1.33 x 10^{15}	218.5	
Combustion	$R_2 = A_2 exp(-E_2/RT)Y_C^n$	4.85 x 10^{14}	228.6	1.16

Chemkin	$k_{fi} = A_i T_{Bi} exp(-E_i/R_c T)$	A_i	B_i	E_i [cal/mol]
Devolatilization	Char => 66.75C(S) + 6CO + 0.25CH4	1.33E+15	0.0	52187.83
Combustion	C(S) + O2 => CO2	4.85E+14	0.0	54600.17
	FORD/O2 1.16/			
	C(S) + OH = CO + H	5.00E+08	0.0	0.00
	C(S) + O2 = CO2 + O	5.80E+08	0.0	576.00
	CO + O2 = CO2 + O	2.50E+12	0.0	47800.00
		1.50E+14	0.0	23600.00
		4.76E+07	1.228	70.00

Chemical Reaction Mechanisms

- Key Soot Formation Reactions

	$k = AT^B \exp(-E/RT)$ A	B	E
Soot nucleation			
2A4 => 32C(B) + 2OH(S) + 28.72(S)	9.00E+09	0.5	0.0
C(S) => C(B)	3.40E+02	0.0	0.0
2CO => C(B) + O2	2.00E+09	0.0	0.0
Soot mass growth reactions			
H + H(S) => (S) + H2	4.20E+13	0.0	13000.0
H2 + (S) => H(S) + H	3.90E+12	0.0	9320.0
H + (S) => H(S)	2.00E+13	0.0	0.0
H(S) + OH => H2O + (S)	1.00E+10	0.734	1430.0
H2O + (S) => OH + H(S)	3.68E+08	1.139	17100.0
C2H2 + (S) => H(S) + 2C(B) + H	8.00E+07	1.56	3800.0
PAH condensation			
A1 + 6H(S) => 6C(B) + 6(S) + 6H2	0.20	0.0	0.0
A2 + 16H(S) => 10C(B) + 16(S) + 12H2	0.10	0.0	0.0
A3 + 20H(S) => 14C(B) + 20(S) + 15H2	0.10	0.0	0.0
A4 + 20H(S) => 16C(B) + 20(S) + 15H2	0.10	0.0	0.0
Soot oxidation reaction			
OH + (S) + C(B) => CO + H(S)	0.20	0.0	8000.0

A1 = benzene, A2 = naphthalene, A3 = anthracene, A4 = pyrene
C(B) = soot, (S) = sites

Adiabatic temperature and molar conversions during lignin and biogas combustion (41% excess O_2, 800°C inlet temperature, CHEMKIN PLUG model)

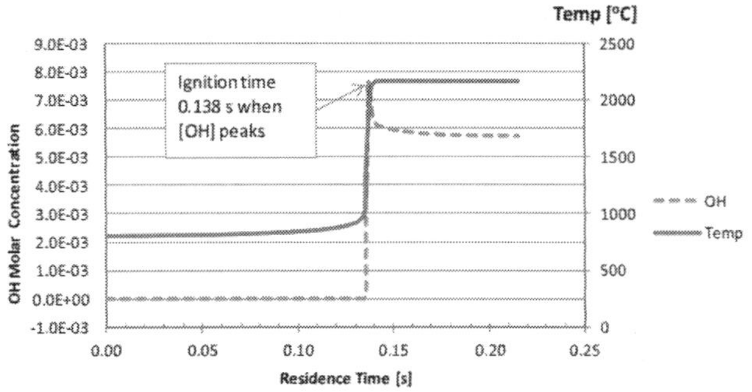

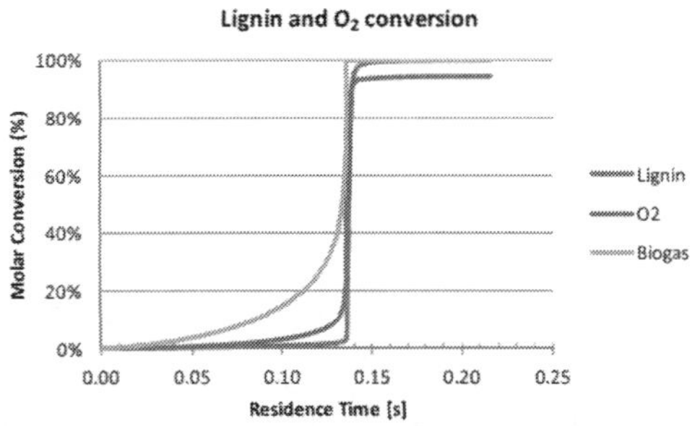

Devolatilization released species during lignin and biogas combustion (41% excess O_2, 800°C inlet temperature, CHEMKIN PLUG model)

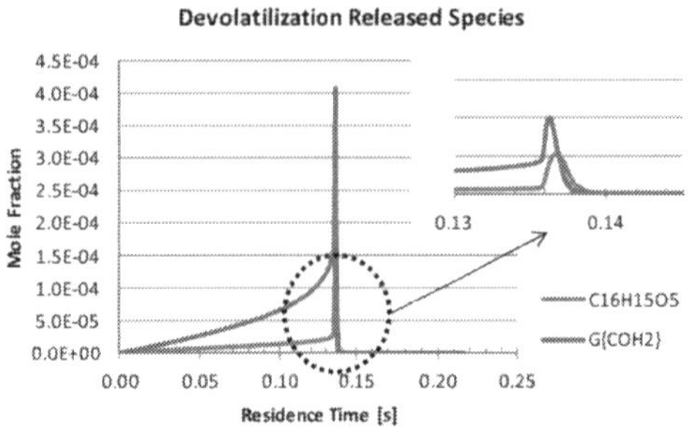

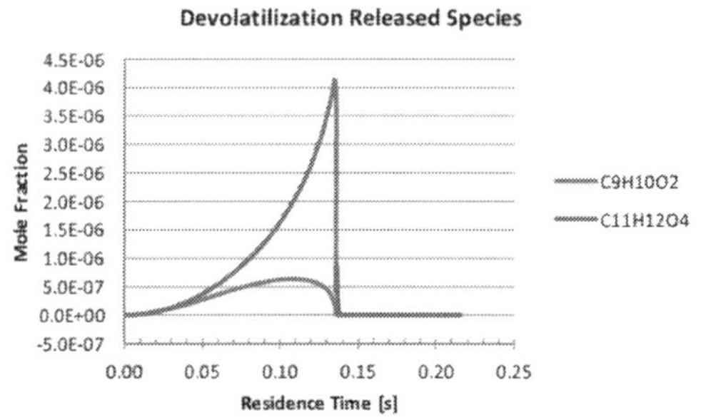

Combustion products and air pollutants during lignin and biogas combustion (41% excess O_2, 800°C inlet temperature, CHEMKIN PLUG model)

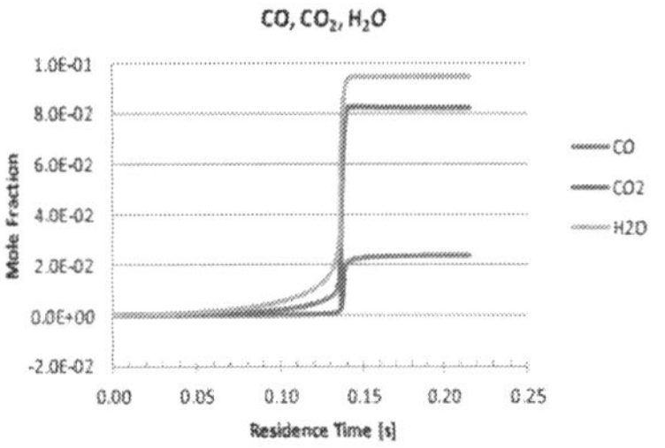

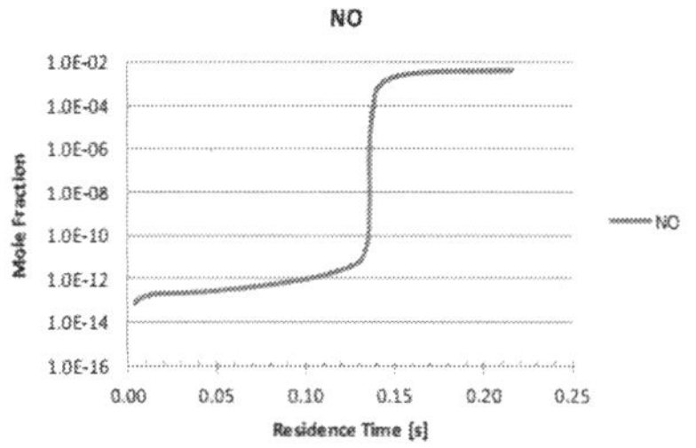

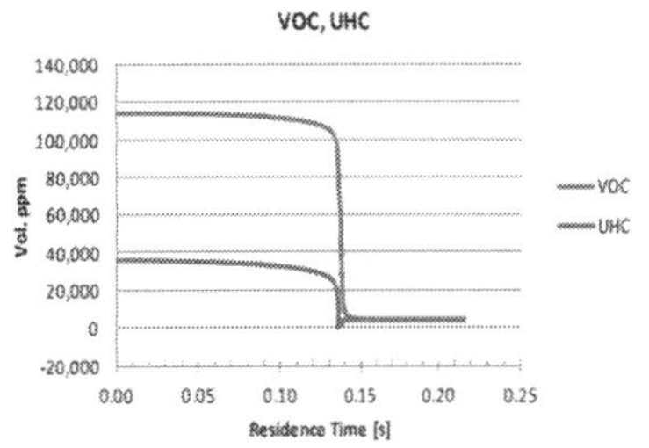

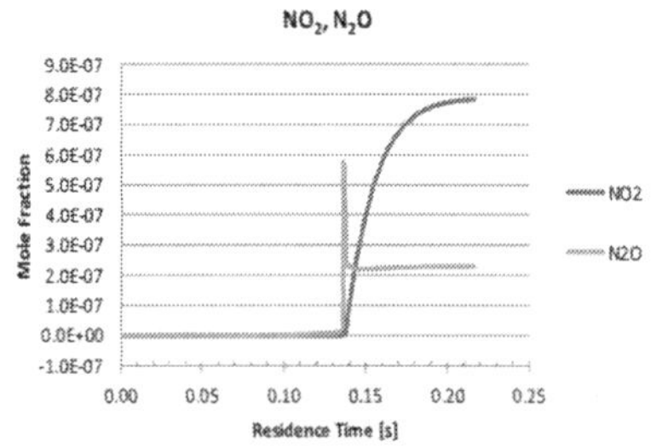

Effect of residence time on the formation of air pollutants (800℃ inlet temperature, 41% excess O_2, CHEMKIN PSR model).

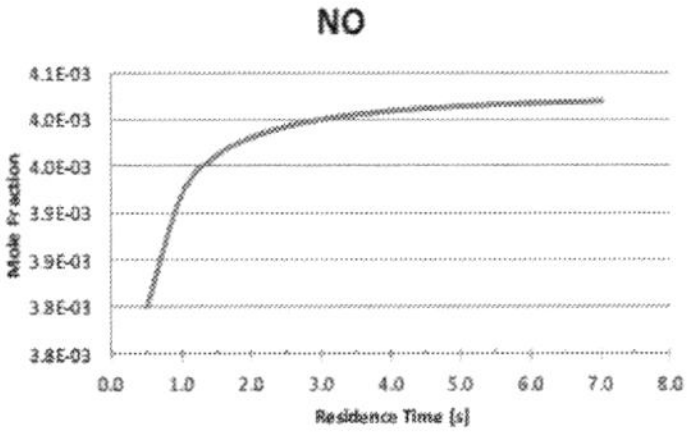

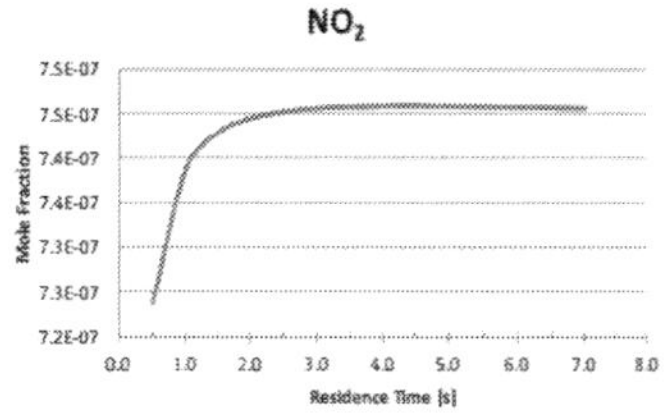

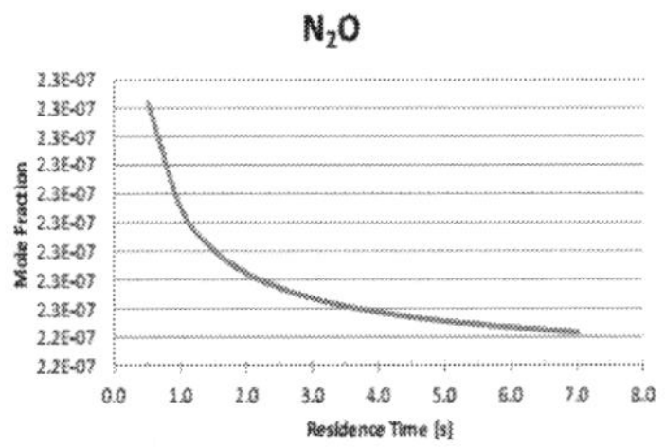

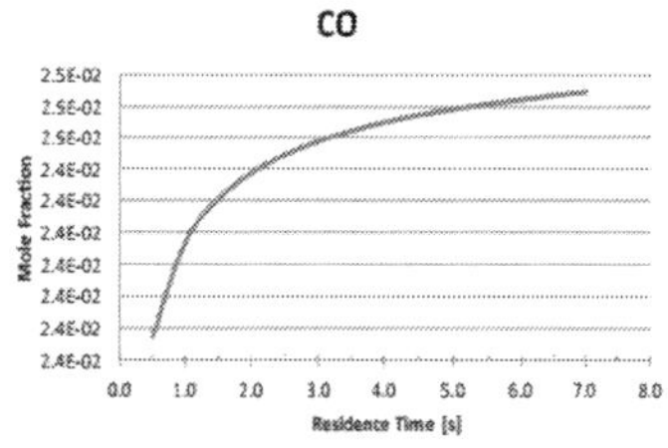

Typical simulation results corresponding to combustion char (CHEMKIN PLUG model)

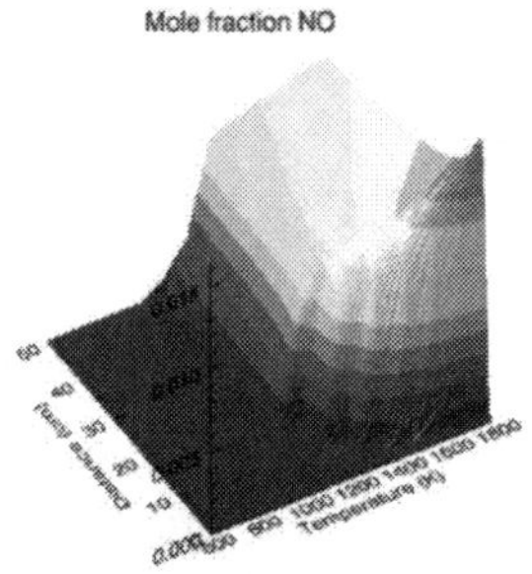

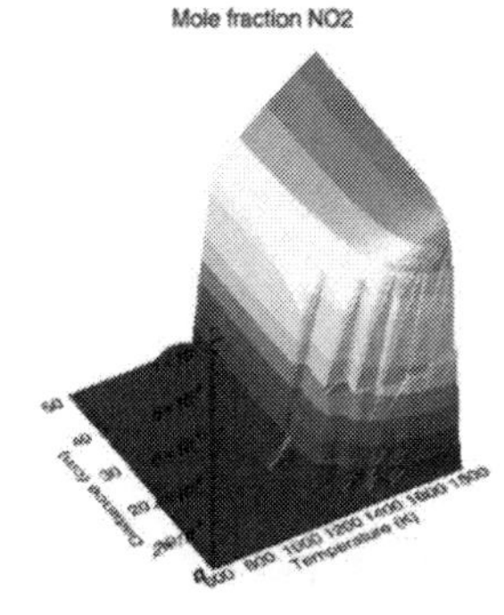

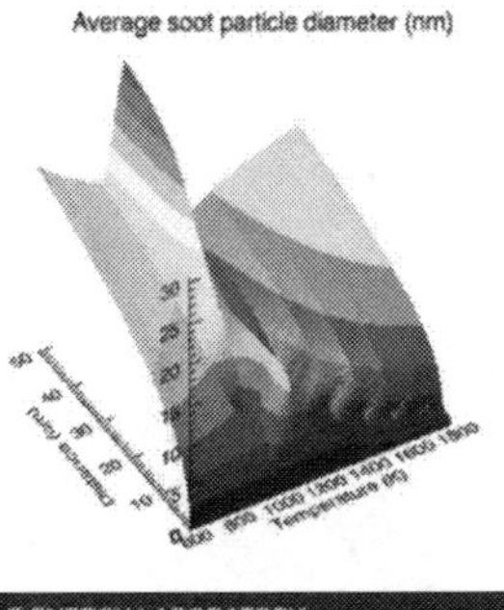

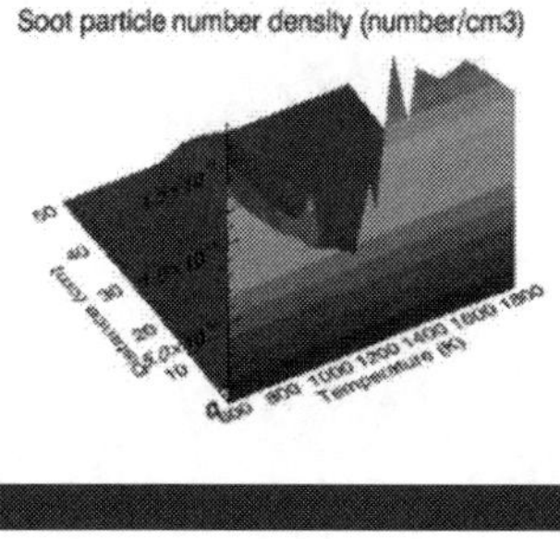

Summary

- We have attempted to use detailed kinetic modeling approach for improved estimation of combustion air pollutant emissions from biorefinery.
- We have developed a preliminary detailed reaction mechanism for biomass combustion. Lignin is the only biomass component included in the current mechanism and methane is used as the biogas surrogate.
- The model is capable of predicting the combustion emissions of greenhouse gases (CO_2, N_2O, CH_4) and criteria air pollutants (NO, NO_2, CO). The results are yet to be compared with the experimental data.
- The current model is still in its early stages of development. Given the acknowledged complexity of biomass oxidation, as well as the components in the feed to the combustor, obviously the modeling approach and the chemistry set discussed here may undergo revision, extension, and further validation in the future.

Acknowledgements

NREL Colleagues
Danny Inman
Yimin Zhang
Garvin Heath

DOE's Bioenergy Technologies Office (BETO)
http://www.eere.energy.gov/biomass

Speaker information:
Eric.Tan@nrel.gov
http://www.nrel.gov/biomass

Thank you for your attention!

434138 Analysis of Environmental Impacts Derived from Burning of the Sugar Cane Bagasse

Abstract:

Among the gaseous emissions from an ethanol distillery, that generated by burning sugarcane bagasse at the distillery boiler is one of the most important. So is justified a study to assess the environmental impacts caused by burning sugarcane bagasse and its relation with the amount of steam produced for maintenance of the process. Once optimization methods may reduce the amount of steam used in the process, then it is possible measure the amount of gaseous emissions that are released into the atmosphere. This work aims to assess the economic relationship between the amount of steam produced and the amount of emissions generated for its production in terms of ecopoints. Analysing the gases generated by burning the sugar cane bagasse in a boiler of frequent use in ethanol distilleries will be used the *Ecoinvent* database for calculating the amounts of emissions reduced from burning bagasse as fuel. The results show that is extremely advantageous economically to the distillery reduce their gas emissions due to the reduction of steam production, not only because the reduction of bagasse used as fuel happens, but also because some changes in the process can require less mechanical effort in the equipments and extending the life span of the plant. It was concluded that it is possible to relate the process variables under study with the reduce of environmental impacts in the utilities sector of this process.

FUNDAMENTAL CHARACTERISTICS OF FILTRATION AND CLASSIFICATION OF FINE PARTICLES IN LIQUID BY ELECTROSPUN NANOFIBER MEMBRANE

Yasuhito Mukai, Nagoya University, Nagoya, Japan

Introduction

Recently, the nanofiber is attracting attention in the textile industry and being developed for various applications in such widespread fields as environment, power energy, electronics, medicine and biotechnology. One of the typical applications of the nanofiber material is a nanofiber membrane, and it is considered to be promising as an alternative to conventional membrane. The nanofiber membrane is expected to be a new filter for environmental purification with high separation performance and various separation functions since it has a nanosized effect, an extremely high specific surface area and a high porosity [1]. In this study, we focused on the application of an electrospun nanofiber membrane to 1. filtration and 2. classification of fine particles dispersed in liquid.

Application to Filtration

The nanofibers are applied to gas or air treatment such as an exhaust gas filter and a chemical and biological protective suit. For liquid treatment, however, the separation characteristics and mechanisms of the nanofiber membrane are poorly understood because few researches have been conducted [2, 3]. In this study, the nanofiber membranes with various fiber diameters and deposition weights were prepared, and the water permeation and particle filtration tests were performed to clarify such fundamental separation characteristics as fluid permeability, pore diameter and particle retention.

Preparation of Nanofiber Membrane

Figure 1 shows the photographs of nanofiber membranes used in this study by a scanning electron microscope (SEM). The nanofiber membranes were prepared by depositing nylon 6 nanofibers produced by electrospinning method on the surface of the base material. Nanofibers with different diameters were produced by controlling the concentration of the nylon 6 solution, and the mean fiber diameter D_f was estimated from the observed data of diameters of randomly-selected 100 fibers projected by SEM. The mean diameter D_f increased with increasing the nylon 6 solution concentration and ranged from 126 to 162 nm for the concentration of 12.5 to 25 wt%. This is attributed to an increase in viscosity under higher concentration. Also, the deposition weight w of nanofiber per unit base area was adjusted to 1.8, 2.6, 3.5 g/m^2 for each nanofiber membrane. These corresponded to the thicknesses of 29, 42, 56 μm. For reference, the mixed cellulose ester microfiltration (MF) membrane with nominal pore size of 1.0 μm was also used.

Permeation and Filtration Tests

A schematic layout of the experimental setup is shown in **Fig. 2**. The feed fluid was poured into the pressurized tank, and the membrane was set to the filter holder with a filtration area of 3.8 cm^2. The

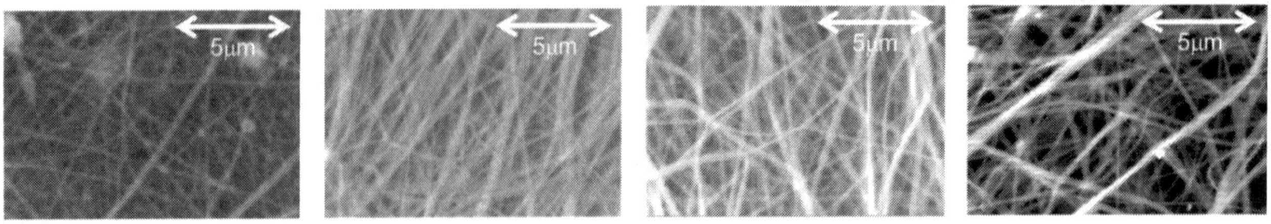

(a) c = 12.5 wt% (b) c = 20 wt% (c) c = 15 wt% (d) c = 25 wt%

Figure 1. Photomicrographs of electrospun nanofiber membrane

water permeation and particle filtration tests for electrospun nanofiber membranes and conventional MF membranes were performed under the constant pressure p by applying compressed nitrogen gas.

In the permeation test, the ultrapure water passing through the membrane was collected in a flask and weighed in an electronic balance connected to a personal computer. Because the permeate volume v per unit membrane area was proportional to the permeation time t, the permeation rate q was obtained from the slope of the plot of v versus t.

In the filtration test, the mono-dispersed polystyrene latex (PSL) particles with uniform diameter d_p of 0.048, 0.100, 0.143 or 0.196 µm were employed. The test suspensions with the concentration s by weight of 2 ppm were prepared by dispersing PSL particles in the ultrapure water. The time variations of the PSL concentration s_f in the filtrate were measured by reading the absorbance to estimate the PSL rejection R.

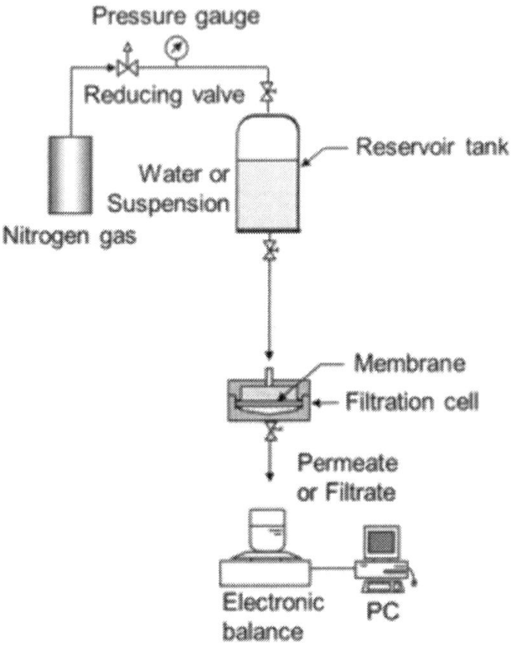

Figure 2. Experimental apparatus

Evaluation of Water Permeation Performance

The permeation tests using ultrapure water were carried out for the nanofiber membrane with D_f of 147 nm and w of 2.6 g/m^2 and the conventional MF membrane with d_m of 1.0 µm, and the permeation rates q were measured for various applied pressures p. The relationship between q and p is expressed by Darcy's equation (1).

$$q = K\frac{p}{\mu L} \tag{1}$$

where K is Darcy's permeability, μ is the viscosity of water, and L is the thickness of the membrane layer. Darcy's permeability K can be calculated from the slope of the plot of q versus p based on Eq. (1). Moreover, the pore size associated with the separation performance of membrane can be estimated as a hydraulic equivalent diameter d_e from the value of K using Eq. (2) [4].

$$d_e = 8.9\sqrt{\frac{K}{\varepsilon}} \tag{2}$$

where ε is the porosity of the membrane layer, and ε of the nanofiber layer can be estimated from the measured values of w and L.

In **Table 1**, the values of L, ε, q/p, K and d_e are summarized to compare the nanofiber membrane with the MF membrane. The distinctive characteristic of the nanofiber membrane is to be extremely thin and porous as is obvious from their values of L and ε. The permeation rate q of the nanofiber membrane is almost four times higher than that of the MF membrane although two membranes have almost the same hydraulic equivalent diameter d_e. This is also one of the major features of the nanofiber membrane, but the permeability K of the nanofiber membrane is not so large as we expected, which indicates that the main reason for the high permeation rate is the thinness of the nanofiber layer.

Table 1. Physical properties and water permeation properties of membranes

Membrane	Nanofiber membrane (D_f = 147 nm, w = 2.6 g/m^2)	Microfiltration membrane (d_m = 1.0 μm)
Material	Nylon 6	Mixed cellulose ester
L [μm]	42	150
ε [–]	0.95	0.80
q/p [m/(Pa·s)]	20.8×10^{-8}	5.1×10^{-8}
K [m^2]	8.8×10^{-15}	7.7×10^{-15}
d_e [μm]	0.86	0.87

Evaluation of Particle Retention Performance

The filtration tests for two membranes compared in the previous section were performed using the PSL suspensions under constant pressure p of 50 kPa. In **Fig. 3**, the rejections R of the PSL particles with different diameters of 0.100, 0.143 and 0.196 μm are plotted against the filtrate volume v per unit

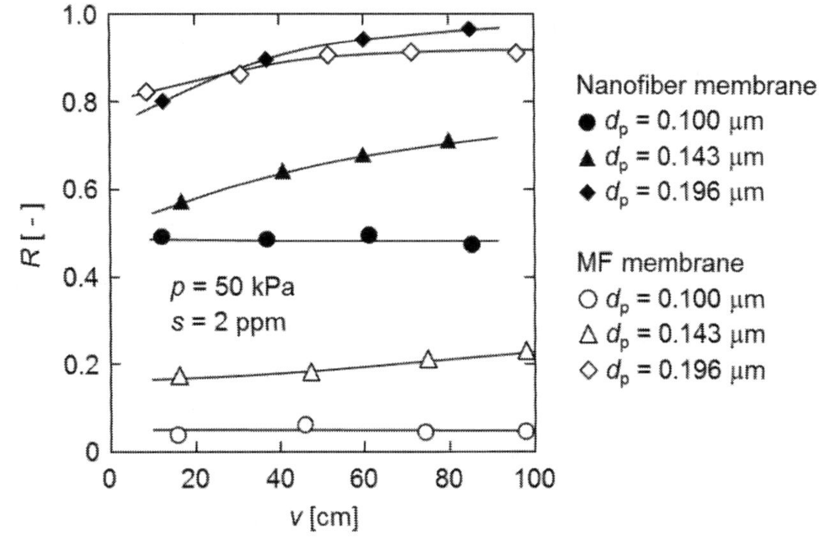

Figure 3. Time variations of PSL rejection by two membranes

membrane area. The rejection R was calculated from the PSL concentration s and s_f in the feed fluid and the filtrate by Eq. (3).

$$R = 1 - \frac{s_f}{s} \qquad (3)$$

Both membranes are reasonably retentive for particles of 0.196 μm, but partially permeable for particles of 0.100 and 0.143 μm, which indicates that the critical particle diameter for these membranes can be estimated to be near 0.2 μm. This diameter is equivalent to one-fifth to one-quarter of their hydraulic equivalent diameter. In the SEM images of the nanofiber membrane during the filtration process shown in **Fig. 4**, it is observed that considerably smaller particles than the pore size are retained inside the pores. Thus, the particle retention by the nanofiber membrane progresses inside the nanofiber layer with the interruption by complex 3D network, the adhesion by extremely large surface area, and the cross-linking between particles.

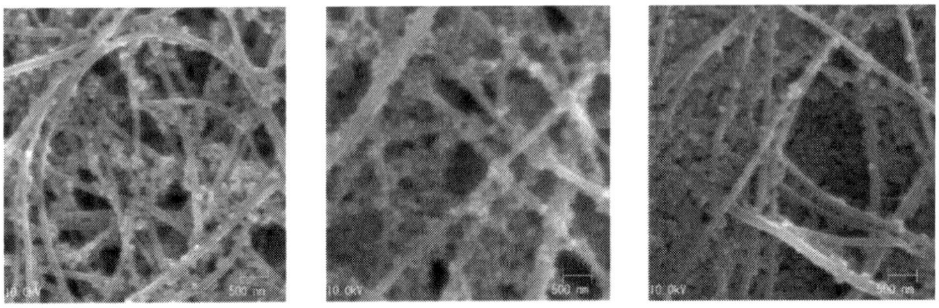

(a) v = 20 cm (b) v = 40 cm (c) v = 60 cm

Figure 4. SEM images of nanofiber membrane during filtration process

Effect of Preparation Conditions of Nanofiber Membrane

The water permeation tests were carried out using the nanofiber membranes with different fiber diameter D_f and deposition weight w. **Figure 5** shows the hydraulic equivalent diameter d_e calculated from Eq. (2) for various nanofiber membranes. The values of d_e in this figure are in the range of 0.7 to 1.2 μm and equal to about six times the fiber diameter D_f. Moreover, this figure clearly demonstrates that d_e has a tendency to decrease with decreasing D_f, which is due to an increase in the specific surface area. On the other hand, d_e is less affected by the deposition weight w of nanofiber.

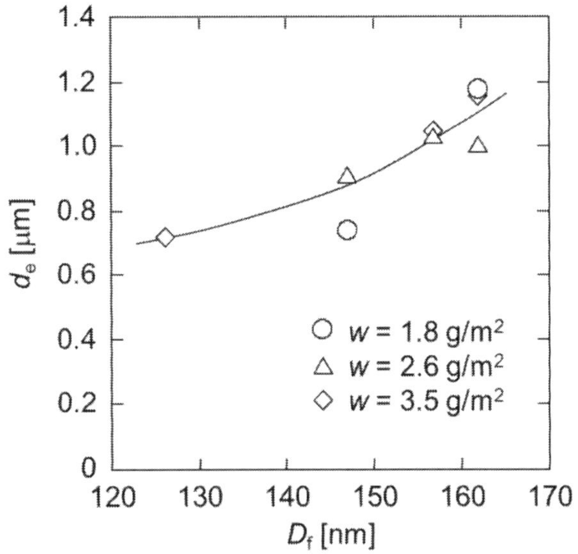

Figure 5. Relation between preparation conditions of nanofiber membrane and hydraulic equivalent diameter

In **Fig. 6**, the rejection R (the steady value at v of 100 cm) of PSL particles with d_p of 0.143 μm is plotted against the fiber diameter D_f. It is expected from Fig. 5 that the particle retention performance is improved by a reduction in the fiber diameter due to a decrease in the pore diameter. In fact, the PSL rejection tends to increase with decreasing D_f as shown in Fig. 6. On the other hand, the dependence of the particle rejection on the deposition weight w of nanofiber is not so significant as compared with the fiber diameter.

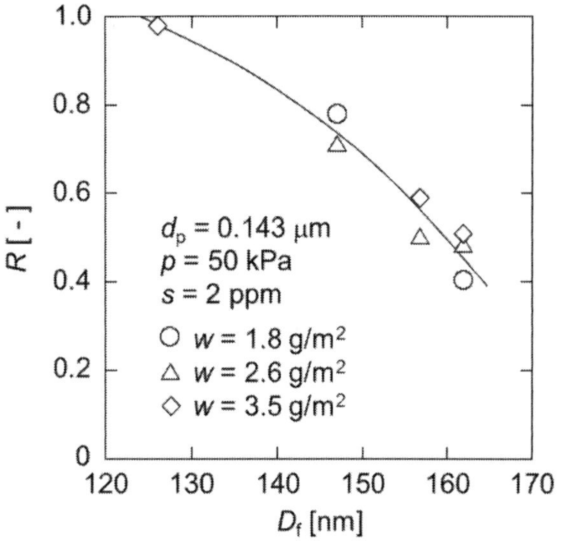

Figure 6. Relation between preparation conditions of nanofiber membrane and PSL rejection

Application to Classification

It is of importance in widely diversified fields to uniformize the particle size of particulate material, and therefore various studies on the classification methods have been conducted. One of promising methods is wet classification by using a porous membrane since the particles larger than the pore size are completely captured on the membrane and only smaller particles can pass through the membrane pores [5]. Actually, the polymer membrane is being applied to the classification of CMP slurry [6]. However, the excessive deposition of larger particles on the membrane retards the passage of smaller particles, resulting in a decline in classification efficiency. In this study, we focused on the nanofiber membrane which has complex 3D network structure and higher porosity as an alternative to conventional porous membrane. When the nanofiber membrane is applied to wet classification, it is expected that the larger particles are retained by the 3D nanofiber network and the smaller particles can pass smoothly through extremely porous space, resulting in high classification efficiency. In order to provide a fundamental understanding of the classification characteristics of fine particles through the nanofiber membrane, the fractionation experiments of the binary mixture of large and small particles were conducted under the various conditions of pH and particle concentrations.

Materials and membrane

Cross-linked polymethyl methacrylate (PMMA) particles with a diameter of 0.5 μm and 1.0 μm and an electrospun nanofiber membrane made of polyacrylonitrile with a fiber diameter of 400 nm were used for the fractionation experiment. The mean pore size of this nanofiber membrane was 1.4 μm, which was calculated from the permeation rate in the water permeation test. The feed suspension was prepared at the given small and large particle concentrations. The pH value of the feed suspension was not adjusted (pH 5.5) or adjusted to pH 11 by the addition of NaOH.

Experimental apparatus and procedure

The experimental apparatus shown in Fig. 2 was also used in this experiment. The experiment was performed under the constant pressure condition of 20 kPa. The time variations of small and large particle concentrations in the filtrate were measured.

Filtration characteristics of single component system

The time variation of the particle concentration in the filtrate obtained in filtration of each single component system was measured before the experiments of the binary mixture system. As a result, the large particles were completely retained from the beginning of filtration by the nanofiber membrane, and almost all small particles passed through the nanofiber membrane through the course of filtration.

Effect of pH on fractionation characteristics

The fractionation experiments of the binary mixture system were conducted at pH 5.5 and pH 11. **Figure 7** shows the transmittance T_p of each particle against the filtrate volume v per unit filtration area, respectively. The transmittance T_p of large or small particles is defined as the ratio of large or small particle concentration $s_{l,f}$ or $s_{s,f}$ in the filtrate to each particle concentration s_l or s_s in the feed suspension (i.e., $T_p = 1 - R$). The large particles were completely captured through the course of filtration at both pH conditions, while the transmittance of small particles at pH 5.5 tended to decrease with progression of filtration, but that at pH 11 kept almost one.

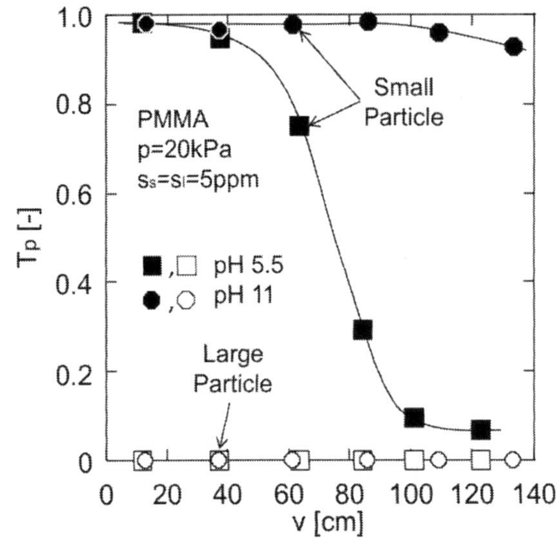

Figure 7. Transmittance of large and small particles

Figure 8 demonstrates the zeta potential of the nanofiber membrane and two kinds of PMMA particles. The zeta potential was measured by a streaming potential method for the nanofiber membrane and by an electrophoretic method for PMMA particles. Both the membrane and PMMA particles are negatively charged, and the absolute values of these zeta potentials tend to increase with an increase in pH. The difference of the fractionation performances between pH 5.5 and pH 11 can be explained based on the electrostatic interactions caused by these zeta potentials. The repulsive forces at pH 11 between similar or dissimilar particles and between the particles and the membrane become larger than those at pH 5.5, and consequently the retention of small particles by the large particle layer and the nanofiber membrane becomes less. In the following fractionation experiments, pH 11 was chosen because of better condition for fractionation.

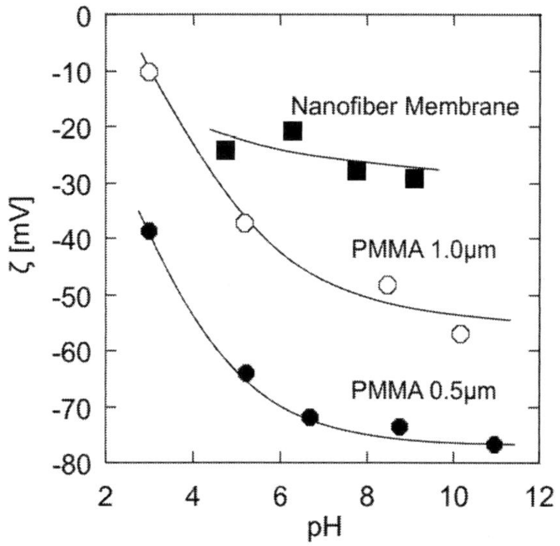

Figure 8. Zeta potential of nanofiber membrane and PMMA particles

Effect of concentration on fractionation characteristics

Figure 9 shows the time variations of the transmittance of small particles at different concentrations of large particles. The transmittance of small particles had a tendency to decline earlier with an increase in the large particle concentration. This indicates that the decline of the transmittance of smaller particles is associated with the amount of large particles trapped by nanofiber membrane.

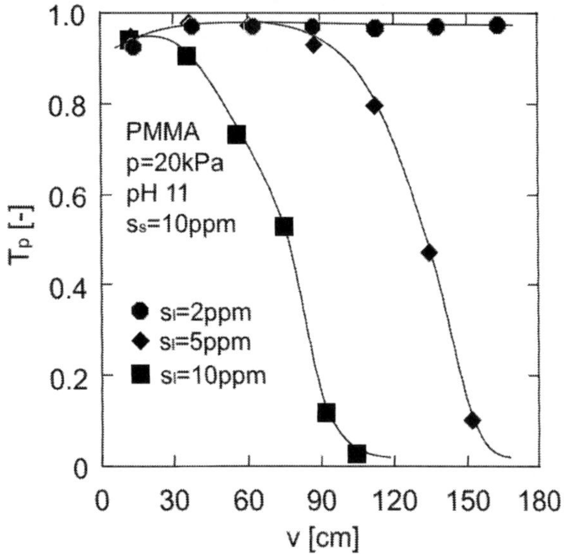

Figure 9. Transmittance of smaller particles at different concentrations of large particles

Conclusions

The electrospun nanofiber membranes were prepared and the water permeation and particle filtration tests were conducted to evaluate its separation performance. As a result, it was revealed that the nanofiber membrane has extremely high water permeation rate and is effective in the separation of the submicron-sized particles in liquid. In addition, the pore size of the nanofiber membrane decreased with a decrease in the fiber diameter, resulting in the improvement of the particle retention performance. Furthermore, it was also demonstrated that the classification performance of the fine particles by an electrospun nanofiber membrane was significantly affected by such conditions of the suspension as pH and particle concentration. It is of practical importance to exclude some factors interfering with the passage of smaller particles through the nanofiber membrane.

Acknowledgements

This work was supported in part by JSPS KAKENHI Grant Number 25420801. The author would like to express the sincere gratitude for the financial support.

References

1. Hongu, T. (2006), "Nanofiber Technology," Nikkan Kogyo Shinbun, Tokyo, pp. 26-27.
2. Yoon, K., B. S. Hsiao and B. Chu (2008), "Functional Nanofibers for Environmental Applications," *J. Mater. Chem.*, **18**, pp. 5326-5334.
3. Feng, C., K. C. Khulbe, T. Matsuura, S. Tabe and A. F. Ismail (2013), "Preparation and Characterization of Electro-spun Nanofiber Membranes and Their Possible Applications in Water Treatment," *Sep. Purif. Technol.*, **102**, pp. 118-135.
4. Furuta, M., Y. Mukai, E. Iritani and H. Nakakura (2004), "Estimation of Pore Size of Microfiltration Membrane Based on Pure Water Permeation Test," *Kagaku Kogaku Ronbunshu*, **30**, pp. 611-614.
5. Ma, W. and Y. Mukai (2012), "Basic Study on Wet Classification of Nanoparticles by Polymer Membrane," *Proc. 44th Autumn Meeting of the Society of Chemical Engineers*, Sendai, Japan, XA2P29.
6. Sumiya, M. (2008), "Improvement in Reliability of Polishing by Filtration of CMP Slurry," in *The Tribology*, Vol. 22, No. 11, Shinjusha, Tokyo, pp. 24-27.

414102 The Effect of Marangoni Currents and Gravitational Forces on the Separation of Two Different Sized Particles during Droplet Evaporation

Abstract:

When a homogenously mixed droplet with two different sized particles evaporates, the particles can separate by size into different rings with one ring for each particle size. This work shows that particle separation is significantly affected by Marangoni currents and by gravitation forces. The finite element method was used to create a transport model that tracks each particle concentration, fluid velocity, pressure, temperature, and vapor concentration. The model created is two dimensional with an axisymmetric assumption. The computational model shows that a majority of particles remain suspended in the droplet due to Marangoni currents. This is in contrast to previously published isothermal models which showed particles rapidly accumulating at the contact line. Experiments using 1 μm and 3 μm polystyrene particles support the computational results. Additional experiments with pendant and sessile droplets show that previous assumptions to neglect gravitational effects on 1 μm polystyrene spheres are incorrect. Finally, the importance of both thermal effects and gravitational forces on the separation of two different sized particles into rings is shown. This is achieved by comparing the depositions profiles for 1 μm and 3 μm particles with and without thermal and gravitational effects.

423355 Determination of Effective Diffusion Coefficient in Divergent Channel: Area Averaging Approach

Abstract:

In this contribution, the long term goal includes the use of the an area-averaging approach to obtain effective transport coefficients for the long-time regime, i..e the Taylor-Aris dispersion in Poiseuille flow in a divergent microchannel while the walls of channel are assumed to have an electrostatic potential. During the current work, electromigration coupled with diffusion and convection is study in a divergent system whereas the fluid in the geometry is assumed to be Newtonian, incompressible, and under steady state conditions. The system used here has been described by the convective-diffusive transport differential model coupled with the Laplace Equation for the electrostatic potential. Preliminary results will be presented and discussed.

413471 Characterizing the Scouring Efficiency of Fluidized Granular Activated Carbon (GAC) Particles in Membrane Fouling Mitigation Via Wavelet Decomposition of Accelerometer Signals

Abstract:

Particle fluidization, a promising unsteady-state shear means of mitigating membrane fouling in order to lower energy requirement and prolong membrane lifespan [1, 2], has garnered increasing attention in membrane-based water treatment processes. In particular, the fluidized particles facilitate back-transport in the polarization layer [3] and also play the role of mechanical scouring agents [4], the latter of which is dominant for millimeter-sized particles. To date, the hydrodynamics of the liquid-solid fluidization of the millimeter-sized GAC particles is not understood, although the understanding of the physical behavior of the liquid and solid phases is key to the fouling mitigation phenomenon. Hence, the current effort aims at bridging the gap in the knowledgebase of the dynamics of the solid particles in effecting the scouring mechanism on membranes to mitigate fouling.

In this effort, an accelerometer was successfully used to reveal the liquid-dolid fluidization dynamics of three GAC particle diameters (namely, 1.20 mm, 1.85 mm, and 2.18 mm) [5]. Specifically, because both liquid and solid phases contribute to the accelerometer signal, wavelet decomposition was used to extract information in the frequency ranges corresponding only to the solid phase [6]. The energy contained in the accelerometer signal corresponding to the solid phase was consistent with the extent of fouling mitigation in the filtration tests. Results indicate that the smallest particle diameter of 1.20 mm conferred the least scouring efficiency, while both the larger particle diameters of 1.85 mm and 2.18 mm provided similar scouring efficiency. The impact of particle diameter is due to the balance between particle inertia, and the difference between superficial liquid velocity and minimum fluidization velocity. Calculations of energy requirement indicate that, the energy requirement of all three particle diameters (d_p) were similar at lower scouring efficiency, whereas the energy requirement of the smallest $d_p$of 1.20 mm was the greatest at higher scouring efficiency.

References

1. Bixler, H.J. and G.C. Rappe, *Ultrafiltration Process.* United States Patent Office, 1970. **3,541,006**.

2. Zamani, F., et al., *Unsteady-state shear strategies to enhance mass-transfer for the implementation of ultrapermeable membranes in reverse osmosis: A review.* Desalination, 2015. **356**: p. 328-348.

3. Davis, R.H., *Modeling of Fouling of Cross-Flow Microfiltration Membranes.* Separation and Purification Methods, 1992. **21**(2): p. 75-126.

4. Meier, J., *Mechanical influence of PAC particles on membrane processes.* Journal of Membrane Science, 2010. **360**(1-2): p. 404-409.

5. Boyd, J.W.R. and J. Varley, *The uses of passive measurement of acoustic emissions from chemical engineering processes.* Chemical Engineering Science, 2001. **56**(5): p. 1749-1767.

6. Ren, J.Q., et al., *Wavelet analysis of dynamic behavior in fluidized beds.* Chemical Engineering Science, 2001. **56**(3): p. 981-988.

420630 Formation of N-Nitrosamines from Amino Acids in Post-Combustion Carbon Dioxide Capture Systems

Abstract:

Amine scrubbing is the most mature technology for post-combustion CO_2 capture. It employs an amine solvent to absorb CO_2 from the flue gas of fossil fuel-fired power plants. Amino acids have been proposed as alternatives to amines for CO_2 absorption due to their high reactivity with CO_2 and low vapor pressure. When selecting amino acids to improve the performance of CO_2 capture systems, the potential of these compounds to form carcinogenic *N*-nitrosamines should be considered to prevent unintended environmental consequences.

In this study, amino acids glycine (Gly), sarcosine (Sar), taurine (Tau), proline (Pro), alanine (Ala), β-alanine, (β-Ala), serine (Ser), phenylalanine (Phe), γ-aminobutyric acid (γ-ABA), glutamic acid (Glu) and aspartic acid (Asp) were selected as model compounds. The amino acids were mixed with strong inorganic bases (e.g., potassium hydroxide) or organic bases (amines) to achieve equimolar concentration of amino acids and bases, and subject to conditions simulating CO_2 capture systems. The effects of amino acid structural characteristics and the selection of bases on the formation of *N*-nitrosamines under absorber conditions were evaluated. The order of amino acid was shown to be a principal factor determining the accumulation of *N*-nitrosamines. Secondary amino acids exhibited the highest potential to form *N*-nitrosamines, an order of magnitude higher than that of primary amino acids. This trend is similar to amines. The potential dealkylation of primary amino acid upon nitrosation was examined. Varying alkyl chain length of amino acid (Gly, β-Ala, and γ-ABA) did not significantly impact *N*-nitrosamine formation. Decarboxylation of amino acid SAR was observed despite the moderate absorber temperature (40 °C), and a volatile nitrosamine *N*-nitrosodimethylamine was observed. Considering that the stability of carboxyl group is critical to the ionic nature of amino acids as CO_2 capture solvent, the decarboxylation mechanism is investigated. Our systematic investigation of *N*-nitrosamines formed from amino acids provides a feasible way to benefit environment and human health for amino acid-based CO_2 capture.

422834 Investigation of Aminopolymer Loading on the MCM-36 Crystal for CO2 Capture

Abstract:

Current research on amine impregnated adsorbents for carbon dioxide capture has focused on the attempt to increase amine loading onto the solid support in order to improve overall capture performance. The hope with this method is to drastically improve the selectivity and capture capacity of the material by shifting it from a physical to chemical adsorbent through the introduction of amine sites, which actively attach to entering carbon dioxide in a 2 to 1 amine to CO_2 molar ratio. However, this methodology so far has missed an opportunity for further optimization of these materials by the careful control of the diffusion and kinetic changes that occur upon impregnation of the amine molecules into the solid support. In our work, we have investigated these changes by the introduction of a branched polyethyleneimine amino-polymer into the large pores of the MCM-36 support.

MCM-36 was chosen for this work because of the large literature background on the synthesis and characterization of this material, as well as the interest within the carbon dioxide capture community on the potential implementation of this adsorbent for capture applications. MCM-36 was first reported in the literature by Corma et al. from the swelling and pillaring of the MCM-22-Precursor crystal (MCM-22-P). This solid is composed of approximately 2 nm microporous layers of aluminosilicate charge balanced one on top of the other by the presence of hexamethyleneimine cations in the interlayer space. This yields a crystal with the porosity of zeolites and the lamellar structure of clays, allowing for the implementation of clay chemistries to swell and pillar the precursor to form the fully inorganic MCM-36 structure, which is composed of microporous layers bridged by SiO_2 pillars, which generate a large pore space between the layers. After synthesis this material can be impregnated with amino-polymers to generate an adsorbent with open CO_2 channels within the layers and large pore spaces filled with amines for chemical adsorption of the CO_2molecules.

The capture capacity, diffusion of entering molecules, and kinetics of carbon dioxide capture were investigated in this work for various amine loadings on the MCM-36 support to attempt to find a method to further optimize polymer impregnated adsorbents for carbon dioxide capture. We have found that increased amine loading within the material past a critical point can only further decrease the capture performance, suggesting that the optimum loading conditions are at a lower weight percent of loaded amines than that currently suggested in the literature. Furthermore this work has found that the use of 2-D porous sorbents may be ultimately hindered for carbon dioxide capture because of the tradeoff between ease of diffusion and capture kinetics upon amine impregnation, and that a new sort of amine impregnated solid must be developed to overcome these challenges if this carbon dioxide capture method is to ever be implemented in practical applications.

429738 Tetraethylenepentamine Supported with Kaolin for CO2 Capture

Abstract:

The increase in greenhouse gas emission due to the consumption of vast amounts of fossil-based energy resources is most likely associated with climate changes. Among the major greenhouse gases is CO_2, which is responsible for approximately 64% of the greenhouse effects observed to date. Therefore, it is essential to improve current CO_2 capture technologies for cost-effective CO_2 capture. In this work, a CO_2 adsorbent was prepared by immobilizing tetraethylenepentamine (TEPA) onto kaolin, an inexpensive and widely available clay. The prepared sorbents were characterized using X-ray diffraction (XRD), scanning electron microscopy (SEM), and Fourier transform infrared spectroscopy (FTIR). At the optimal TEPA loading (30 wt%) on the kaolin, the CO_2 sorption capacity of the sorbent reached 2.13 mmol CO_2/g-sorbent at 70°C with the presence of moisture (ca.10 vol %). The regeneration of the adsorbent can be achieved using nitrogen at 100 °C. This work shows that this polymer composite has the potential to be applied as a solid sorbent for CO_2 capture.

435184 Detailed Analysis of Foam Bioreactor Using Escherichia coli Expressing Carbonic Anhydrase for Optimization of CO2 Capture

Abstract:

The foam bioreactor consisting of the recombinant *E. coli* expressing carbonic anhydrase (CA) in tiny aqueous foam bubbles has demonstrated high CO_2 capture at short gas retention time (up to 93% and 9570 g CO_2 per $m^3{}_{bioreactor}$ h^{-1}, 24 s gas retention time, cell conc. of 4 g_{dw} L^{-1} at 4% inlet CO_2). The present study showed the detailed analysis of alternative medium, mass transfer, biokinetic, characterization of carbonic anhydrase and operating strategies. The glucose-based auto inducible medium resulted in the specific growth rate of 0.082 h^{-1} and CA activity of 1022 mU $OD_{600nm}{}^{-1}$ mL^{-1} at 24 h which was four to fourteen times higher than those by the conventional medium (LB-IPTG). The CA in the *E. coil* grown on the glucose-based auto inducible medium showed highest activity at pH 8.5 while it exhibited high stability up to 40 °C. The foam bioreactor provided high mass transfer (K_La of 80 - 210 h^{-1} at 1.3 - 2.3 x 10^{-5} m^3/s) due to the large gas-liquid interfacial area in the foam bioreactor. The K_La in the bioreactor was also linked with the foam structure such as foam size, liquid film thickness of foam and gas holdup. The biokinetic equation for the CO_2 capture followed the conventional Monod-type equation, but it also showed the drastic substrate inhibition with the gaseous CO_2 concentration above 12%. On the other hand, our study showed of significant inhibition of carbonic anhydrase by the produced bicarbonate in the medium which was 50% rate reduction at 40 mM bicarbonate. However, the periodic removal of bicarbonate from the culture medium was implemented and was shown to allow sustained high and stable CO_2 removal (~ 92%) with a high rate of mineralization. Besides, the long term operation of the foam bioreactor revealed limitation of glucose (for cell growth), lactose (for enzyme expression) and nitrogen source, thus feeding of glucose, lactose and ammonium chloride at regular interval (dilution rate of 2 d^{-1}) helped to sustain stable CO_2 removal (> 90-92%) for 8 d. In short, the foam bioreactor using the *E. coli* displaying carbonic anhydrase would be a cost-effective process to effective CO_2 capture at short gas contact time and mild operating conditions.

426397 Development of a New Type of Accelerated Mineral Carbonation Process

Abstract:

Accelerated mineral carbonation (AMC) method has been recognized as a promising measure for carbon dioxide fixation. The drawback of the AMC is the cost and power consumption for the acceleration of the carbonation reactions. Uses of chemicals such as acids, alkalis or some special reagents have been proposed for the acceleration reaction, mainly for the extraction of alkali earth metals from the sources. These chemicals should be recycled or reused; otherwise the emission amount of carbon dioxide associated with the production of these chemicals should outnumber the amount fixed in the process. The effective recycling or reusing method should be essential for the AMC as a practical option for carbon dioxide fixation.

The authors have proposed a new type of AMC process by using acids as an extraction reagent of calcium or magnesium from the sources and alkali metal hydroxide solutions as a capturing reagent of carbon dioxide in the flue gas. The solution of extracted magnesium or calcium will be mixed with the solution of alkali metal carbonate after capturing carbon dioxide to precipitate magnesium or calcium carbonates. The remained solution will then be treated with the electrodialysis process to recover the acid and the alkali.

We have examined the feasibility of the process by laboratory experimental studies. Since the capturing of carbon dioxide with alkali metal hydroxide solution and the precipitation of carbonates are well known processes, we concentrated to the extraction step and acid-alkali recovery steps. For the extraction step, we have examined the combination of solid sources such as wollastonite, serpentine, and waste cement and acid extractants such as nitric acid, hydrochloric acid, and other weak acids. The extraction reactions were found to progress rather rapidly, even for the cases using the weak acids, and the extraction of impurities in the sources such as iron, silicon could be minimized under a condition of proper solid-liquid ratios.

The recovery step with electrodialysis can be achieved by using the bipolar membrane electrodialysis (BMED). The alkali-acid pair was recovered for all the combinations studied. However, both the recovery rate and the energy-efficiency were significantly higher for the cases using the combination of weak acid and calcium or magnesium.

In conclusion, the use of weak acid as a extractant for the alkali earth metals is preferable for the process efficiency of the carbon dioxide fixation.

427482 Investigation of the CO2 Capture of Flue Gas By Ca-Based Sorbents from Cement Industrial

Abstract:

The CO_2 emissions of cement industry accounts for 15% of the whole industrial CO_2 emission in China. China is a big country for the production and consumption of cement, and the cement production of China has been ranked first for 24 years continuously in the world. Meanwhile, Chinese cement production and its CO_2 emissions is growing year after year. It is calculated that 0.815 tons of CO_2 emissions were produced directly from the production of every ton of cement clinker. Therefore, abatement of CO_2 emission is becoming the focus of government attention and academic research. CCCR technology (a Ca-based sorbents Cycling Calcination/Carbonation Reaction technology) is a kind of new and effective method to capture CO_2 in flue gas from industry.

Based on the cement industry, this paper studied the carbonation of a Ca-based absorbents and its stability in cycling carbonation reaction under high CO_2 concentration (C_{CO2}>30%), commonly found in cement industry. The orthogonal tests were carried out to determine the appropriate parameters for the CO_2 capture process. Composition of Ca-based absorbents, the size of particle, calcining temperature, carbonating temperature, carbonating reaction time for the cycling carbonation rate of absorbents (X_N) as well as the kinetics of limestone thermal decomposition and CaO carbonation were studied.

The results of orthogonal experiments show that, Ca-based sorbents can keep fine pore structure and able to resist sintering when contains a certain amount of dolomite. As a consequence, the X_N of sorbents was improved. And we chose the calcination temperature of 850°C, carbonation temperature of 700°C and the CO_2 concentration of C_{CO2}=35% for the subsequent experiments.

The investigation were carried out under a high CO_2 concentration condition (≥30%) to study the kinetics of limestone thermal decomposition and CaO carbonation in a stacked state and a simulated suspension state respectively. Under a stacked state or a simulated suspension state, limestone thermal decomposition process could be both modelled as a randomly nucleation, accompanying by particle size growth. The reaction order of limestone thermal decomposition varied from 2/5 to 3/4 under different CO_2 concentrations. The higher the CO_2 concentration was, the higher activation energy E and pre-exponential factor ln$A$ could be. The activation energy varied from 958.28 to 2556.10 kJ/mol and the pre-exponential factor ln$A$ changed from 103.67 to 242.44. In addition, the activation energy E rose exponentially with an increase of CO_2 concentration in a stacked state, followed a relationship as.

On the basis of these studies,we found that the CO_2 capture and sequestration (CCS) capability of limestone in flue gas was dramatically decline after several cycles due to the surface"sintering". In order to improve the efficiency of recirculated limestone, we modified the limestone with four kinds of materials such as carbon materials, metal oxide and natural mineral (attapulgite and vermiculite).

A variety of carbon, such as carbon black, graphite, bamboo woods and coconut woods were studied. Our results show that coconut woods is the best carbon source and it has the performance of 13% increase of efficiency with 2 wt% adding amount.

we used some metal oxides such as MgO,SiO_2 and Fe_2O_3 to modify the morphology of the limestone particles to increase the CO_2 capture capability of limestone. The efficiency can be increased 7%`8% after adding 1wt% MgO particles.

The limestone modified with attapulgite, a kind of mineral with nano-fibrous structure, can reduce the sintering phenomenon between the limestone particles. So the carbonation rate during the CCS process can be improved effectively. The best performance of 13.11% increase of efficiency with 5wt% adding amount.

We used natural vermiculite which has unique piece structure and high temperature resistant properties to modify the limestone surface in order to delay the limestone particles sintering. The ability of CO_2 capture, the morphology changes have been tested by TGA, XRD, FTIR and SEM. The efficiency can be increased 8.12% by adding 1wt% vermiculite.

424646 Understanding Mechanisms of CO2 Capture into Monoethanolamine Solution with Different CO2 Loading

Abstract:

Due to the high reactivity with CO_2, monoethanolamine (MEA) has been used in industrial processes to capture CO_2 for many years. The mechanism of CO_2 absorption into MEA solution has been widely studied, but it is still in dispute. There is few literature reported regarding to the detailed reaction processes of CO_2 absorption into MEA solution with high CO_2 loading and desorption from the CO_2-saturated MEA solution.

To get a clear picture of the process mechanism, ^{13}C nuclear magnetic resonance (NMR) was used to analyze the species of MEA solution with different CO_2 loading, and the detailed absorption and desorption mechanism of MEA-CO_2 system was evaluated in this work. The results demonstrated that the absorption reaction was started with chemical reactions of MEA with CO_2 to form carbamate according the zwitterion mechanism, followed by the hydration of CO_2 to form HCO_3^-/CO_3^{2-}, and accompanied with the hydrolysis of carbamate. Base on our results by ^{13}C and experimental observations, the reaction was found to vary under different CO_2 loading. At low CO_2 loading, CO_2 absortion into MEA was an exothermic reaction and the reaction rate was fast. In this stage, the absorption capacity of the solvent was found to be 0.40 mol CO_2/ mol MEA, which was comparatively comformable to the zwitterion mechanism. At high CO_2 loading, the hydration of CO_2into solution was the main reaction and was an endothermic process.

It is interesting to find that the existence of carbamate will be influenced by CO_2 loading, and it is rather unstable at high CO_2 loading. At low CO_2 loading, carbamate is formed fast by the reaction of CO_2 and MEA. But at high CO_2 loading, it will be formed by the reaction of CO_3^-/CO_3^{2-} with MEA, and be easily hydrolyzed by H^+. The Hydrolysis rate was faster than the formation rate, and the concentration of carbamate decreased as the absorption carried out at high CO_2loading.

Moreover, CO_2 desorption from the CO_2-saturated MEA solution was proved to be a reverse process of the absorption. Initially, part of HCO_3^- were heated to release CO_2 and part of HCO_3^- reacted with carbamic acid ($MEAH^+$) to form carbamate, then the carbamate decomposed to MEA and CO_2.

The results indicated that the reaction of CO_2 capture into MEA solution varied with different CO_2loading, and the process mechanisms would more accurate reflect the reaction.

---

*Corresponding author. Tel: +86-592-6166216; Fax: +86-592-6162300;

E-mail address: zhoujing@hqu.edu.cn (G. H. Jing)

429152 Increase the Feeding Value of Corn Ethanol Coproducts Via Genetic Engineering of Yeast

Abstract:

With fast growth in recent several years, the U.S. corn ethanol industry provides enormous amount of co-products, for instance, distiller dried grains with solubles (DDGS), to the feed market, making the ethanol industry as one of the largest contributors to the US feed supply. Although DDGS is a good source for protein, the low content and low digestibility of several essential amino acids (lysine, methionine, threonine, arginine and tryptophan) is a limiting factor when considering the inclusion of DDGS in animal diets and its impact on animal performance. High inclusion of DDGS in animal diets causes amino acid imbalances with subsequent consequences on animal performance and excess excretion of nitrogen into the environment. Addition of synthetic sources of amino acids can alleviate amino acid imbalances and result in significant decreases in the nitrogen discharge and manure odor. Commercially available synthetic amino acids for feed supplementation are only available for some amino acids while costs can fluctuate if supplies are short. Many amino acids remain too costly for commercial feeding purposes. Regardless of the availability of amino acid supplements, the value of an ingredient such as DDGS is improved if the content and digestibility of limiting amino acids such as lysine, tryptophan and arginine are increased. *Saccharomyces cerevisiae* has been considered as a natural food product. Non-living dry yeast powder can often be used as a food supplement for human nutritional and pharmaceutical purposes. In corn ethanol co-products, yeast biomass is an important composition and is expected to provide more of these limiting amino acids. Although genetic modification of yeast has been intensively carried out to improve the baker's yeast for ethanol production and some of the engineered strains have been implemented and commercialized for industrial ethanol production, no genetic engineering research has been specifically focused on the animal feed improvement for the corn ethanol coproducts. Using genetic manipulation to engineer yeast strains is considerably easier than engineering feed grains while engineered yeast is also more likely to be adopted by the industry. The goal of this study is to significantly increase the feeding value of corn ethanol coproducts through genetically engineering the yeast strain to enrich several key amino acids (first limiting lysine, commercial costly tryptophan and arginine) during the fermentation so that the yeast biomass can significantly improve the nutritional value of DDGS. Specifically, the metabolic pathway of lysine, arginine and tryptophan in *S. cerevisiae* is genetic engineered to generate the biomass with more of these amino acids. Quite different from the current mainstream emphasis on the ethanol production, this research focuses on the improvements of animal feeds generated from corn ethanol industry and will potentially benefit all sides involved in this industry, including the animal feeds and the corn farmers, and potentially decrease the environmental impacts of the whole corn ethanol industry.

432905 Towards Biofuel Production in Synechocystis Sp. PCC 6803: Expanding the Molecular Biology Toolbox for Pathway Engineering

Abstract:

Biofuel production in microorganisms has great potential to fulfill this nation's need for fungible "green" transportation fuels. Because of the simplicity of bacterial cells, photosynthetic cyanobacteria make good candidates as a platform for biofuel production. *Synechocystis* sp. PCC 6803 is an attractive production platform due to rapid doubling times, a sequenced genome and an established molecular biology toolbox. This molecular biology toolbox was established to probe the biology of photosynthesis and has not been optimized for pathway engineering which is necessary for biofuels production. This presentation will discuss our efforts to develop an alternative counter selection marker and to develop and characterize promoters for gene expression under varying conditions. The use of these tools to study and optimize biofuel production in Synechocystis will also be discussed.

(466c) Effect of Different Feeding Strategies of Glucose and Short-Chain Carboxylic Acids on the Production of Natamycin By Streptomyces Natalensis

Elsayed, E. A., National Research Centre
Farid, M., National Research Centre
Wadaan, M., King Saud University
EL Enshasy, H., Universiti Teknologi Malaysia
Originally presented on: 11/11/2015 9:06:00 - 9:24:00

Natamycin is an industrially important tetraene antibiotic produced in submerged culture by different strains of *Streptomyces* sp. Natamycin has a wide spectrum activity against molds and yeasts with low toxicity against mammalian cells. It is used in the treatment of bronchopulmonary aspergillosis, mycotic keratitis, yeast keratomycosis and candidiasis. Natamycin is used in many food industries for food preservation and has been granted the GRAS status (Generally Regarded As a Safe) by FDA. This present work was carried out to investigate the optimization of natamycin production through different feeding strategies of glucose as a limiting substrate as well as short-chain carboxylic acids (acetic, propionic and butyric) and their effect on cell growth and natamycin biosynthesis. The optimization was performed in shake-flask level as well as 2-L stirred-tank bioreactor. Cultivations have been evaluated under different feeding strategies of glucose as a limiting substrate, as well as different short-chain carboxylic acids (acetate,propionate, and butyrate) . The obtained results showed that the highest natamycin level was obtained at 20 $g.L^{-1}$ glucose. Moreover, increasing glucose concentration resulted in a significant decrease in natamycin level. In addition, natamycin decomposition was observed at lower glucose concentrations (10-20 $g.L^{-1}$) after glucose depletion. The addition of acetic and propionic acids at concentrations below 2 $g.L^{-1}$ at cultivation start showed stimulatory effects on natamycin production. Furthermore, the addition of acetic and propionic acids in a mixture (7:1) at a total concentration of 2 $g.L^{-1}$, increased antibiotic production significantly. Conclusively, the addition of short chain carboxylic acids increased antibiotic yield and decreased the production time.

433138 Systematic Characterization of a Panel of Strong Promoter/Terminator Pairs to Facilitate Scheffersomyces Stipitis Engineering

Abstract:

Systematic Characterization of a Panel of Strong Promoter/Terminator Pairs to Facilitate *Scheffersomyces stipitis* Engineering

Scheffersomyces stipitis is an unconventional yeast that has the highest native capability for xylose fermentation, therefore holding a lot of potentials in lignocellulosic biomass conversion. However, its broad application as a popular host in biorenewables field is limited by the lack of genetic manipulation tools. Here, we report the identification and characterization of a panel of strong constitutive promoters and terminators. Based on RNA-seq analysis of *S. stipitis* growing in both mixed sugars and single sugar conditions, nine promoters were selected for further comprehensive characterization under conditions with various sugar and oxygen levels. The strengths of majority of these promoters were more than 10-fold higher than that of the commonly used TEF1 promoter, representing a collection of strongest Scheffersomyces promoters reported in literature. In parallel, the transcriptional termination efficiency of the selected terminators was studied and compared to the tight TEF1 terminator. Such a systematic characterization of potent promoters and terminators not only paves the way for manipulating pathways in *S. stipitis* to produce industry-relevant high-value biochemicals from mixed sugars, but also provides a foundation for future mechanism study of xylose uptake and conversion in the native host, which will benefit the entire biorenewables field.

424924 Isotopically Nonstationary 13C Flux Analysis of Isobutyraldehyde Production in Synechococcus Elongatus

Abstract:

Recent studies have demonstrated the feasibility of converting energy from sunlight and carbon from CO_2 directly into biofuels using photosynthetic microorganisms. Despite the advances made in cyanobacterial biofuels production, the productivities achieved are currently too low for industrial feasibility and few tools are available that specifically address the challenges of determining and redirecting metabolic flux in photosynthetic microbes. Our group is developing novel approaches that use ^{13}C metabolic flux analysis (MFA) to quantitatively assess in vivo metabolic phenotypes of photoautotrophs. Previously, we mapped carbon fluxes in the cyanobacterium *Synechocystis* sp. PCC 6803 by applying isotopically nonstationary MFA (INST-MFA), which involves model-based regression of transient ^{13}C-labeling patterns of intracellular metabolites. The flux analysis revealed unanticipated photosynthetic inefficiencies tied to oxidative metabolic pathways, despite minimal photorespiration.

We have now applied a similar modeling approach to mapping photoautotrophic metabolism in another strain of cyanobacteria, *Synechocococcus elongatus* PCC 7942. Metabolism was quantified in the wild-type strain of *S. elongatus,* as well as a strain engineered to produce isobutyraldehyde (IBA, a direct precursor of isobutanol). GC-MS profiling of isotopically labeled intracellular metabolites, growth, and IBA production were used to create flux maps of central carbon metabolism in the WT and IBA-producing strains. The flux analysis results quantitatively describe increased flux through phosphoenolpyruvate carboxylase (PEPC) rather than pyruvate kinase (PK) in both studied strains, with an increased overall PEPC flux in the IBA-producing strain. This pointed to an alternative carbon route to pyruvate formation, a major precursor to IBA (phosphoenolpyurvate → oxaloacetate → malate → pyruvate → → → IBA).

Based on these results, we generated three cyanobacterial single gene overexpression strains in the IBA-producing background for reactions involved in the pyruvate formation bypass: IBA/PEPC^{ox}, IBA/MDH^{ox}, and IBA/ME^{ox}. We found that the IBA/ME^{ox} strain showed a significant improvement in IBA production, while maintaining a growth rate comparable to its IBA-producing parental strain. We also overexpressed pyruvate kinase in the IBA-producing background strain, but found that IBA production did not increase and the growth rate actually decreased.

437376 Scale-up 101: What the Novice Should Know about Batch Scale-up

Abstract:

Scaling-up batch chemical processes involves much more than deciding the size of the reactor according to the quantity of product desired. Scale-up affects heat transfer and mixing which in turn have an impact on many reactions. In the best case, inattention to scale-up may result in substantially lower yield or purity. In the worst case, poor scale-up may result in a runaway reaction opening relief devices, possibly causing a fire or other spectacular decomposition. This paper provides an overview of the issues and provides rules of thumb useful to the novice.

436547 Criticality Reduction through Alternative Nitration Conditions

Abstract:

Development of novel explosives always poses challenges to safe scaleup. Nitrations can result in unstable reaction mixtures with resulting high criticality classes. Nalas Engineering evaluated potential processing options to produce 2,6-diamino-3,5-dinitropyrazine-1-oxide (LLM-105) from 2,6-dimethoxy-pyrazine (DMP).

Previous investigations by Nalas revealed a strong decomposition of the desired intermediate 2,6-dimethoxy-3,5-dinitropyrazine (DMDNP) in the reaction mixture over time. This instability was evident and prevented further scale-up. Identification of process conditions that reduced the criticality class or afforded alternative manufacturing conditions suitable to produce DMDNP were investigated to include continuous processing options.

The objective of this effort was to compare the time to maximum rate (TMR) and thermal stability of the reaction mixture (TD_{24}) as a function of nitric acid stoichiometry as well alternative nitrating media. This information afforded a reduced criticality class and higher level of overall safety for the nitration.

436663 Minimizing Risk at Scale-up through Process Modeling

Abstract:

Scale up of a new chemical processes to larger production often involves changes in equipment and processing time as compared to that of the smaller scale development. These process changes increases the potential risk to a successful scale up. Through a systematic approach using modeling of the chemical and process related rates it becomes possible to manage the risk associated with scale up. This approach utilizes appropriate small scale data to allow the models to make projections to the large scale so as to ensure a seamless scale up of the process. This systematic approach will be elaborated and demonstrated with numerous case studies.

Challenges in Relief Design for Pilot Plants

Mike Trainor, P.Eng.
Project Engineer

ZETON
PILOT PLANT TECHNOLOGY

Zeton Inc., 740 Oval Court, Burlington, Ontario, L7L 6A9, Tel: 905-632-3123, E-mail: info@zeton.com

Pilot Plant Challenges

- Smaller equipment = higher area/volume ratios
- Pool fire = all-enveloping
- Commercial availability of small ASME relief devices limited
- Unique challenges associated with smaller equipment, not as big an issue at commercial scale (i.e. shell & tube exchangers)

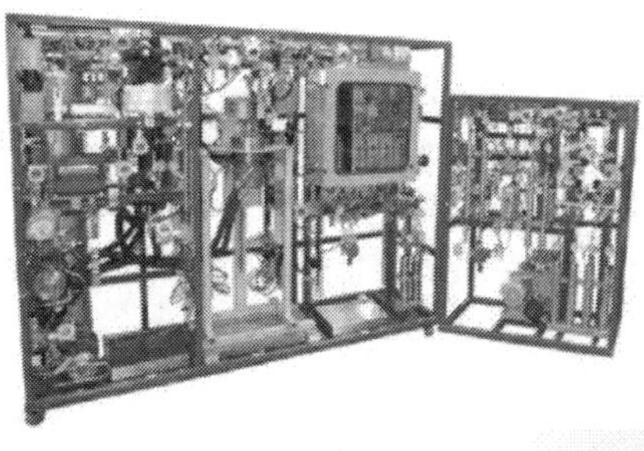

Code Requirements

Commercial equipment, and most pilot-scale equipment designed to ASME VIII Div 1, requiring certified relief valves. Outside scope of ASME VIII Div 1:

- parts of rotating equipment (i.e. pumps, compressors)
- vessels with design pressure not exceeding 15 psig
- vessels with a diameter less than 6", with no limitation on length or pressure (6" schedule 80 pipe and smaller)

Outside of ASME VIII Div 1, system will likely be designed according to ASME B31.3 (piping).

Code Requirements

From ASME B31.3, 2014 ed.:

322.6.3 Pressure-Relieving Devices

(a) Pressure-relieving devices required by para. 301.2.2(a) shall be in accordance with the BPV Code, Section VIII, Division 1, UG-125(c), UG-126, UG-127, and UG-132 through UG-136, excluding UG-135(e) and UG-136(c). The terms *design pressure*[10] and *piping system* shall be substituted for *maximum allowable working pressure* and *vessel*, respectively, in these paragraphs. The required relieving capacity of any pressure-relieving device shall include consideration of all piping systems that it protects.

Code Requirements

For systems designed to ASME B31.3 (piping, small vessels, rotating equipment), per section 322.6.3 excluded sections:

- UG-128: liquid service pressure relief valves may be less than NPS ½
- UG-129: less stringent marking requirements
- UG-130: certification mark not required
- UG-131: capacity certification not required

Basically, allows use of non-certified relief devices, i.e. proportional relief valves (Hoke, Circle Seal, Swagelok, Nupro). Sizing via Cv calculations for liquid/gas flow.

Code Requirements

For ASME-stamped vessels, require certified valves. Generally required orifice areas are many orders of magnitude smaller than those commercially available.

Flowsafe F84/F85	A-G Series 81/83	Farris 2700	Crosby 900
0.003 in^2	0.012 in^2	0.038 in^2	0.085 in^2
0.015 in^2	0.028 in^2	0.068 in^2	
0.034 in^2	0.049 in^2	-	
0.065 in^2	-	-	

Red – long delivery, typically a factory special

Black – readily available, short delivery

Difficult Scenarios – Pool Fire

Typical pool fire considers a 2,500 ft^2 zone (30 ft radius), up to an elevation of 25 ft above the source of the flame

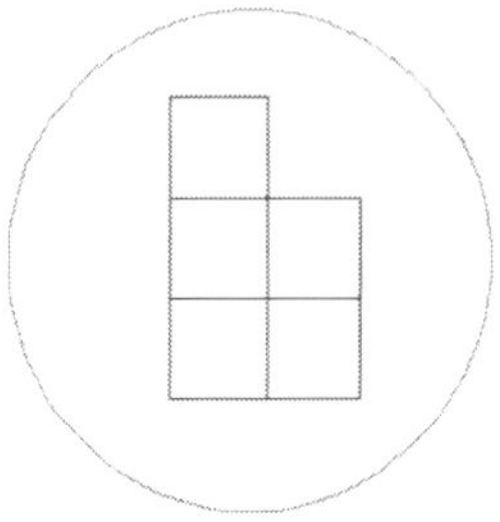

60 ft Circle,
Qty. 5 - 12 ft Modules

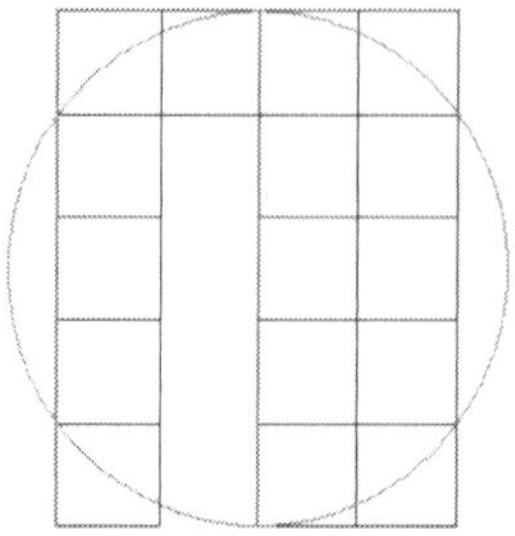

60 ft Circle,
Qty. 16 - 12 ft Modules

Problem: on a typical pilot plant, that could be every valve on the plant

Difficult Scenarios – Pool Fire

Entire plant considered in pool fire

Modularized design, only consider fire up to 25 ft elevation?
Depends on layout, client...

Difficult Scenarios – Pool Fire

Entire plant will fit within one fire circle, resulting in coincident pool fire relief of all vessels on the plant.

As well, the larger surface area to unit volume ratio associated with pilot plant scale vessels leads to disproportionally large relief valve sizes compared with full scale plant vessels.

So what are
the options?

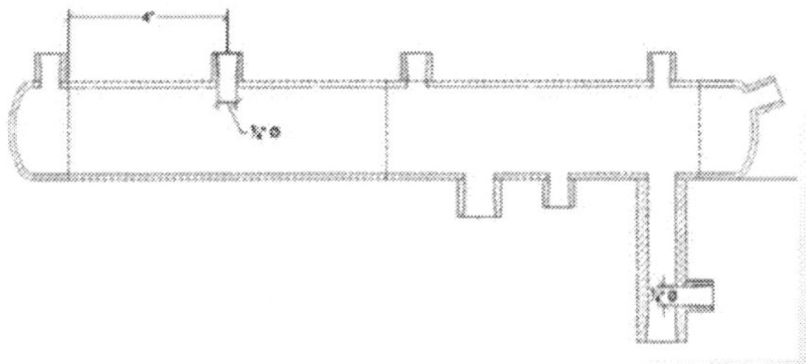

Difficult Scenarios – Pool Fire

Option: Coincident Fire – All Vessels in Zone

- most conservative
- leads to "huge" relief header (8" and beyond) ... eats up a lot of space on pilot plant skids
- site disposal requirements, existing flares may be inadequate to handle instantaneous design load
- can potentially reduce loading by using fireproofing insulation on vessels, utilizing environment factor "F" in API 521
- disproportionately sized flare KOs (5' diameter)

Difficult Scenarios – Pool Fire

Left:
65 lb/hr Production Rate
38 Relief Valves in Pool Fire
39,000 lb/hr relief load
8" Header, 6' KO Pot

Left:
1,600 lb/hr Production Rate
9 Relief Valves in Pool Fire
16,000 lb/hr relief load
6" Header, 5' KO Pot

Right:
2,900 lb/hr Production Rate
15 Relief Valves in Pool Fire
17,400 lb/hr relief load (all at once)
7,900 lb/hr relief load (staged relief)
6" Header, 5' KO Pot

Difficult Scenarios – Pool Fire

Option – Detailed Analysis

Option 1 - For pilot and demonstration plants, the volume of flammable liquids stored within all the tanks can be low enough such that the pool fire scenario can be eliminated, or the fire circle reduced (however, beware of nearby tank farms and equipment which can store flammable liquids).

Option 2 - If not eliminated, number of coincident reliefs during a pool fire scenario can be minimized through detailed analysis. Differences in vessel size, set-pressure differences, insulation thickness etc. can be used for a dynamic analysis. Reduce peak load since not all relief valves will pop a the same time.

Difficult Scenarios – Pool Fire

Option 3 – "Beyond-Design" Strategies

A reasonable approach for lower capital cost pilot and demonstration plants, if client is OK with it. These approaches allow for vessel damage, but protect personnel by keeping containment of process fluids.

Vessel wall stresses at fire relieving temperature evaluated relative to:

a) the ultimate tensile stress (UTS) with an adequate safety factor
b) lower time creep rupture stress values (i.e. 100 hour creep rupture stress)

Difficult Scenarios – Guillotine Tube Failure

Double-guillotine failure from a single tube results in highly disproportional relief sizes relative to the exchanger size at pilot scale.

Typical pilot scale exchanger could have as few as 10 tubes, compared to thousands of tubes at the commercial scale. Required flowrate to relieve the same regardless of scale. Double-guillotine less likely at pilot scale? Smaller diameter tubes with relatively short pitch.

Difficult Scenarios – Guillotine Tube Failure

Typical pilot plant tubesheets...

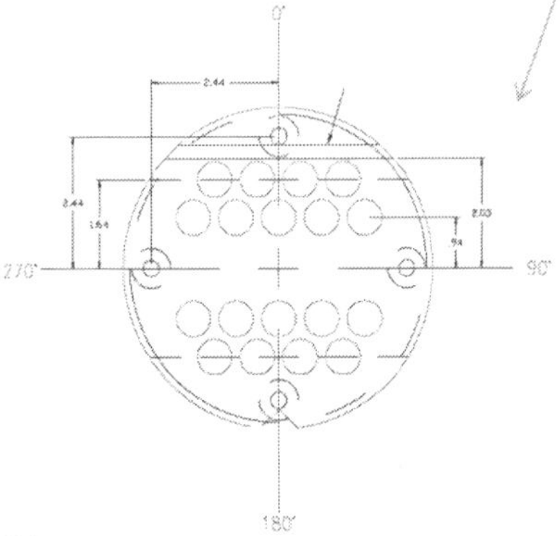

Commercial tubesheet

Difficult Scenarios – Guillotine Tube Failure

Mitigate the scenario:

- utilize 10/13 rule for design of low pressure side (including nearby piping)
- consider cost of pilot scale equipment, look at "beyond design" strategies to yield, but not lose containment
- consideration of pinhole leaks as opposed to double-guillotine, typically Zeton would consider a 1/16" pinhole with a 0.62 orifice coefficient, since a leak of this size is usually detectable

Difficult Scenarios – Guillotine Tube Failure

Recent project, nominal flow on tube side 1,650 lb/hr. Required flow for double-guillotine: 216,540 lb/hr.

Would have required a tee in the exchanger shell, with rupture disc the same diameter as shell.

Problems?

- premature failures during normal operation (MTBF rates higher for rupture discs than for guillotine scenario)
- may not even be able to react in sufficient time, pressure shock wave will travel at speed of sound in the fluid

The 3% Rule and Chatter

Non-recoverable frictional pressure drop in inlet piping allowed up to 3% of relief valve set pressure.

3% rule stems from concern with valve chattering (rapid opening and closing) that can:

1. Reduce the valve capacity as the stagnation pressure at the valve inlet drops below the blowdown pressure
2. Lead to valve body damage and possible loss of containment

Can occur if valve is grossly oversized, very common at pilot plant scale.

The 3% Rule and Chatter

At pilot scale, some key considerations:

- low inventories in small vessels are far below what would be expected to be required for valve body damage
- lower spring constants and associated forces with smaller relief valves, less likely to damage (testing required)
- valve lift in smaller valves is a fraction of that of large valves

Manufacturer	Orifice Area	Lift
Flowsafe F84/F85	0.015 in^2	0.075 in
Farris 2700	0.038 in^2	0.050 in
Crosby 900	0.085 in^2	0.106 in
Farris 2600	12.27in^2	1.186 in

Summary

- rationalize relief scenarios for evaluation while going through relief contingency table
- recognize lower risk associated with pilot scale equipment
- appreciate unique challenges at pilot scale

(492a) Creating More Value from Asphaltenes Using Continuous Oxy- solubilisation Process

Ashtari, M., University of Calgary

Originally presented on: 11/11/2015 8:30:00 - 8:55:00

A Northern Alberta facility for bitumen upgrading processes 4000 tons of asphaltenic pitch every day via solvent deasphalting[1]. Increased asphaltene production in northern Alberta facilities, leads to searching for processes for better utilization of that material. Asphaltene presence in petroleum has negative effects on viscosity, density and elemental composition of the crude oil. Due to the problems that asphaltene particles can cause along production lines like well bore plugging, fouling and coke formation, dealing with their separation and trying to extract valuable products from them by further processing is of significant importance [5, 6].

Existing technologies for processing asphaltenes such as delayed coking or gasification only add a small amount of value and need high capital investment. So, With regard to big problems which asphaltene could create in oil, new technologies are needed that can directly transform asphaltenes using fewer processing steps and low Capital Intensity infrastructure, and can produce increased yields of high economic value products, such as transportation fuels or petrochemicals, while reducing the environmental pollutions [6].

Exploring alternative extensive uses of asphaltene pitch in search of potential valuable products led us to a variation of the Wet Air Oxidation (WAO) process commonly used for organic wastes processing. As so far known, WAO is a destructive technology based on the oxidation of wastes at high temperature and high pressure in liquid phase. In that system, molecular oxygen dissolved in the wastewater reacts with hydrocarbons and carbohydrates. The oxidation process depends on the solubility of oxygen at high temperature and high pressure. High temperature could increase reaction rates and production of free radicals [7, 8].

The concept proposed here is that by controllably oxidizing asphalthene particles in water, the inter-core links between asphalthene molecules can be broken and simultaneously partial oxidation -Oxycracking- can release small molecular weight oxidized molecules while decreasing the size of aromatic associations.

According to our chemical studies, the material produced from this process is of similar characteristics to fulvic and humic acids which could be used in agricultural applications as fertilizers, sewage purification, radionuclide, CO2 sequestration and surfactants [9].

This work focuses on wet air oxidation of asphaltene particles as potential new technology for producing fulvic or humic acids which could be used as fertilizers commercially. For this purpose, at first the optimum operating conditions had obtained in batch mode and then continuous pilot set up was designed, built and tested based on batch experiments results to evaluate maximum solubilisation of asphaltene particles in water.

425792 Design and Analysis of the Natural Gas Liquefaction Optimization Process- Energy Storage of Cryogenic Carbon Capture (CCC-ES)

Abstract:

The cryogenic carbon capture (CCC) process provides energy- and cost-efficient carbon capture and can be configured to provide an energy storage system using an open loop natural gas (NG) refrigeration system. This system stores energy during non-peak times by liquefying and storing natural gas to be used as a refrigerant. When the energy is at a premium during peak demand, the stored liquefied natural gas carries out the cryogenic separation process relieving the parasitic load on the power plant. The natural gas either returns to the pipeline or combusts in a simple natural gas turbine after the gas has been vaporized and warmed, increasing the power output of the plant during peak demand by reducing the parasitic losses associated with carbon capture.

This investigation compares four different natural gas liquefaction processes simulated by Aspen HYSYS as incorporated as part of the CCC-ES process. In these processes, LNG vaporizes in the CCC process and the cold vapors return through the LNG heat exchangers exchanging sensible heat with the incoming flows. Aside from this difference, this investigation uses process designs similar to traditional LNG processes. The simulations meaningfully compare these alternative liquefaction options, eliminating differences in assumptions and process details inherent in comparing processes simulated by different authors or different codes. The comparisons include costs and energy performance with individually optimized processes, each operating at three operating conditions: energy storage, energy recovery, and balanced operation. The results indicate that turbomachinery efficiency influences overall performance. Given similar quality turbomachinery, efficient heat exchangers in particular reduce energy input requirements and maximize energy savings and capital costs, including heat exchangers used to cool compressed gases.

429045 Process Synthesis of Mixed Refrigerant Cascade System for Ethylene Plants

Abstract:

Refrigeration system generally works by indirectly transferring the heat from low-temperature sources to high-temperature sinks at the expense of electricity or compression work. Refrigeration system is one of the most important and critical operating system in chemical and petrochemical industries. It must satisfy different cooling tasks from different units working at different temperatures/pressures with a wide range; meanwhile it has to be integrated with other utilities (e.g., steam and cooling water) to accomplish an optimal plant-wide energy saving system. The traditional cascade refrigeration system (CRS) used in ethylene plants employs multiple refrigerants, each of which is circulated in a closed cycle and operated at multiple temperature and pressure levels. Unfortunately, the use of the multiple refrigerants require a very number of separate compressors, heat exchangers, pumps and associated pipes for each temperature and pressure level. Recently, mixed refrigerant cascade system (MRCS) has been introduced, where the pure working fluids are replaced by mixed working fluids. MRCS has many significant advantages than the tradition single-component refrigeration system. It needs fewer process units and fewer maintenances and thus less expenditures compared with CRS. The system operates within smaller temperature differences at the lower temperature which leads to a smaller increase in entropy and consequently a smaller work consumption. In this paper, process synthesis model and operational conditions are developed in which the pure ethylene cycle of the conventional cascade refrigeration system is replaced by a mixed refrigerant. Meanwhile, a general MINLP (mixed-integer linear programming) model for the optimal synthesis of MRCS in ethylene plants has been solved. A case study about a mixed refrigerant cascade system used in a typical ethylene plant is employed to demonstrate the efficacy of the developed methodology. Comprehensive thermodynamic analysis is also conducted to give deep understandings of the optimization results.

426768 Key Technologies to Produce Bio-Ethylene and Vam

Abstract:

Bio-manufacture is expected as a solution to deal with challenges decreasing of petroleum resources and increasing environmental protection. A packaged industrial technology to produce bio-Ethylene and Vinyl Acetate Monomer (VAM) from bio-ethanol has been developed by Tianjin University. A 100,000 MTA VAM plant was built using this technology in 2012, which is the first VAM industrial plant in China.

This process comprises of sections of dehydration of bio-ethanol to produce bio-ethylene and acetoxylation of bio-ethylene to produce VAM. In this advanced process, a series of innovations have been made including a high efficiency process with adiabatic bed reactor was invented for ethanol dehydration to ethylene; the deactivation mechanism of the catalyst by CO was established via the DFT calculations, based on which a low energy consuming process for VAM production with high CO tolerance was developed; a dedicated database for VAM production was set up based on the combination of quantum mechanics and GEMC method; the Mechanical Vapour Recompressor (MVR) technology was adopted to recover the low-rank reaction heat in the start-up period; a high efficiency VAM reactor was designed via the investigation of hydrodynamics properties of both single-phase and multiphase flow with the assistance of computational fluid dynamics.

A novel approach to synthesis ethoxylated alkyl sulfonates in microemulsion microreactor

Ji Chen, Ming Yang, Yun Fang*

The Key Laboratory of Food Colloids and Biotechnology, Ministry of Education; School of Chemical & Material Engineering, Jiangnan University, 1800 Lihu Avenue, Wuxi, Jiangsu 214122, P. R. China

Abstract: Alkyl ethoxylated sulfates (AES) is one of the commercial anionic surfactants with excellent properties in wetting ability, dispersiing ability, emulsification and electrolyte resistance. However, the unstable sulfate ester group in the molecule of AES makes it easy to decompose at high temperatures or in strong acidic or alkaline condition, which hinders it being used more widely in certain extreme environments [1,2]. Ethoxylated alkyl sulfonates (EAS) contain a chemical stable sulfonate group in molecule in stead of unstable sulfate group in AES, which can be widely used in various harsh environments such as high temperature, high salinity, strong acids or strong alkalis. In this work, the nucleophilic substitution reaction between AES and Na_2SO_3, to produce EAS was run in micelle and microemulsion microreactor, respectively. The effects of time, temperature, the initial surfactant concentration, the initial sulfite concentration, the initial mole ratio of the oil/water on the conversion of AES to EAS were determined. By comparison, an optimum surfactant concentration was found at the maximum sulfonation yield of EAS, which obtained in microemulsion microreactors is 82.1%. This is because that the distance of the adjacent AES ionic head becomes larger in microemulsion microreactor than it in micelle microreactor due to existing co-surfactant and oil phase [3,4]. Therefore, the electrostatic repulsion between AES ionic head and SO_3^{2-} becomes weaker in microemulsion microreactor, which will improve the sulfonation yield of EAS.

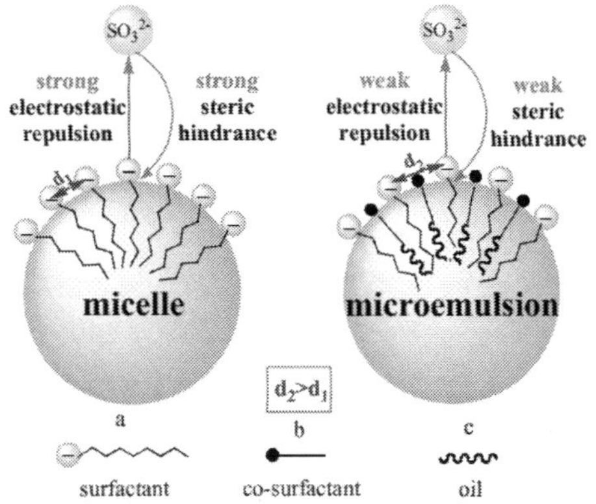

Key word: ethoxylated alkyl sulfonates, alkyl ethoxylated sulfates, micelle, microemulsion, sulfonation

REFERENCES

[1] Attaphong C, Do L, Sabatini D A. Vegetable oil-based microemulsions using carboxylate-based extended surfactants and their potential as an alternative renewable biofuel. Fuel. 2012; 94: 606-613.

[2] Klaus A, Tiddy, G T, Solans C, Harrar A, Touraud D, Kunz W. Effect of Salts on the Phase Behavior and the Stability of Nano-Emulsions with Rapeseed Oil and an Extended Surfactant. Langmuir. 2012; 28(22): 8318-8328.

[3] Husein M M, Weber M E, Vera J H. Nucleophilic Substitution Sulfonation in Microemulsions and Emulsions. Langmuir. 2000; 16(24): 9159-9167.

[4] Fredrik C, Martin A, Krister H. Oxidation of Self-Organized Nonionic Surfactants. Langmuir. 2004; 20(10):3835-3837.

437114 Taking the Whole Family Overseas: Experiences and Tips for Survival

Abstract:

Moving a large family from the U.S. to Australia brought challenges and taught lessons that may be valuable for others considering international assignments. Issues like education, housing, food, pets, finding friends for children, transportation, healthcare, and other issues need to be considered and managed. A variety of personal stories and their lessons will be shared.

437953 Learning to Be Flexible

Abstract:

The speaker will discuss ways in which chemical engineers can overcome cultural and regional barriers to find success in international assignments, based on work experiences involving Middle Eastern, Asian, and North American companies.

421116 Managing Innovation with Partners and Teams from India

Abstract:

India is a land with rich opportunities for chemical engineers, but proper preparation and knowledge is needed for Westerners to work most effectively in India or to maximize success with teams in India. Vital tips to prepare for working in India or collaborating with teams in India will be shared, including guidance on how to maximize innovation success.

421089 Innovation with Chinese Teams: What Western Companies Need to Know for Success

Abstract:

In the West, there is a common but incorrect stereoptype about the lack of innovation and innovation skills in China. In reality, Chinese engineers, scientists, and entrepreneurs can innovate and invent remarkably well, though there are some cultural barriers that can reinforce old stereotypes and hinder productivity when Western managers work with Chinese teams. Here we will share experiences in innovation and provide critical guidance to help Western companies and managers tap the innovation power of Chinese partners and associates.

CHEMICAL ENGINEERING & THE CHALLENGE OF INDUSTRIALIZATION OF AFRICA.

DR. N. JOHN ERINNE, FNSChE, FNSE

PAST PRESIDENT, NIGERIAN SOCIETY OF CHEMICAL ENGINEERS (NSChE)

&

MANAGING CONSULTANT, CHEX & ASSOCIATES, LAGOS, NIGERIA

PRESENTED AT THE

2015 AIChE ANNUAL MEETING

SALT LAKE CITY, UTAH, USA

NOV. 11, 2015

ABSTRACT

Most of the African continent is today faced with the dilemma of massive unemployment and excessive dependence on imports, even in the face of abundant primary materials. This paper explores the prospects for deploying Chemical Engineering skills as a tool for spearheading the industrialization of the continent and resolving this dilemma. It also proposes arrangements and structures for strengthening the profession and practice of Chemical Engineering in Africa, in order to position it to play the anticipated role and achieve its potentials.

Keywords: Africa, Chemical Engineering, industrialization.

1. AFRICA: HISTORICAL PERSPECTIVES

The continent of Africa, with a very sizeable population of over one billion, consists presently of 54 independent states and two partially recognized/unrecognized states **(APP. I)**[1]. As is well known, most of the continent came under colonial rule by the great powers of Europe (U.K, France, Belgium, Portugal, Spain, Netherlands, etc.) by the turn of the last century. In the aftermath of the Second World War however the clamour for decolonization of the continent began to gain momentum, eventually culminating in the wave of independence of African State in the 1960s.

The post-independence history of the continent has nevertheless been characterized mostly by instability, sometimes bordening on chaos and anarchy, as the various states grappled with the grave challenges of self-governance and nation building, with very limited back-ground experience. This naturally also had a negative impact on the economic development of the continent. After 40-50 years however, many of the states are now settling into stable, democratic structures, more amendable to economic and social development.

It is pertinent nevertheless to observe that there are notable exceptions to this historical pattern in the continent. Whereas countries such as Egypt and Ethiopia never came under colonial rule at any stage, colonization of South Africa developed through a completely different trajectory.

2. AFRICA: RESOURCES & ECONOMY

The continent of Africa is indeed very richly endowed with abundance of natural resources, in form of:

- Petroleum (oil & gas)
- Solid Minerals
- Agricultural Products

i. Petroleum

Africa has been producing oil & gas actively since the late 1950s and early 1960s. Most of the activities were in North Africa (Algeria, Libya and Egypt) and on the West Coast (Nigeria, Cameroun, Equitorial Guinea, Gabon, Republic of Congo and Angola) for several decades. Lately, several other countries have also joined the train and are actively producing oil & gas. These include: Sudan, South Sudan, Ghana, Ivory Coast, Chad, Namibia, Tunisia, DR. Congo, Mauritania, etc. Others have as well recently struck oil and/or gas and are in the process of developing the reserves for production (Uganda, Kenya, Tanzania, Mozambique, Morocco, Sierra Leone, etc.).

Libya, with proven reserves of 47 billion barrels of crude oil, has the largest reserve in Africa, while Nigeria is the leading producer, with output of about 2.5 million barrels per day **(APP. II)**[2].

With a total proven reserves of crude oil of 165 billion barrels in 2010, Africa is estimated to have nearly 12% of the total world reserves of 1,383 billion barrels[3]. Besides, Africa accounts for about 10 million barrels per day (bpd) of crude oil production, representing about 11% of total world production of 90 million bpd[3].

Similarly, of the total world proven reserves of natural gas of 6,609 trillion cubit feet (tcf) in 2010 Africa had 520 tcf, a very significant 8% share[3].

In particular, Nigeria and Algeria, with reserves of about 175tcf and 154tcf respectively, have the largest reserves **(APP.III)**[4] and rank 8th and 9th globally.

Unfortunately however the total oil refining capacity of Africa is only about 3.1 million bpd **(APP. IV)**[5], representing barely 30% of its crude oil output and a paltry 3.6% of global capacity[6].

ii. **Solid Minerals**

The range of endowment of the continent in solid minerals consists of:

- Coal

- Other Non-Metallic Minerals: limestone, kaolin, talc, diamond, phosphate, bentonite, barytes, etc.
- Metallic Minerals: gold, copper, iron ore, bauxite, tin, zinc, platinum, cobalt, etc.

For example, South Africa with proven reserves of about 48,000 million tonnes of coal and annual production of 270 million tonnes ranks 6th in the world, **(APP. V)**[7].

iii. **Agricultural Products**

Ivory Cost is the world leader in the production of cocoa, with output of 1.4 million tonnes in 2012, followed by Ghana with 870,000 tonnes **(APP.VI)**[8]. The effect is that Africa accounts for a very impressive 70% of total world cocoa production of 4,000 million tonnes per annum **(APP.VII)**[8].

However, again, Africa lags in respect of processing capacity, with only a 16.5% share of world capacity **(APP.VIII)**[8].

On the other hand, for coffee, though less impressive, Africa still contributes about 14% of estimated world output of 7.4 million tonnes. Ethiopia, Ivory Coast and Uganda rank 5th, 11th and 12th respectively among the producing nations[9].

These further illustrate the huge endowment and potential in Africa in respect of primary materials production and processing. Indeed, in addition to cocoa and coffee, the continent is an important producer of a wide range of other agricultural crops, such as: cotton, tea, sugarcane, rubber, palm, groundnuts, cassava, sisal, maize, millet, sorghum, rice, tobacco, etc.

iv. **Water Resources**

Furthermore, Africa is very well watered by several rivers which provide water for agriculture and for power. Notable among these are such rivers as the Nile, Niger, Senegal, Congo, Orange, Limpopo and Zambezi.

v. **Colonial Economy**

In the colonial era, the economy of Africa was essentially set up to obtain these primary materials required to sustain the industrial societies of the European powers. By this very uneven division of responsibility, Africa was reduced to merely growing primary crops or extracting vital minerals while industrial activities were domiciled in Europe. Coal obtained from Africa assisted in sustaining electric power generation in Europe, just like cocoa, coffee and tea grown in African sustained the beverage and confectionary industry. Similarly, African cotton was the mainstay of the textile industry while palm produce and groundnuts provided vegetable oil used in soap making and for food.

The gap between resource availability and processing capacity for a number of commodities highlighted above clearly underlines this phenomenon.

Notably too, the finished products from these industries ultimately found their way back to the African market, mostly at prices disproportionate to the African primary materials prices.

vi. **Post-Colonial Economy**

This economic pattern has persisted, long after the decolonization of the continent. Even with the advent of petroleum, the pattern has remained unaltered. The result is that today, Africa exports crude oil to the world market and imports refined petroleum products. What an absurdity!!

If the colonial powers were however to blame for the initial aberration, they cannot be held accountable for the sustenance of the same pattern of trade several decades after independence. We in Africa must therefore accept responsibility substantially for the perpetuation of poverty inherent in this lopsided pattern of trade.

vii **Poverty**

To buttress the issue of poverty in Africa, the relative GDP per capita of Africa and the other developing regions of the world are presented in **APP.IX**[10]. Clearly, Africa is at the bottom of the ladder, despite its rich endowment with natural resources, as illustrated earlier. Even the current average annual growth rate of about 5% in the continent, as impressive as it is, is not expected to close this gap in the foreseeable future.

3. IMPERATIVES OF CHANGE

In the light of the foregoing, it becomes imperative in this 21st century that Africa must change this economic order, for her survival. With increasing number of African countries finding oil & gas in their territories, Africa's crude oil must be refined for both local consumption and export to the world. Similarly, before we export natural gas overseas, we must use our gas to generate electricity to support local industry, as well as feedstock for petrochemical industries making vital chemicals for both African consumption and export to world markets. Indeed, Africa can readily become a hub for such vital industries as petrochemicals, textiles/garments, fertilizers, aluminum products, chocolates/beverages, rubber products, etc. it is only by this that the critical developmental issues of massive unemployment, value addition to primary materials, creation of wealth and reduction of poverty can begin to be effectively addressed. It is indeed by this that the relatively high growth rates currently prevailing in the continent can be sustained and have true meaning in the lives of the ordinary people of Africa.

Attainment of such lofty goals however cannot be easy. It will certainly require massive mobilization and deployment of vital input and precursors necessary for gainful exploitation of the abundant primary resources. These inputs include:

- Finance/Funding
- Technology/Know-How
- Human Capital

4. PUBLIC INFRASTRUCTURE

It is also pertinent to note that envisaged industrialization of the continent cannot indeed be reality without the simultaneous development of public infrastructure to support it. This will be in such vital areas as:

- Electric Power
- Fuels & Energy Supply
- Water Supply
- Telecommunications
- Information Technology
- Ports/Harbours
- Airports
- Railways
- Roads/Highways

- Sewage/Waste Management
- Environmental Protection Systems
- etc.

Whereas impressive progress may have been made lately across the continent in such areas as telecommunications and information technology, critical gaps are identifiable in most other areas, between the colonial/post-colonial infrastructure available and the requirements of modern infrastructure to support industrialization. Massive deployment of resources is therefore envisaged in this area, along with the focus on industrialization.

5. ENTER CHEMICAL ENGINEERING

Chemical Engineering, as we know full well, may be defined as that aspect of endeavor concerned with the transformation of primary materials into finished products required for consumption. By this, Chemical Engineers are essentially involved in development, design, construction, operation, maintenance and optimization of processes for transformation of materials from the raw stage to finished products. Indeed, it may be pertinent to note that such processes do not always necessarily entail chemical reaction or chemical changes. Often they are essentially physical processes. Separation of air to obtain oxygen and nitrogen or extraction of vegetable oil from oil seeds readily come to mind in this regard. There are of course numerous others. This however clearly illustrates why Chemical Engineers are often referred interchangeably as Process Engineers. It also underscores the versatility of the profession.

In view of this, it is evident that Process Engineers have the unique training and background to drive the much desired economic emancipation and industrial development of the African continent. I am yet to see or know any other profession that is as uniquely positioned in this regard.

Though Chemical Engineering may not be able to provide answers to the question of financing, it certainly can drive technology and know-how, in most respects. It can also provide the vital technical manpower required in most of the sectors.

Without doubt, Process Engineers are best positioned to drive the refining of crude oil into petroleum products or processing of natural gas for use in power generation or even transforming natural gas into petrochemical products and agricultural fertilizers. A few more strategic examples will suffice to fully buttress the point:

- Production of sugar from sugarcane.
- Production of cement from limestone
- Smelting of aluminum from bauxite
- Extraction of vegetable oils from oil seeds and refining of same.

The list may as well be indeed endless.

Besides, with their training and skills Process Engineers are also eminently positioned to assist African nations structure optimum resource development and utilization strategies and plans for a very broad range of primary materials.

In addition, Chemical Engineers are also equipped to contribute in varying degrees to the development of the public infrastructure required to underpin these resource improvement initiatives, as highlighted earlier. In particular, Chemical Engineers have a strong role to play in such areas as fuels & energy supply, thermal power generation, alternative energy development, water treatment, urban waste management and environmental protection systems.

6. CHEMICAL ENGINEERING IN AFRICA

Internationally, Chemical Engineering is a late comer as a distinct Engineering discipline, relative to the more traditional branches of Engineering: Civil, Mechanical and Electrical. It's existence dates back to only a little over a century ago when it emerged and was recognized as a distinct professional discipline in USA and UK.

The pattern is also somewhat reflected in Africa. This is underscored by the earlier emphasis on basic public infrastructure development in the colonial and post-colonial periods in the continent. The focus was on hydro-power, transportation (harbors/airports/railways/roads) and communications infrastructure, etc, just barely adequate to support the colonial economy which was centered on export of primary commodities, with little or no interest in processing and local value addition. This clearly culminated in a subdued role for Chemical Engineering in the development process in most parts of the continent.

However with the compelling dynamics of change in the continent clearly pointing in the direction of local value addition and processing of primary materials, a major boost is expected in the role of Process Engineers in the continent in the coming years. This is also further buttressed by the increasing role of Process Engineers in modern public infrastructure development.

The highest concentration of Chemical Engineers in Africa presently are in South Africa, Egypt and Nigeria, expectedly coinciding with the three largest economies of the continent. A distinct and vibrant Chemical Engineering society exists in each of these countries. In the rest of the continent, Chemical Engineers do not appear to have the 'critical mass' for independent or semi-independent existence yet and are therefore understandably subsumed in the wider Engineering societies.

It is expected that the changes and developments anticipated in the economies of African nations should help to further energize the Chemical Engineering formations across the continent, both in countries where Chemical Engineering associations exist and those where they do not exist, presently. In this regard, it is imperative that Chemical Engineering is properly positioned to play the expected role in the industrial development of the continent.

7. RECOMMENDATIONS

A few recommendations are proferred, which may assist in realization of these lofty expectations outlined above.

i. **Development Strategy**

African nations are enjoined to more than ever before design their development strategies to leverage on the abundant primary materials to grow manufacturing industry, in order to combat the menace of poverty and unemployment in the continent.

ii. **Education/Training**

African nations are also encouraged to establish Chemical Engineering schools in their Universities, tailored to their needs for resource improvement and industrial development. Special attention should be paid to the curricula, to ensure that they strongly reflect the specific needs of their economies and not merely designed along the lines of the broad, classical Chemical Engineering curriculum. Besides, such schools must be well resourced with faculty and facilities, to ensure quality training. Indeed, quality is perhaps considered more important than quantity with regards to training of future Chemical Engineers in the continent, for fairly obvious reasons.

iii. **Professional Development**

In as much as current thinking clearly favors private sector lead in industrial development, it is necessary for African countries to have the mechanism in place to ensure that young Chemical Engineers and other related technical professionals have adequate opportunity for professional development and career growth. This should be considered to be an important part of governmental regulatory function which is a vital ingredient for success of the free enterprise system. It is indeed essentially the means to ensure that these professionals mature in their chosen fields in due course and are able to play the roles expected of them.

iv. **Industry Support**

The emerging industrial sector in Africa is also urged to strongly support governmental efforts in education, training and professional development of Chemical Engineers and allied technical cadres, as indeed an exercise in enlightened self-interest.

v. **African Chemical Engineers Forum**

There is compelling need for Chemical Engineers in the continent to organize themselves into a forum or association which will serve as a platform for regular exchange of ideas and experience. Such a body will also facilitate the much desired robust contribution of Chemical Engineers to the industrial development and economic integration of Africa. The existing strong Chemical Engineering bodies in South Africa, Egypt and Nigeria readily provide the nucleus and platform for such an organization. The responsibility therefore clearly devolves on them to get together and lead the way in this regard. The Nigerian Chemical Engineers in NSChE are willing and ready. Are our brothers and sisters in SAIChE and ESChE (Egyptian Society of Chemical Engineering) also similarly willing and ready? This is a call to service for mother Africa. May we indeed not be found wanting!

8. CONCLUSION

As Africa transits to the stage of rapid industrial development using its vast raw materials as feedstock, a central role is foreseen for Chemical Engineers as key drivers of the process. It is therefore critical that all stakeholders take the necessary steps to ensure that this role is fulfilled. In particular,

special attention to the education, training and professional development of Chemical Engineers is called for, from African governments. Professional Chemical Engineers in the continent are also specially called upon to organize themselves more effectively, to position them to play the expected role.

REFERENCES

1. CIA Directorate of Intelligence, The world fact book: "Africa", 2012.
2. US Energy Information Administration/International Energy Statistics, Oil & Gas Journal, 2011.
3. Christina Katsouris, Energy Intelligence, Africa's Oil & Gas Potential, 2012.
4. OPEC, Natural Gas Proven Reserves by Country, 2013
5. Energy Information Administration (EIA), Form EIA-820 "Annual Refinery Report", 2014
6. British Petroleum, BP Statistical Review of World Energy, 2013.
7. British Petroleum, BP Statistical Review of World Energy, 2008
8. ICCO, USDA, Reuters, 'LMC Report', 2012
9. USDA, Hawaii Farm Facts, 2012.
10. World Economic Forum, The Global Competitiveness Report, 2013.

APP. I

AFRICAN NATIONS & POPULATIONS

S/NO	COUNTRY	POPULATION	%OF POPULATION
1	Algeria	33,333.267	3.25
2	Angola	16,941,000	1.65
3	Benin	8,439,000	0.82
4	Botswana	1,839,833	0.18
5	Burkina Faso	13,228,000	1.29
6	Burundi	7,548,000	0.74
7	Cameroon	17,795,000	1.73
8	Cape Verde	420,979	0.04
9	Central African Republic	4,216,666	0.41
10	Chad	10,146,000	0.99
11	Comoros	798,000	0.08
12	Democratic Republic of Congo	75,507,308	7.36
13	Djibouti	906,000	0.09
14	Egypt	84,550,000	8.24
15	Equatorial Guinea	504,000	0.05
16	Eritrea	5,880,000	0.57
17	Ethiopia	85,237,338	8.31
18	Gabon	1,384,000	0.13
19	Gambia	1,517,000	0.15
20	Ghana	23,000,000	2.24
21	Guinea	10,057,975	0.98
22	Guinea-Bissau	1,586,000	0.15
23	Ivory Coast	17,654,843	1.72
24	Kenya	34,707,817	3.38
25	Lesotho	2,067,000	0.20
26	Liberia	4,128,572	0.40
27	Libya	6,036,914	0.59
28	Madagascar	18,606,000	1.81
29	Malawi	12,884,000	1.26
30	Mali	13,518,000	1.32

31	Mauritania	3,069,000	0.30
32	Mauritius	1,219,220	0.12
33	Morocco	35,757,175	3.49
34	Mozambique	20,366,795	1.99
35	Namibia	2,031,000	0.20
36	Niger	13,957,000	1.36
37	Nigeria	174,507,539	17.01
38	Republic of Congo	4,012,809	0.39
39	Rwanda	7,600,000	0.74
40	Sao Tome & Principe	183,176	0.02
41	Senegal	11,658,000	1.14
42	Seychelles	80,654	0.01
43	Sierra Leone	6,144,562	0.60
44	Somalia	9,832,017	0.96
45	South Africa	52,981,991	5.17
46	South Sudan	8,260,490	0.81
47	Sudan	36,787,012	3.59
48	Swaziland	1,032,000	0.10
49	Tanzania	44,929,002	4.38
50	Togo	7,154,237	0.70
51	Tunisia	10,102,000	0.98
52	Uganda	27,616,000	2.69
53	Zambia	14,668,000	1.43
54	Zimbabwe	13,010,000	1.27
		1,021,398,191	99.58
	Partially Recognized/ Unrecognized States		
55	Somaliland	4,000,000	0.39
56	Sahrawi Republic	266,000	0.03
		1,025,664,191	100.

APP. II

AFRICA PROVEN RESERVES OF CRUDE OIL & PRODUCTION - 2010

	COUNTRY	RESERVES (billion barrels)	PRODUCTION ('000 barrels per day)
1.	LIBYA	47	1,750
2.	NIGERIA	37	2,520
3.	ANGOLA	13.5	1,940
4.	ALGERIA	13.4	1,700
5.	SUDAN	6.4	112
6.	EGYPT	3.7	680
7.	GABON	2.2	242
8.	EQUATORIAL GUINEA	1.7	350
9.	CONGO REPUBLIC	1.8	275
10.	DR CONGO	0.8	115
11.	CHAD	1.2	17
12.	MAURITANIA	1.5	16
13.	CAMEROON	0.7	77
14.	OTHERS	2.5	180
	TOTAL	165	9,974

• includes South Sudan

APP. III

TOP TEN WORLD PROVEN RESERVES OF NATURAL GAS - 2010

RANK	COUNTRY	RESERVES (TCF)
1	RUSSIA	1,623
2	IRAN	1,120
3	QATAR	837
4	TURKMENISTAN	583
5	USA	315
6	SAUDI ARABIA	273
7	VENEZUELA	184
8	NIGERIA	175
9	ALGERIA	150
10	AUSTRALIA	128

APP. IV

AFRICA OIL REFINING CAPACITY 2010

RANK	COUNTRY	CAPACITY ('000
1	EGYPT	680
2	SOUTH AFRICA	490
3	ALGERIA	450
4	NIGERIA	445
5	LIBYA	340
6	MOROCCO	155
7	IVORY COAST	70
8	KENYA	65
9	SUDAN	50
10	ANGOLA	45
11	CAMEROON	45
12	GHANA	45
13	TUNISIA	35
14	SENEGAL	25
15	ZAMBIA	25
16	GABON	20
17	CONGO REPUBLIC	20
18	CHAD	20
19	NIGER	20
20	ERITERIA	18
21	TANZANIA	17
22	MADAGASCAR	12
23	MAURITANIA	10
		3,102

APP. V

LEADING WORLD RESERVE & PRODUCTION OF COAL 2007

PROVEN RECOVERABLE RESERVES (million tonnes)			PRODUCTION (million tonnes per annum)		
RANK	COUNTRY	RESERVES	RANK	COUNTRY	OUTPUT
1	USA	246,643	1	CHINA	6,396
2	RUSSIA	157,010	2	USA	2,537
3	CHINA	114,500	3	INDIA	478
4	INDIA	92,445	4	AUSTALIA	394
5	AUSTRALIA	78,500	5	RUSSIA	314
6	SOUTH AFRICA	48,750	6	SOUTH AFRICA	269
7	UKRAINE	34,153	7	GERMANY	202
8	KAZAKHSTAN	31,279	8	INDONESIA	175
9	POLAND	14,000	9	POLAND	146
10	BRAZIL	10,113	10	KAZAKHSTAN	94

APP. VI

WORLD PRODUCTION OF COCOA 2012

REGION	COUNTRY	PRODUCTION ('000 tonnes)
AFRICA	Cameroon	220
	Ghana	870
	Ivory Coast	1,400
	Nigeria	230
	Other Africa	81
TOTAL AFRICA		**2,801**
ASIA/OCEANIA	Indonesia	500
	Malaysia	18
	Other Asia	-
TOTAL ASIA/OCEANIA		**623**
AMERICAS	Brazil	185
	Ecuador	170
	Other Latin America	208
TOTAL AMERICAS		563
TOTAL WORLD		**3,987**

APP. VII

REGIONAL SHARES OF COCOA PROCESSING (GRINDING) 2012

REGION	OUTPUT ('000 tonnes)	% SHARE
EUROPE	1,554	39.1
AMERICAS	865	21.8
ASIA & OCEANIA	897	22.6
AFRICA	657	16.5
	3,973	100.0

APP. VIII

REGIONAL SHARES OF COCOA PROCESSING - 2012

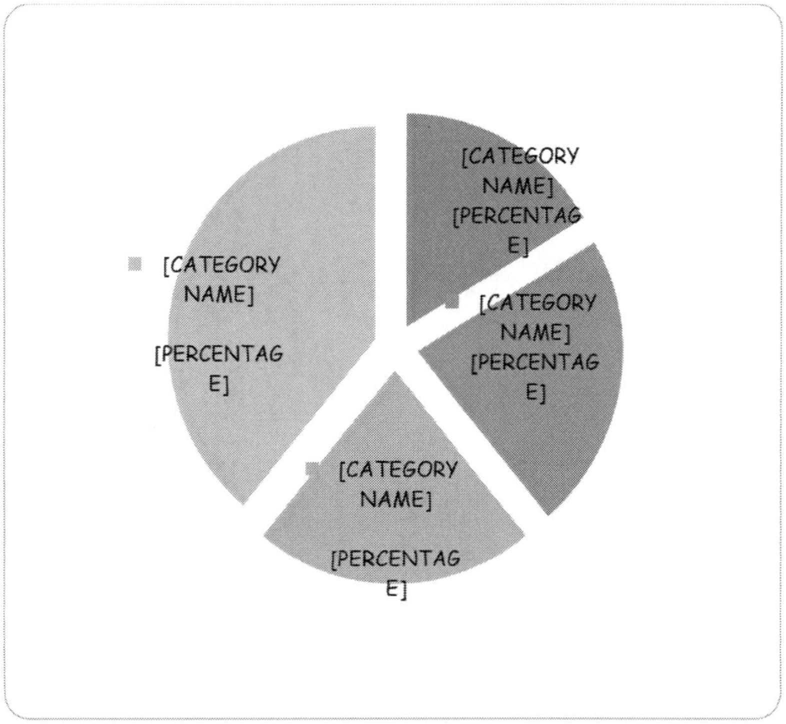

APP. IX

GDP's OF WORLD DEVELOPING REGIONS 2012

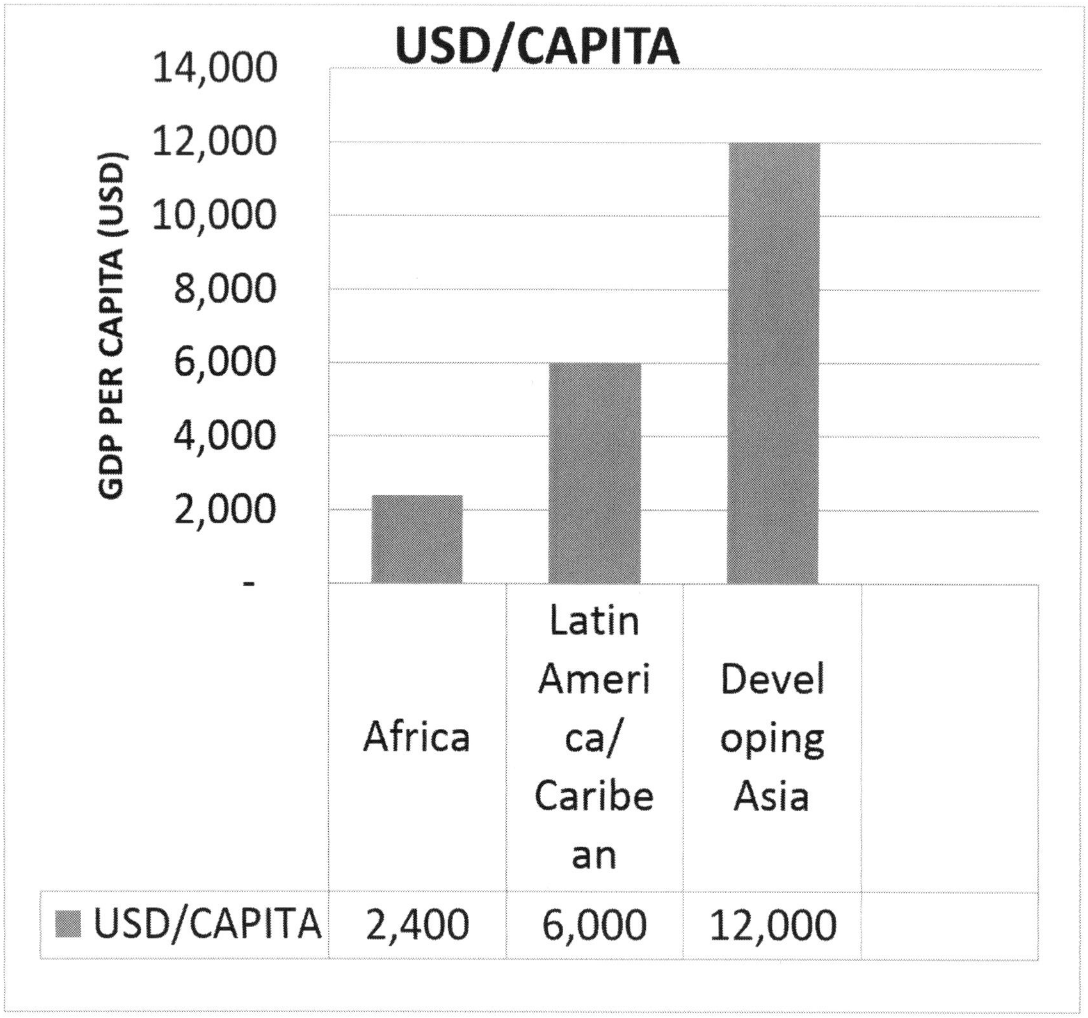

Figure 1USD/CAPITA

433229 The Research on a Polyamidoamine-Based Demulsifier for Oil/Water Separation

Abstract:

Oil-in-water emulsions widely exist in the waste water resulted from polymer flooding oil recovery process. Because there are large amounts of interfacial active substance and ultrafine oil droplets in this kind of emulsion, it is very hard to treat with traditional methods. To effectively deal with this kind of emulsion with ultrafine oil droplets and highly concentrated surfactants, one kind of chemical method, demulsification is adopted. In this thesis, the synthesized 3rd generation polyamidoamine (PAMAM) was studied as a typical demulsifier. FTIR and MNR were applied to analyze the molecular structure of the demulsifier. The influences of temperature, settling time, and concentration of the demulsifier used on the efficiency of the demulsifier were investigated in detail. At the optimal operating condition (demulsification time: 120 min; demulsification temperature: 30 °C; demulsifier concentration: 1500 mg L-1), the oil removal rate of the demulsifier could reach more than 90%. The surface tension and interfacial tension at the diesel–water interface were also measured to give a basic understanding of the demulsification mechanism.

436869 Three Types of Filter Media Produced from Electrospun Cellulose Acetate (Cac)-Polystyrene (PS) Composite Membrane for Separating Oil-Water Mixtures

Abstract:

Three Types of Filter Media Produced from

Electrospun Cellulose Acetate (Cac)-Polystyrene (Ps)

Composite Membrane for Separating Oil-Water Mixtures

L. Baghernejad[1] Chem., E. Iski[2] Chem.

Department of Chemistry and Biochemistry

O. Shoham[3], Department of Petroleum Engineering

R. Mohan[4], Department of Mechanical Engineering

S. Odueyungbo[5], Chevron Energy Technology Company

The University of Tulsa

Tulsa, Oklahoma 74104

(918) 812-7956

lida-baghernejad@utulsa.edu

ABSTRACT

(261 Words)

Oil-water mixtures are encountered in many domestic, industrial and environmental activities and are often undesired and problematic. Particularly, oil-spills such as the sudden Mexican Gulf catastrophe have attracted tremendous interest towards the separation of oil-water mixtures. Membrane filtration is one of the most effective methods so far in separating oil-water mixtures. In this project, Cellulose Acetate (CAc) fibers were reinforced by Poly (styrene) (PS) fibers via co-electrospinning CAc and PS from one solution to produce a highly porous, nanofibrous and flexible composite membrane. Three types of filter media were developed from the electrospun CAc-PS membranes: 1) a hydrophobic-oleophilic filter medium (Filter #1), which separated oil from oil-water mixtures, 2) a hydrophilic-oleophobic filter medium (Filter #2), which separated water from oil-water mixtures and 3) a hydrophobic-oleophobic filter medium (Filter #3), which is expected to separate water and oil droplets from wet gas. These filter media were produced by heating and/or treating the electrospun CAc-PS membrane with two commercially available surfactants named PHOBOL-CP® and ULTRAPHIL® from the Huntsman-DuPont Textile Company. The filter media were characterized physically, mechanically and chemically. The filtration performances of the first two filter media were tested in a lab-scale batch filtration setup, while the performance of the third filter medium remains to be tested. These filter media are novel with respect to their material, structural topography and morphology, surface roughness and filtration performance. The uniqueness of these novel filter media can be attributed to their high reproducibility, scalability, low basis weight, flexibility, high separation efficiency (> 90%) and cost-effectiveness due to their abundant and cheap starting materials. In such terms, these filter media can compete with conventional filter media.

436152 Reducing Drilling Fluid Losses Using a Lab Scale Shale Shaker

Abstract:

The recovery of drilling fluid is essential in the oil and gas drilling under the environmental and economic point of view. These fluids are basically colloidal suspensions classified according to their base components in aqueous and non-aqueous fluids. The equipment typically utilized in solid-liquid separation units are divided into three groups sequentially arranged: shale shakers, hidrociclones (desander and desilter) and centrifugal decanter. Understanding the dynamic behavior of the shale shaker is very important to reduce the process load applied to the subsequent equipment. This paper presented a contribution on the experimental research on employment of g-force control loop dimensionless number applied to a vibrating screen prototype. The study was conducted based on the method traditionally employed in solids control units. The manipulated process variable is the rotation imposed on motor vibrators installed in the vibrating screen. The results obtained in the experiments indicate an appropriate and consistent operation of the proposed control loop. The dimensionless number g-force was maintained close to the set-point value, ensuring repeatability for the experiments.
The use of higher dimensionless values of g-force does not necessarily assured greater removing of residual liquid entrained with the solid material retained by the sieve screen. One possible reason for this behavior is the fact that the greater intensity of vibration provided by motor vibrators helped to increase the solid transport speed on the display screen, reducing the effect of elimination of retained fluid.

404743 Destabilization and Treatment of Produced Oil-Water Emulsions from EOR Application Using Polyacrylamides

Abstract:

Surfactant and polymer flooding technology can greatly enhance the oil recovery through the expansion of sweeping and displacing efficiency. The recovered oil from surfactant and polymer flooding emulsifies the residual chemical, which makes the separation of water from oil quite difficult, yet the impact of the enhanced oil recovery (EOR) chemicals on the produced water cycle is generally neglected in chemically-based EOR studies. This includes compatibility of EOR chemicals with the additives used to pre-treat the injected water or change reservoir wettability and result in producing oil/water emulsion after EOR breakthrough.

The largest waste produced in oil and gas industries is believed to be the produced water, as it contains different sort of organic and inorganic admixture. The discharging of produced water contaminates not only the water, both surface and underground, but also the soil. There are a number of treatment methods available for produced water. To separate water from oil in a much efficient manner and to reach the emission standard, a new class of water soluble polymer of polyacrylamides (PAMs) was used as destabilizing agents for water-oil emulsions, which have been stabilized by surfactant (Talowamine Acetate).

The impact of the surface charge form, the density of polyacrylamides in turbidity reduction , zeta potential, COD, viscosity of separated water were explored in this study. Different anionic polyacrylamide of different surface charge density and molecular weight were evaluated. Different anionic polyacrylamides were utilized, and at optimum dosage, anionic AN94 PAM at its optimum concentration was proved as the best way to reduce the residual turbidity compared with other PAMs mentioned in this research. The effects of electrolytes such as aluminum sulphates on produced water degree of flocculation in the existence of PAM were also investigated in terms of turbidity reduction and volume of separated water after jar test. The results showed that the volume of separated water increased more than 25% compared when only PAMs were used, and the turbidity of separated water improved significantly.

431325 Destabilization and Treatment of Produced Water-Oil Emulsions Using Anionic Polyacrylamide and Electrolyte of Aluminum Sulphate and Ferrous Sulphate

Abstract:

Surfactant and polymer flooding technology can greatly enhance the oil recovery through the expansion of sweeping and displacing efficiency. The recovered oil from surfactant and polymer flooding emulsifies the residual chemical, which makes the separation of water from oil quite difficult, yet the impact of the enhanced oil recovery (EOR) chemicals on the produced water cycle is generally neglected in chemically-based EOR studies. This includes compatibility of EOR chemicals with the additives used to pre-treat the injected water or change reservoir wettability and result in producing oil/water emulsion after EOR breakthrough.

The largest waste produced in oil and gas industries is believed to be the produced water, as it contains different sort of organic and inorganic admixture. There are a number of treatment methods available for produced water. To separate water from oil in a much efficient manner and to reach the emission standard, a new class of water soluble polymer of polyacrylamides (PAMs) was used as destabilizing agents for water-oil emulsions, which have been stabilized by surfactant.

The impact of the surface charge form, the density of polyacrylamides in turbidity reduction, COD, viscosity of volume separated water were explored in this study. The effects of electrolytes such as aluminum sulphates and ferrous sulphate on produced water degree of flocculation in the existence of anionic polyacrylamide were investigated in terms of turbidity reduction and volume of separated water after jar test. Different concentrations of both sulphates added into optimum concentration polyacrylamide selected from jar test were utilized, and at optimum dosage, anionic AN 934 PAM with aluminum sulphate at its optimum concentration was proved as the best way to reduce the residual turbidity compared with other additives mentioned in this research. The results showed that the volume of separated water increased more than 25% compared when only PAMs were used, and the turbidity and COD reduction of separated water improved significantly.

435174 Reliable Use of a Baghouse in High-SO3 Environment

Abstract:

Recent EPA regulations set very stringent particulate emission standards for coal-fired power plants, as well as limit mercury and acid gas emissions. This has led to an increased use of fabric filter baghouses, often in combination with sorbents for mercury and/or acid gas removal. EPRI and Southern Company have research collaboration around the use of a pilot-scale (20,000 ft3/min slip stream) baghouse on flue-gas from medium-to-high-sulfur fuel. This talk will review recent results around emission control, particulate filtration, pulse-jet cleaning, and baghouse O&M.

410043 Continuous Solid-Phase Crystallization

Abstract:

Continuous crystallization is one of the emerging technologies that has been widely investigated recently. Most of the advancements have happened in the pharmaceutical industry, and for the continuous crystallization of small molecules, all of the platforms have been in the liquid phase. There has not been any work in continuous solid-phase crystallization for complex matrices, when the solid domain (micron sized particles) consists of a small and large molecule mixture.

This work presents a novel technique for continuous solid-phase crystallization of dairy powders, as a model system, which contains lactose (small molecule, initially amorphous), proteins (large molecules, which have few effects on the crystallization process through different molecular interaction with lactose and water), and minerals. Experimental results and mathematical modeling of the process demonstrate the performance of the system for a wide range of processing conditions and lactose/protein ratios. The ability of the technique to manipulate the particles' micro-structure, and to design the particles' physico-chemical properties, by controlled and partial crystallization of the powders, will also be discussed.

The challenges for powder processing at highly critical conditions, crystallization within a complex matrix, and implementation of the technique for the industrial applications, will be presented. The approaches to overcome these challenges, as well as the gap in the knowledge and potentialities for further improvements will also be discussed.

The wide applications of the technique are in the Pharmaceutical, Food, and BioNanotechnology industries for solid-phase and solvent-exchange crystallization, and conditioning of the particles' structure.

423627 Think Small, Think Flexible: The Expansion Strategy of a Continuously Operated Modular Multi-Product Plant

Abstract:

As a result of product diversification and increasing volatility of markets, the capability to adapt the production output of chemical plants becomes crucial for economic success. The plant design for pharmaceutical products is determined by a shorter time-to-market and uncertain market sales. Therefore a product launch requires a production of small amounts during the market entry. Thus, pharmaceutical chemicals are conventionally produced batch-wise in a multi-product plant that is flexible regarding production rates. However, a batch-wise processing is inefficient in terms of resource consumption, batch failures and cleaning effort. Due to an uncertain market demand, for the production of increased product quantities a more efficient and flexible plant concept is required [1].

In this context, a continuously operated modular multi-product plant offers the solution of an efficient and flexible production. Continuously operated processes enable an increase of efficiency in production. The modular design enables a short time-to-market and a flexible adaptation of the production capacity. Thus, an efficient early market entry is possible due to the design of modular production line with a fixed design capacity for small product amounts. When market demand increases a stepwise numbering-up of the modular production line allows a fast and flexible adaptation of production capacities by reuse of engineering. The smaller the scale of the modular production line is, the more expansion steps are necessary to supply the increasing market demand. As a consequence, this capacity flexibility results in additional costs incurred by the loss of economy of scale. Furthermore, a flexible reallocation of the production of different products within the production capacity is possible. This planning flexibility also results in additional costs for setup changes and storage of the products. The increased flexibility provided by this modular plant concept has to be balanced with the consequent costs.

In this work the modular plant concept regarding capacity flexibility and planning flexibility is evaluated. Regarding an uncertain market development, the sequential expansion by numbering-up of a modular production line is investigated. Therefore, a long-term production planning for lot-sizing and scheduling of a multi-product production is applied. Thus, necessary expansion steps are identified when the market demand exceeds the production capacity. An economic evaluation is conducted to consider additional cost by the sequential investment and the production planning. A decision tree analysis is applied to balance the added value of flexibility with the additional costs. Therefore, the obtained flexibility of different scales of a modular production line is evaluated.

[1] T. Bieringer et al., Chemical Engineering & Technology, 36 (6), 2013, 900–910

410571 Membrane Enhanced Peptide Synthesis (MEPS) – Integrating Organic Solvent Nanofiltration (OSN) into Iterative Chemical Reactions

Abstract:

In the pharmaceutical industry, the high potency and specificity of peptides make them more attractive than traditional small-molecule drugs. The chemical synthesis of peptides involves the repetitive addition of amino acid, in between which the intermediate products have to be isolated for purification purposes. It is of paramount importance to achieve high yield for the isolation of each intermediate product or the cumulative mass loss will be detrimental to overall yield of the synthesis. The conventional peptide synthesis methods, namely Liquid Phase Peptide Synthesis (LPPS) and Solid Phase Peptide Synthesis (SPPS), have been used in the industry over the past few decades, but innovation is still required to improve process performance. Membrane Enhanced Peptide Synthesis (MEPS) integrates Organic Solvent Nanofiltration (OSN) into LPPS. A hexapeptide (Pyr-Ser(Bzl)-Ala-Phe-Asp-Leu-NH2) was synthesised successfully via MEPS at a scale of 33.65 mmol and a soluble anchor concentration of 10.4 wt% in the starting solution. Before the cleavage and global deprotection of peptide, MEPS achieved a high purity (88.1%) and a decent overall yield (71.2%), similar to LPPS (100.0% and 72.4% respectively), in which the intermediate products were precipitated and purified. Consequently, the performance of MEPS was better than LPPS in terms of material cost, process time, volumetric efficiency and E-factor, showing that OSN is indeed a more attractive option than precipitation for the isolation of intermediate products in liquid phase iterative chemical reactions.

432830 Prediction of Intracellular Nucleotide Sugar Variation Impact on Monoclonal Antibodies (mAbs) Glycan Distribution Using Glycosylation Reaction Network Models

Abstract:

Prediction of intracellular nucleotide sugar variation impact on monoclonal antibodies (mAbs) glycan distribution using glycosylation reaction network models

Sha Sha[a] and Seongkyu Yoon[a]

[a]Department of Chemical Engineering, University of Massachusetts, Lowell, MA, USA

Glycosylation is a critical post-translational process on monoclonal antibodies (mAbs) produced by mammalian cells as the glycan profile on mAbs plays significant impact on the pharmaceuticals' stability, efficacy and half-life. Glycosylation central reaction network (CRN) in cellular *Golgi* compartment is the branching paths for mAb to be attached with various intermediate or terminal glycan structures via the mediation of a series of enzymes. The recent development of reaction network models is an approach to simulate glycosylation process in *Golgi* CRN and predict the distribution of glycan on final monoclonal antibodies (including a number of intermediate or terminal glycan structures) through simulation using the model, and thus are potentially of great value in the application in the industrial bioprocess. Reaction network models are formulated by describing the kinetics of reactions in the network and involve various intracellular factors associated with glycosylation as model parameters. The capability of models being built to reflect key intercellular factor variation is one of the priorities for models to ensure proper performance.

However, the performance of model prediction for glycan distribution out from intracellular factor (e.g. nucleotide sugar) variation has not been sufficiently evaluated in the past studies. In this work, a reaction network model was developed, describing the CRN by resembling Golgi as a four-compartment continuous stirred-tank reactor (CSTR) and the glycan distribution was estimated by solving the model based on mass balance of all glycan structures in the network at a steady state. The impact of intracellular nucleotide sugar concentration variation on the glycan distribution output was then simulated using the model and further compared with nucleotide sugar- glycan variation correlation being examined by published literature. The model was then optimized to minimize the gap between model prediction and experiments-derived data, therefore becoming a more reliable tool to be utilized in the future for glycosylation prediction.

(552e) Understanding the Effect of High Gas Entrance Velocity on Chinese Hamster Ovary (CHO) Cell Culture Process

Chaudhary, G., Boehringer Ingelheim Fremont Inc.
Guillen, N., Boehringer Ingelheim Fremont Inc.
Karim, K.,
Whiteley, E.,
Lin, H., Boehringer Ingelheim
Khetan, A., Boehringer Ingelheim Fremont, Inc.
Originally presented on: 11/11/2015 13:54:00 - 14:15:00

Growing demand for biologics requires high intensity unit operations such as high cell density fed-batch processes. High yielding cell culture process requires higher aeration/oxygenation to satisfy high oxygen demand and for carbon dioxide removal in a bioreactor. Higher aeration/oxygenation through sparger can result in higher gas entrance velocity (GEV), which can lead to cellular damage and can adversely impact the cell culture performance. A study was performed to evaluate the effect of high GEV on CHO cell culture process using spargers of different number/size of holes. Our results demonstrate that high GEV imparts significant negative effect on cell growth, viability, and productivity. Similar effects were observed in large scale production bioreactors. In addition, the extent of cell damage (necrosis or apoptosis) under high GEV was examined by lactate dehydrogenase measurement and flow cytometry analysis. Insights into the cell condition under high GEV will be presented and discussed.

435020 Efficient Optimization of Tablet Coating Operation Using Hybrid CFD-DEM Simulations

Abstract:

Tablet coating is a common unit operation in pharmaceutical manufacturing. The coating process depends on a number of process variables – spray parameters, coater geometry, mass loading, drying air flow etc. Process engineers in the pharmaceutical industry are interested in optimizing coater operation from both process and product perspectives. Performing physical experiments with coaters to determine the optimal set of process parameters is, more often than not, impractical. Digital experiments using computer simulations of tablet coating offer a viable alternative to experiments. We present here an efficient hybrid CFD-DEM model framework that enables a relatively rapid and efficient optimization of the coating process. The approach described here is an effective alternative to full scale CFD/DEM co-simulation and is designed to provide useful design guidance to process engineers at a significantly reduced computational expense.

437563 Multistage Continuous MSMPR Crystallizer with Solids Recycle

Abstract:

Continuous manufacturing has been well developed in food and chemical industries, however, batch operation is still the most common method used in the pharmaceutical industry [1]. For crystallization, continuous processing has the advantages of enhanced reproducibility of results, reduced cost, improved process efficiency and flexibility in production capacity [2]. Due to these reasons, continuous crystallization has obtained great interest in both industry and academia in recent years [3].

The major challenges associated with the transition from batch to continuous crystallization process are to improve the yield of active pharmaceutical ingredients (APIS) and the purity of products. Batch process can obtain a higher yield by pushing the process to equilibrium while a continuous process is operated at steady state [4]. An appropriate recycle stream can be applied in continuous process to boost the yield. A study of cooling crystallization of cyclosporine showed a 91.8% yield by using a single-stage mixed-suspension, mixed-product removal (MSMPR) crystallizer with concentrated mother liquor recycle system [5]. However, previous research shows that for MSMPR, the mean crystal size and the purity decrease as improving the yield by recycling the mother liquor [6]. This is because mother liquor recycle will make the steady state with low supersaturation and high impurity/API ratio: the slow growth caused by low supersaturation will lead small crystals and the high impurity/API ratio causes more impurity incorporation.

A common approach of improving the yield of continuous crystallization is to lower the operating temperature and thus decrease the solubility of API. However, for some APIs, e.g. cyclosporine, whose growth kinetics are very sensitive to temperature, the growth rate is very low (~0.1 μm/min) at low temperature. Thus, low operation temperature will cause low yield or it is required a much longer process residence time. The goal of this work is to enhance the yield under low temperature and push the steady state close to equilibrium while keep the residence time short and remain the purity of the crystal. To accomplish this goal, a two-stage MSMPR crystallizer with concentrated slurry recycle stream is established for cooling crystallization of cyclosporine. By recycling the crystals back to crystallizer, the total growth surface area increases so that more crystal growth occurs even if the liner growth rate is low. Thus, the yield can be improved. As the residence time of the crystals increases by recycling, crystals can grow more and have larger crystal size. Since the residence time of the mother liquor is shorter than that of the crystals, the impurity in the mother liquor will not accumulate in the crystallizer and the crystal purity can remain.

Moreover, a population balance model in conjunction with experimentally determined crystallization kinetics and distribution coefficients is developed. The model can simulate n-stage MSMPR crystallizer with different number of slurry recycle or mother liquor recycle to any stage(s). Using this model, the optimal stage temperatures, residence time, feed concentration, and recycle ratios can be obtained. Experiments guided by the model is used to obtain higher experimental yield and further valid the model.

This work presents an alternative idea to apply recycle stream to improve the yield and purity and control the crystal size in a continuous process.

References:

[1] Parikh, D. M. Handbook of Pharmaceutical Granulation Technology; Marcel Dekker: New York, 1997

[2] Vervaet, C.; Remon, J. P. Chem. Eng. Sci. 2005, 60, 3949–3957.

[3] Poechlauer, P.; Manley, J.; Broxterman, R.; Gregertsen, B.; Ridemark, M. Org. Process Res. Dev. 2012, 16, 1586−1590.

[4] Chen, J.; Sarma, B.; Evans, J. M. B.; Myerson, A. S. Cryst. Growth Des. 2011, 4, 887−895.

[5] Wong, S. Y.; Tatusko, A.; Trout, B. L.; Myerson, A. S. Cryst. Growth Des. 2012, 12, 5701-517.

[6] Alvarez, A. J;Singh, A.; Myerson, A. S. Cryst. Growth Des. 2011, 11, 4392-4400.

427189 Separation of Fine Oil Droplets from Oil-in-Water Mixtures By Dissolved Air Flotation

Abstract:

Dissolved air flotation (DAF) is a separation technique, often used after a primary gravity separator or filters, to enhance the quality of the wastewater before it can be released to streams, rivers and the sea. The main aim of the DAF investigations reported here was to measure the removal efficiency of oil droplets in the ranges of 15-80 μm from the oil-in-water mixtures. Several operating parameters were varied and two kinds of oil were used. It was decided that vegetable oil should be tested as it is relevant to the food industries. To compare the vegetable oil droplet removal efficiencies, investigations were also carried out using a mineral oil. Saline water in this experiment was made by adding 3.5% by concentration of NaCl, to mimic the average salinity of water produced from the oil wells. An oil-in-water measuring technique (FastHEX) and a droplet counting method (Coulter Counter) were used to measure the concentration and numbers of oil droplets in the water, respectively. The best oil droplet removal efficiency obtained from these experiments was found to be 96 % and it was for the removal of vegetable oil droplets from a saline oil-in-water mixture. The vegetable oil has a large and positive spreading coefficient (14.6×10^{-3} N/m), which facilitates the attachment of bubble and oil droplet and therefore produced higher removal efficiency than lamp oil that has a negative spreading coefficient (-7.11×10^{-3} N/m). The experimental results were subjected to dimensional analysis and the removal efficiency was found to be a function of several dimension and dimensionless groups. The experimental data tabulated in dimensionless form has been subjected to multivariable linear regression. The resulting correlation was found to have a root mean square error of 11.1 % but predicted oil droplet removal efficiencies greater than one and less than zero. An alternative mathematical formulation was devised that cannot predict efficiencies outside the range. Regression of this formulation, which had the same number of adjustable parameters as the linear regression, was successful with a root mean square error of 10.9 %.

419313 Modeling of Complex Bio Systems Using Batchreactor

Abstract:

Process simulation is widely used by process engineers in the design of new units as well as in operations of existing plants for process optimization, units troubleshooting or debottlenecking, plants revamping, performing front-end engineering analysis… It is used in nearly all process industries: chemicals, pharmaceuticals, petrochemicals, oil and gas, refining, specialty chemicals… However, bio-industries still do not use intensively process simulation. In these industries, microorganisms (fermentation), enzymatic reactions or reactive transformation of the plant material are involved. They are characterized by a great complexity of the raw material and the phenomena which are taking place in reactive unit operations. The number of components (often non-exhaustive for a plant material) and the number of reactions (several thousand for a given microorganism) limit the number of robust reactional models available in literature. As they most often operate in batch mode, bio-reactions are rarely presents and used in process simulation software.

BatchReactor® is a software dedicated to the simulation of batch chemical reactors [1], which provides the user with a detailed modeling of the reactor (heating/cooling system, condenser, mixing device…) and a reliable description of the production recipe. BatchReactor® allows modeling of complex systems by taking advantage of the power of Simulis® Thermodynamics (thermodynamic properties server) which offers an extensive set of thermodynamic models with a pure components database of more than 2,000 pure components (AIChE's DIPPR® database [2]). Additional new components can be easily added in database as for example microorganisms represented with a CHONPS formula. Moreover, thanks to the user "interpreted" kinetic rate model available through Simulis® Reaction (chemical reaction server), any kind of specific kinetic models can be used in BatchReactor®. This is essential to represent kinetic bio-reactions with often non-classical laws [3]. So, in BatchReactor®, stoechio-kinetic model can be coupled with thermodynamic and heat transfer models already available in the software to simulate the bio-reactor.

Several applications from various industries have been successfully simulated: food industry (tomato juice oxidation, brewing of beer, alcohol fermentation), pharmaceutical (production of thiolutine) and white biotechnology (production of polyhydroxybutyrate). A stoechio-kinetic model was necessary for all these applications but, when it was known, the strengths of BatchReactor appeared: accurate thermodynamic and transfer phenomena representation, recipe description, user-friendly interface.

References

[1] www.prosim.net

[2] R.L. Rowley, W.V. Wilding, J.L. Oscarson, N.F. Giles, DIPPR® Data Compilation of Pure Chemical Properties, Design Institute for Physical Properties, AIChE, New York, NY (2011)

[3] Ramon Portugal F., Fillon M., Meyer X.M., Pingaud H., Strehiano P., I.S.B.N. 2-87805-022-3, 159-170 (1997)

424888 Oxygen Transfer Rate Effect on Pigment Production By Penicillium Purpurogenum GH2 in Shake Flask Cultures

Abstract:

Fermentation processes have become a popular alternative to obtain natural food ingredients such as natural colorants. Filamentous fungi seem to be an attractive source of bio-pigments due to the wide range of colors that are able to produce at high yields. Production of pigments by filamentous fungi in submerged fermentation is affected by a large number of operating variables like pH, temperature, agitation and aeration rate that may interact strongly between themselves and with the specific characteristics and properties of growth medium and microorganisms, creating a unique environment resulting in a particular final yield of the process. Industrial application of fungal pigments will depend on the successful scaling up of the process. In aerobic fermentation, oxygen transfer rate (OTR) is a key parameter utilized on the design and scale up of microbial processes aiming for the highest productivity.

The presented study aimed to characterize the oxygen transfer conditions in shake flask culture and to analyze its effect on growth and pigment production by *Penicillium purpurogenum* GH2.

Fermentation experiments were carried out in Erlenmeyer flasks (125 mL) for 8 days at 30 °C. Oxygen transfer was controlled by varying agitation speed (200 to 250 rpm) and the medium volume in flask (25, 50 and 75 mL). Oxygen transfer coefficient (k_La) was determined without biological consumption of oxygen using the dynamic method. OTR at the evaluated agitation speeds and liquid volumes was estimated by OTR = k_La (C^*-C_L) using the obtained k_La values, where (C^*-C_L) varied approximately from 7.0-7.5 mg/L at the range of conditions studied.

Pigment production and growth were affected by the conditions tested. Pigment production increased as liquid volume decreased from 75 to 25 mL at both agitation speeds (200-250). The highest pigment production obtained at 250 rpm with 25 mL of liquid volume was 14.42 OD_{500nm} which was 23% higher than the obtained with 25 mL at 200 rpm. Meanwhile the maximum biomass growth was obtained at 50 mL (3.36 g/L) at 200 and at 75 mL (3.37 g/L) at 250 rpm. The oxygen transfer coefficient k_La decreased with the liquid volume and increased with the agitation speed (11.02-17.60 h^{-1}). Pigment production showed a linear correlation with k_La (Pigment = 2.17k_La-23.73, R^2=0.98) and with OTR (Pigment = 0.32OTR-26.67, R^2=0.90). Biomass growth increased by decreasing k_La from 17.60-11.59 h^{-1} reached a maximum biomass 3.37 gL^{-1} and decreased at a k_La value of 11.02 h^{-1}.

The above results indicated that oxygen transfer rate in shake flasks depends on agitation speed and working volume used in submerged fermentation. The production of pigments by *Penicillium purpurogenum* GH2 is strongly affected by the oxygen supply in the medium. A direct correlation was found between pigment production and k_La and OTR which represent a major finding as reference for future scale up of the process from shake flask to bioreactor.

435117 Experimental Orders on an Entrepreneurial Pace: Dairy Wastewater Characterization Control Using Sequenced Blending

Abstract:

A creative mindset must be adopted in the effort to create sustainable processes. Whether it be a process improvement to an existing technology or a novel application of sustainable processing, screening the effects through experimental orders often must be done in a fast pace environment with limited information. The case study presented is dairy wastewater control applying an entrepreneurial mindset to practical experimental orders.

Discharge specifications for effluents to the POTW have pH as a significant regulatory compliance component. Many small to medium cheese plant operations discharge directly to the POTW. Usually, cheese and milk products are produced in the same plant. There is both acidic and basic cleaning solution discharge streams. By using a selective blending control system the effluent from the plant to the POTW can be modified in process and in some cases water can be recovered for other operations. Cycling of the cleaning operations is not practical with the pressures upon production. The installation and use of pH monitoring and flow control systems can temper the blended discharge. pH curves can be determined to assess the capacities of both the basic and acidic fluids to accept volumes of the other for in process neutralization.

Cheese whey is acidic. Samples are in the pH 4 – 5 range. The high pH cleaning residue can accept cheese whey as an effluent constituent. Whey residue contains calcium which will precipitate as the pH is raised. The baseline pH variability of the current effluent treatment system can be established and translated to a bench lab titration. The cheese whey was characterized and added in a mass ratio to the available effluents then titrated similarly to the current system. The pH neutralization curves were established for each of the three loading levels. The family of curves were used to establish the mass balance expected at full scale Sample analysis was conducted at the nominal operating temperatures.

Typically high pH characteristic flow can accept up a mass of the cheese whey without dropping the pH below 10. This condition could occur frequently. A sequenced blending control scheme would control of the cheese whey addition should be programmed to occur only when the waste tank material has the capacity to accept the cheese whey. A second waste tank characteristic with a lower pH in the range of 7 – 9 was also observed. It did not have capacity to neutralize the added cheese whey. The sequenced blending control scheme will be needed to avoid this possible out of specification condition.

The addition of cheese whey may require additional caustic neutralization. This may add additional precipitates to the existing waste stream. The pH neutralization kinetics could be effected by the formation of additional precipitates. The mixing parameters can be validated by lab work in sample jars and beakers. These indicative results can then be run in a scalable mix system. Effects of mixing on the reaction kinetics can be difficult to observe in small lab volume samples.

428251 Chemical Mapping of Sweeteners and Additives in Commercial Chewing Gums Using Imaging FTIR Spectroscopy

Abstract:

Chemical Mapping of Sweeteners and Additives in Commercial Chewing Gums using imaging FTIR Spectroscopy

Ramazan KIZIL,

Istanbul Technical University, College of Chemical and Metallurgical Engineering, Chemical Eng. Dept. Maslak, 34469 Istanbul-Turkey. Contact: kizilr@itu.edu.tr

Determination of the structural organizations of chemical constituents, such as lipids, proteins and carbohydrates, within a food at the microscopic level is expected to significantly contribute to the modern food technology whose prime interest lies in development of controlled processes and operations to preserve, transform and create or destroy food structure (Aguilera, 2001). Food microstructure analysis has an ultimate goal of understanding structural hierarchies, forming naturally or manufactured, from images that depict spatial arrangements and interactions of chemical compounds that form flavor, color, texture and nutritional value. Developing new tool to correlation microstructure with food texture perception is expected to contribute to food industry in production of textually attractive food products (Wilkinson et al. 2000).

Vibrational spectroscopic tool such as Raman or infrared spectroscopy coupled with an optical microscope can be used to visualize structures in a form of image produced from chemical composition contrast. The technique is known to be chemical imaging and it is becoming a promising modality for diagnosis of cancer as a complimentary molecular histopathology tool. Another interesting application field of chemical imaging is pharmaceutical industry for which visualization of the distribution of active ingredients and excipients is prime interest.

Analyzing molecular vibrational transitions of atoms of a compound as a result of absorbance of IR radiation provides a strong basis for chemical structure determination of organic or inorganic substances. This analysis can be best performed by implementing a multiplexing detection of all wavelengths at the same time using a focal plane or linear array detectors. A classical FTIR system coupled to an optical microscope with a motion controlled stage and a multiplexing detector collects spectrum at a resolution set by the diffraction limit, which is around 5.5 micron for most commercial IR imaging systems.

Chemical imaging is a powerful tool for visualization of structure of a specimen from its functional groups in a form of chemical contrast map, representing the spatial distribution of chemical entities in the sample. FTIR provides the contrast necessary for enabling chemical structural imaging by means of the selectivity of vibrational absorption modes detected only at specific frequencies/energies. The spectrum at each pixel is referred to as fingerprint of chemical composition.

This talk covers experimental studies of determining the spatial distribution of sweeteners and additives in chewing gums to elucidate the chemical nature of developed microstructure in different brands. The samples were investigated in attenuated total internal reflectance (ATR) mode in order to obtain images at pixel resolution of 1.65 micron by oversampling approach and solid immersion lens (SIL) technology (Kazarian and Chan, 2010). Similar to oil immersion optics, ATR uses hemispherical Germanium element which has refractive index of 4.1. This allow to obtain images with 4 times enhanced resolution comparing to that of transmission or transreflectance modes of IR imaging. Three different brand commercial chewing gums were used to study microstructure imaging. ATR imaging necessitates a firm contact of the Ge element with the sample, so a little distortion of images due to contact pressure related stretch. Since all samples will be subjected to a similar stretch condition, the images offer a comparing realistic visualization of the microstructure. Chemical mapping strategy was followed to show the spatial distribution of sweetener and additives in the texture from their own molecular fingerprint. The main matrix was well represented in the images and inhomogeneous distributions of other components were mapped at a pixel resolution of 1.65 micron. The findings of the study and sensory evaluations suggests that controlling the microstructure can help improve the sensory quality and functional characteristics of chewing gums.

References:

Aguilera, J.M. 2005. Why food microstructure. Journal of Food Engineering. 67:3-11.

Kazarian, S.G., Chan, A. 2010. Micro-and macro ATR infrared spectroscopic imaging. Applied Spectroscopy. 64:135A-152A.

Wilkinson, C., Dijksterhuis, B.G., Minekus, M. 2000. From food structure to texture. Trends in Food Science and Technology. 11:442-450.

437187 Temperature Control of Microchannel Reactors Using Bimetallic Thermally-Actuated Valves

Wednesday, November 11, 2015: 12:30 PM
Salon D (Salt Lake Marriott Downtown at City Creek)

Richard Pattison, McKetta Department of Chemical Engineering, University of Texas at Austin, Austin, TX, Akash Gupta, Department of Chemical Engineering, The University of Texas at Austin, Austin, TX, Melissa Donahue, James R. Fair Process Science and Technology Center, Austin, TX and Michael Baldea, McKetta Department of Chemical Engineering, The University of Texas at Austin, Austin, TX

Temperature Control of Microchannel Reactors Using Bimetallic Thermally-Actuated Valves

Richard Pattison, Akash Gupta, Melissa Donahue, and Michael Baldea

McKetta Department of Chemical Engineering

The University of Texas at Austin, 1 University Station C0400, Austin, TX 78712

email: mbaldea@che.utexas.edu

Microchannel reactors coupling exothermic and endothermic reactions are one of the most prominent results of the process intensification paradigm. The high surface-area-to-volume characteristics of CPRs result in a process with minimal transport limitations and consequently, a size and cost that are an order of magnitude less than conventional reactors of the same capacity [1].

While CPRs bring significant potential economic benefits, they also pose significant control and operational challenges. Synchronizing heat generation and consumption in the reactors is particularly challenging, and if not properly addressed, could lead to the formation of temperature hot spots that may damage the catalyst coating of the plates or the reactor structure itself [2-4]. Furthermore, in practical scenarios, the reactors are subject to disturbances in the inlet conditions and potentially unequal distribution of flow to the channels, which may further contribute to the formation of hot spots. The implementation of measurements (temperature, composition, and/or flow rate) at the channel level is practically challenging. As a consequence, thus far, the majority of feedback control systems for CPRs rely on measuring the properties (typically temperature) of the bulk outlet streams, and on boundary actuation, in the sense that the total (rather than channel-wise) flow rate of fuel is adjusted in response to changes in reactor outputs.

In this work, we present a novel approach for microchannel reactor control, that provides actuation and temperature control at the individual channel level. Specifically, we introduce a new class of thermally-actuated valves constructed from bimetallic strips [5]. Bimetallic strips consist of two different metal strips that are rigidly attached. The difference in thermal expansion properties of the two metals causes the strips to deflect upon changes in temperature, transforming temperature variations into mechanical displacement. Thus, affixing bimetallic strips to either side of the combustion channels in microchannel reactors is equivalent to implementing a thermally-actuated valve as seen in Figure 1.

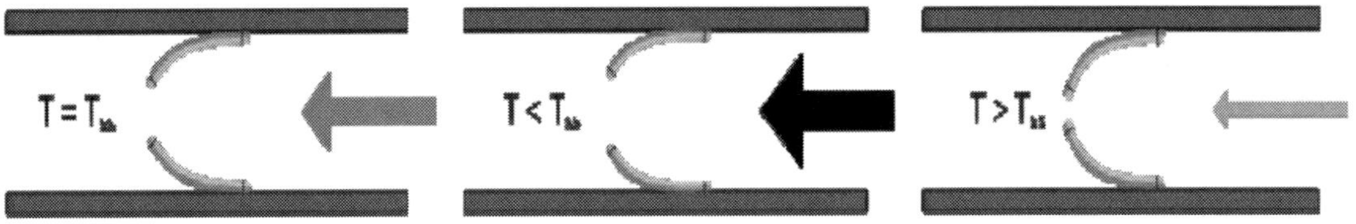

Figure 1: Response of a thermally-actuated valve to temperature changes. Tss represents the design temperature (adapted from [5]).

In this system, the flow rate through the channel is dictated by the pressure drop across the valve. If the operation deviates from the nominal conditions, temperature changes in the reactor will result in a deflection of the strips and a change in the valve position, and consequently an adjustment in the fuel flow rate. For example, in the presence of a disturbance that reduces the amount of heat consumed in the reforming channels, the temperature in the reactor will rise causing the strips to deflect toward the channel center. Consequently, the fuel flow rate will be restricted, and, in turn, the heat generated in the combustion channel will drop, returning reactor temperature towards the nominal value.

Using a detailed model of a steam methane reforming CPR as a case study, we illustrate the considerations involved in the design of thermally-actuated control valves, and demonstrate via dynamic simulations their effectiveness for temperature control when the reactor is subject to flow maldistribution and operational disturbances.

References

[1] Zanfir, M.; Gavriilidis, A. Catalytic combustion assisted methane steam reforming in a catalytic plat reactor. *Chem. Eng. Sci.* **2003,** *58*, 3947-3960.

[2] Baldea, M.; Daoutidis, P. Dynamics and Control of Autothermal Reactors for the Production of Hydrogen. *Chem. Eng. Sci.* **2007,** *62*, 3218-3230.

[3] Zanfir, M.; Gavriilidis, A. Influence of flow arrangement in catalytic plate reactors for methane steam reforming. *Chem. Eng. Res. & Des.* **2004,** *82*, 252-258.

[4] Vaccaro, S.; Malangone, L.; Ciambelli, P. Modelling of a catalytic micro-reactor coupling endothermic methane reforming and combustion. *Int. J. of Chem. Reactor Eng.* **2010**, *8*.

[5] Pattison, R.C., Donahue, M.M., Gupta, A., Baldea, M. Localized Temperature Control in Microchannel Reactors Using Bimetallic Thermally-Actuated Valves, Ind. Eng. Chem. Res., in press

424226 Advanced Control Strategies Towards the Intensification of Monoclonal Antibody Production

Abstract:

Advanced control strategies towards the intensification of monoclonal antibody production

Maria M. Papathanasiou[a,d], Ana Quiroga Campano[a], Fabian Steinebach[b], Thomas Mueller-Spaeth[c], Massimo Morbidelli[b], Athanasios Mantalaris[a], Efstratios N. Pistikopoulos[d]*

[a]*Dept. of Chemical Engineering, Centre for Process Systems Engineering (CPSE), Imperial College London SW7 2AZ, London, U.K*

[b]*Institute for Chemical and Bioengineering, ETH Zurich, Wolfgang-Pauli-Str. 10/HCI F 129, CH-8093 Zurich, Switzerland*

[c]*ChromaCon AG, Technoparkstr. 1, CH-8005 Zurich, Switzerland*

[d]*Artie McFerrin Department of Chemical Engineering, Texas A&M University, College Station TX 77843*

**stratos@tamu.edu*

Keywords: process intensification, continuous processing, monoclonal antibodies, multi-parametric control

Process intensification plays a dominant role in the improvement of sustainability and productivity of biopharmaceutical processes. In particular, during the past few years there has been increasing demand on product quality and higher titers in the production of monoclonal antibodies (mAbs) [1]. The latter consists of two main parts: (1) the upstream processing (USP), where the cells are cultured and the therapeutic agent is produced and (2) the downstream processing (DSP) that involves the isolation/purification of the targeted product. Both the upstream and the downstream processing, however, are affected by various parameters related to the design of the bioreactor, the methods chosen for separation/purification as well as the operation mode (e.g. batch or continuous).

In order to develop environmental friendly, cost effective processes, current industrial trends suggest gradual shift from batch to continuous operation. Along with process intensification efforts, this can be facilitated by the use of advanced computational tools that will derive continuous operating conditions under minimum cost. In this work we present a complete computational study on the development of advanced control strategies for the production and purification of monoclonal antibodies. We follow the PAROC framework and software platform [2, 3] for the development of advanced multi-parametric control strategies that comprises four main steps: (i) development of a high-fidelity process model, (ii) approximation of the complex, process model into a linear state space representation, (iii) design of the multi-parametric controller, (iv) 'closed-loop', in-silico validation of the controller against the process model. The controller development is based on the derivation of multi-parametric Model Predictive Control (mp-MPC) policies that allow the solution of the optimization problem offline, improving the controller performance during online operation.

The examined system is described by (1) the upstream and (2) the downstream processing. The upstream process considers fed-batch culturing of GS-NS0 cells producing a chimeric IgG4 antibody. The mathematical model describing the system is based on multiple energy metabolic pathways and considers standard Monod kinetics. The model monitors the population of viable, early apoptotic and death cells, as well as the consumption/production profile of glucose and lactate, amino acids and the ATP balance. It comprises 16 differential and 23 algebraic equations, 41 variables and 45 parameters. The model is simulated in gPROMS® ModelBuilder v4.0.0 and the results are tested against experimental data. Due to its high complexity, the model cannot be used directly for the design of advanced controllers. It has to be therefore reduced to a linear state space model used for the formulation and solution of the multi-parametric programming problem (MATLAB®).

The purification of the upstream mixture (downstream processing) is performed using the Multicolumn Countercurrent Solvent Gradient Purification (MCSGP) process [4]. Similarly, using the model described by Krättli, et al. [5] we develop an advanced control strategy to ensure maximum recovery yield, while satisfying purity constraints [3, 6]. The 'in-silico' connection of the two systems is realized using the output of the upstream processing as a disturbance for the downstream system.

The results of this study form a solid basis for advanced "in-silico" experiments to determine the optimal operation conditions and evaluate the efficiency of the designed controllers. Additionally, such experiments will highlight the process bottlenecks and will provide the user with the necessary insight on the system dynamics, thus facilitating the realization of continuous monoclonal antibody production.

Acknowledgements

Financial support from the European Commission (OPTICO/G.A. No.280813) is gratefully acknowledged.

References

[1] P. Gronemeyer, R. Ditz, and J. Strube, "Trends in Upstream and Downstream Process Development for Antibody Manufacturing," *Bioengineering,* vol. 1, pp. 188-212, 2014.

[2] E. N. Pistikopoulos, "Perspectives in multiparametric programming and explicit model predictive control," *AIChE Journal,* vol. 55, pp. 1918-1925, 2009.

[3] E. N. Pistikopoulos, N. A. Diangelakis, R. Oberdieck, M. M. Papathanasiou, I. Nascu, and M. Sun, "PAROC

—An integrated framework and software platform for the optimisation and advanced model-based control of process systems," *Chemical Engineering Science,* p. In press.

[4] L. Aumann and M. Morbidelli, "A continuous multicolumn countercurrent solvent gradient purification (MCSGP) process," *Biotechnology and Bioengineering,* vol. 98, pp. 1043-1055, 2007.

[5] M. Krättli, F. Steinebach, and M. Morbidelli, "Online control of the twin-column countercurrent solvent gradient process for biochromatography," *Journal of Chromatography A,* vol. 1293, pp. 51-59, 6/7/ 2013.

[6] M. M. Papathanasiou, F. Steinebach, G. Stroehlein, T. Müller-Späth, I. Nascu, R. Oberdieck, *et al.*, "A control strategy for periodic systems - application to the twin-column MCSGP," in *Computer Aided Chemical Engineering*, ed, 2015, p. accepted for publication.

425075 Modeling and Analysis of Heat Recirculating Microreactor for Catalytic Combustion of Propane

Abstract:

Combustion in catalytic microreactors finds several applications due to the high energy density of hydrocarbon fuels. These include portable power generation, micro-propulsion, decentralised processing of fuels in thermally integrated micro-reactors, etc. Due to high surface area to volume ratio, high rates of catalytic reactions can be obtained compared to conventional devices. However, the large surface area also makes the system susceptible to extinction due to heat losses.

Heat recirculating geometries have shown to increase the range of stable operation of micro-reactors. The working principle for heat recirculating reactors is that the incoming cold reactant stream is preheated by transferring the excess enthalpy of the hot product stream. This involves carrying out reaction in a heat exchanger like assembly, to facilitate heat exchange between hot outlet and cold reactant streams. This can be achieved in simple geometries such as counter-current reactor, or in more complex Swiss-Roll geometries. Swiss-Roll geometry is especially popular in homogeneous and catalytically stabilised combustion in micro-burners.

In this work, we present numerical simulations for a new heat recirculating geometry. Specifically, we modify the Swiss Roll geometry to develop a heat recirculating micro-reactor with a spiral geometry. The inner walls are coated with catalyst. Thorough numerical simulations are presented for understanding the role of heat recirculation on thermal management and stability of the microreactor. The proposed geometry is simpler than the Swiss-Roll, while retaining the main advantage of larger area for heat transfer between the combustion channel and the incoming cold inlet stream. Coupling the micro-reactor with energy generation is also investigated using simple heuristics. We analyse the effect of geometry on stability of propane combustion in the micro-reactor, and compare the reactor with more traditional geometries.

436943 An Integrated, Multi-Stage, Multi-Scale Framework for Achieving Sustainable Process Synthesis-Intensification-Control

Abstract:

The chemical and biochemical industry needs major reductions in energy consumption, waste generation, etc., in order to remain competitive through the design and operation of more sustainable chemical and biochemical processes. These required reductions can be addressed through process synthesis-intensification-control, that is, the efficient use of raw materials (feedstock), the use of sustainable technologies and the design (and control) of processes that directly impact and improves sustainability/LCA factors. The unit operations concept, which has been sufficient until now, is one of the most used for performing process synthesis (and intensification) because it allows the association of tasks (functions) with the processing route to be followed. At the unit operations scale (Jaksland et al., 1995) and task scale (Siirola, 1996) alternatives are limited to existing (well-known) unit operations and therefore, may not be able to generate new integrations/combinations of intensified existing equipment.

However, to find innovative processes designs, extensions of the current concepts are necessary. Here, process synthesis-intensification using a phenomena based process synthesis method (Lutze et al., 2013) can play a major role because it provides the opportunity to perform the same tasks in a more sustainable way, new/novel unit operations can be generated (Lutze et al, 2013) and more sustainable processes can be designed (Babi et al., 2014).

An integrated, multi-stage, multi-scale, computer-aided framework has been developed in order to perform process synthesis-intensification-control. The framework operates at different scales, the unit operations scale, task scale and phenomena scale. In stage 1, process synthesis is performed (at the unit operations scale) using computer-aided flowsheet design (CAFD) (Tula et al, 2015), considering a superstructure of all possible alternatives based on known technologies, in order to generate a base case design. In stage 2, the base case design is decomposed into the smallest constituent units, that is, tasks then the involved phenomena. The base case design is analysed using economic and sustainability analyses in order to identify process limitations (hot-spots) that are translated into intensification design targets. In stage 3, an integrated task-phenomena-based synthesis-intensification method is embedded and applied (Babi et al., 2015) that consists of combining the phenomena to fulfil tasks, which are then translated into unit operations that constitute the (more sustainable) flowsheet alternatives which satisfy the intensification design targets. In this way, truly predictive and innovative solutions are generated much in the same way as atoms are combined to form molecules with desired properties (that is, analogous to computer-aided molecular design). The final stage involves validation and selection. Here detailed model-based calculations and/or experiments are performed to evaluate and compare the new solutions. In stage 4, design-control integration issues are validated/fine-tuned in order to generate the more sustainable controller structure. Note that in stages 1-3 design decisions are structured and made in such a way that the designed process(s) will also have the best opportunity for control (Mansouri et al., 2015).

In this presentation, the integrated process synthesis-intensification-control framework will be presented together with the corresponding databases, computer-aided models and tools needed to achieve sustainable synthesis-intensification. Different examples, related to synthesis, intensification and control will be presented for each stage of the framework.

References

1. Babi, D. K., Holtbruegge, J., Lutze, P., Gorak, A., Woodley, J. M., & Gani, R. (2015). Sustainable Process Synthesis-Intensification. Computers & Chemical Engineering.
2. Babi, D. K., Lutze, P., Woodley, J. M., & Gani, R. (2014). A Process Synthesis-Intensification Framework for the Development of Sustainable Membrane-based Operations. Chemical Engineering and Processing: Process Intensification, 86, 173–195.
3. Jaksland, C. A., Gani, R., & Lien, K. M. (1995). Separation process design and synthesis based on thermodynamic insights. Chemical Engineering Science, 50(3), 511–530.
4. Lutze, P., Babi, D. K., Woodley, J. M., & Gani, R. (2013). Phenomena Based Methodology for Process Synthesis Incorporating Process Intensification. Industrial & Engineering Chemistry Research, 52(22), 7127–7144.
5. Mansuori S, Sales Cruz M, Huusom JK, Gani R. In: Integrated process design and control of reactive distillation processes. Submitted for the International Symposium on Advanced Control of Chemical Processes – ADCHEM 2015; 2014
6. Siirola, J. J. (1996). Strategic process synthesis: Advances in the hierarchical approach. Computers & Chemical Engineering, 20, S1637–S1643.
7. Tula, A. K., Eden, M. R., & Gani, R. (2015). Process synthesis, design and analysis using a process-group contribution method. Computers & Chemical Engineering.

435211 On Quantification of the Attainable Region for Separator Networks with Varying Network Outlet Flowrates

Abstract:

In this work the process network attainable region (AR) concept is explored for separator networks in which the network outlet flow rates are allowed to vary. For separator unit models to which the IDEAS conceptual framework is applicable, it is shown that identification of AR boundary membership is equivalent to feasibility assessment of an infinite linear program (ILP). A number of important AR properties are then theoretically established, including AR convexity for fixed network inlet flowrate ratios, and representation of the AR in a process network outlet state space of reduced dimension. Finite dimensional approximations of the aforementioned ILP are then employed in creating increasingly accurate representations of the variable-flow-rate separator AR. A case study involving construction of the AR for the vapor-liquid equilibrium based separation of a ternary azeotropic mixture will be presented.

430169 Multi-Model Operability Approach As a Tool for Process Intensification: Application to a Membrane Reactor for Direct Methane Aromatization

Abstract:

This presentation introduces a novel multi-model operability-based approach for design and intensification of processes described by high-dimensional and nonlinear models. This novel approach is applied to membrane reactors (MRs) for the direct methane aromatization (DMA) conversion to benzene and hydrogen. In the last decades, MRs have emerged as examples of intensified processes as they combine reaction and separation unit operations in one process improving the system efficiency and lowering the cost[1,2,3]. Specifically, in the DMA-MR process, higher methane conversion than in conventional reactors is enabled due to the selective hydrogen removal from the reaction side through the membrane. This removal shifts the reaction equilibrium towards the products, thus increasing methane conversion and benzene production rate. The nonlinear and multivariable nature of the DMA-MR system associated with the reaction-transport phenomena interplay make this system challenging to design and optimize.

Process operability was developed as an approach for the design and control of complex chemical processes[4]. In particular, process operability enables the verification of a design's ability to achieve the feasible region of operation associated with process specifications, considering the physical limitations of the process design. Operability describes the relationship between the input (manipulated and/or disturbance) and output variables through a linear/nonlinear mapping obtained by employing the process model[5]. This approach has been applied to chemical processes that are described by high-dimensional linear models[6]. For the design of nonlinear systems, a low-dimensional operability implementation for the DMA-MR was analyzed using a subset of the system variables, including: membrane selectivity and permeance, reactor pressure and diameter, methane conversion and benzene production rate[7,8]. The results of this previous study demonstrated the promising capabilities of the operability approach to provide guidelines for experimental research, in terms of determining the set of membrane characteristics that should be obtained to achieve desired process specifications. However, the DMA-MR process intensification analysis using process operability and the full set of process variables has not been addressed, especially due to the limitation of current operability methods for nonlinear systems in terms of the problem size that they can address.

In this research, we first explore the extension of previously formulated operability algorithms for large-scale linear systems[9] to develop a multi-model operability approach for high-dimensional and nonlinear models. This approach employs sensitivity calculations and linear programing tools to quantify the output space as a function of the input space using the minimum number of linearized models that are necessary to adequately describe the nonlinear process operating envelope. The proposed approach is then applied as a tool for process intensification of complex high-dimensional systems. The performed operability calculations consider the multivariable nature of the DMA-MR process, by including the following variables: (i) reactor: diameter, length, temperature, pressure, and input flows; (ii) membrane: area, permeability, and selectivity; and (iii) economics: membrane and catalyst costs. Preliminary results on the application of the operability method as a tool for process intensification show reduction of the DMA-MR footprint (reactor volume and membrane surface area) for an equivalent level of performance. This case study indicates that operability can be a powerful tool for process intensification of membrane reactors and other complex chemical processes.

References

1. Stankiewicz A. I., Moulin, J. A. Process intensification: transforming chemical engineering. *Chem. Eng. Prog.* **2000**, 96(1), 22-34.
2. Drioli E., Stankiewicz A. I., Macedonio F. Membrane engineering in process intensification - An overview. *J. Membr. Sci.* **2011**, 380(1–2), 1–8.
3. Brunetti A., Caravella A., Drioli E., Barbieri G. Process intensification by membrane reactors: high-temperature water gas shift reaction as single stage for syngas upgrading. *Chem. Eng. Technol.* **2012**, 35(7), 1238–1248.
4. Vinson D. R., Georgakis C. A new measure of process output controllability. *J. Proc. Cont.* **2000**, 10(2-3), 185-194.
5. Georgakis C., Uzturk, D., Subramanian S., Vinson D. R. On the operability of continuous processes. *Control Eng. Prac.* **2003**, 11 (8), 859–869.
6. Lima F. V., Georgakis C. Design of output constraints for model-based non-square controllers using interval operability. *J. Proc. Cont.* **2008**, 18(6):610-620.
7. Carrasco J. C., Lima F. V. Nonlinear operability of a membrane reactor for direct methane aromatization. *In Proceedings of the 2015 IFAC ADCHEM Symposium*, British Columbia, Canada, June **2015**.
8. Carrasco J. C., Liu D., Lima, F. V. Modeling and nonlinear operability analysis of a membrane reactor for direct methane aromatization. *Presented at 2014 AIChE Annual Meeting*, Atlanta, Georgia, November **2014**.
9. Lima F. V., Georgakis C. Input-output operability of control systems: the steady-state case. *J. Proc. Cont.* **2010**, 20(6):769-776.

425808 Synthesis of Azeotropic Reactive Distillation Networks Involving Quaternary Mixtures

Abstract:

The synthesis of an isobaric reactive distillation network involving a quaternary azeotropic mixture is investigated in this work. The network is comprised of vapor-liquid equilibrium trays, with liquid holdup that changes from tray to tray. Kinetically and/or equilibrium limited reactions are assumed to occur in the liquid phase.

The Infinite DimEnsionAl State-Space (IDEAS) conceptual framework, is first shown to be applicable to the problem, and is then employed to formulate an infinite linear program (ILP) that can rigorously assess the feasibility of the considered process synthesis problem. The formulated IDEAS ILP's infimum value is approximated by the sequence of optimal values of a number of finite dimensional linear programs (FLP) of ever-increasing size. The method is demonstrated on a case study involving the isobaric reactive distillation of an acetic acid/isopropanol/isopropyl acetate/water azeotropic mixture.

437111 Comparison of Life-Cycle Carbon Footprint of Fischer-Tropsch Diesel Versus Ultra-Low Sulfur Diesel

Abstract:

The increasing availability and competitive prices of shale/natural gas in the United States are spurring much interest in monetization options. One of these options is gas-to-liquids (GTL) technology to produce transportation fuels. Comparisons are commonly made between petroleum-based fuels and Fischer-Tropsch diesel (FTD) produced via GTL based on technical performance and economics. Given the growing interest in environmental and sustainability issues, there is a need to also consider and compare the greenhouse gas (GHG) emissions associated with both petroleum- and GTL-based fuels. In this work, we present a well-to-wheel life cycle assessment (LCA) for GTL diesel for a hypothetical commercial-scale facility to be located in Texas using pipeline-quality natural gas and an LCA for Ultra Low Sulfur Diesel (ULSD) produced at a refinery along the Gulf Coast from crude oil imported from offshore countries. An approach is proposed for determining and allocating GHG for the various steps of the LCA. The appropriate design and environmental tools are discussed and evaluated. We explore the various building blocks from raw material acquisition and transport, conversion, fuel transport and end use using the life cycle framework to compare GTL fuels to ULSD.

432145 Assessment of the Sustainability-Potential of Adipic Acid Processes

Abstract:

The concept of sustainability-potential has been demonstrated to be a valuable tool for considering alternate chemical process schemes. This hierarchical approach is a generalization of the notion of gross profit potential (GPP). As such, it permits assessment of the technoeconomic, environmental, and societal impacts of a process in the early stages of its design. In the present work, the sustainability-potential of adipic acid production from a biorenewable feedstock will be compared with that of conventional adipic acid processes based upon petroleum feedstocks. Consideration of the gross available exergy dissipation for the conventional route (-983 kJ/mole) is much more negative than that for the biorenewable approach (-146 kJ/mole). As such the biorenewable approach is more sustainable from a thermodynamic standpoint. The gross profit potential shows the opposite trend. The GPP for the biorenewable approach is quite low while the conventional adipic acid process has a positive GPP. Arriving at a toxicity index is a much more subjective process. Using measured and estimated values for the fathead minnow LC_{50} as a toxicity index, the two processes are roughly equivalent. Alternate indices for environmental and health considerations will be discussed.

432869 Resource and Techno-Economic Analysis of Biogas from WWTP (WasteWater Treatment Plants) Biosolids Using Anaerobic Digestion

Abstract:

Waste is an under-utilized biomass resource and is high-impact renewable feedstock with collection infrastructure already in place. Thus, Waste-to-energy (WTE) technologies can position at near-term market entry point with wide spectrum of end products (fuels, chemicals and power) while mitigating methane emissions from current waste treatment. Among various wastes, municipal solid waste (MSW) and biosolids have potential to produce about 590 TBtu (equivalent to 5 billion gallons of gasoline or 1/3 of current biofuel production) per year. In the other hand, the United State of environmental protection agency (EPA) has recognized the benefits of promoting net low-carbon fuels derived from biogas, and in a recent rulemaking, EPA classified many sources of biogas from cellulosic feedstock for transportation fuels as part of the Renewable Fuel Standard (RFS) [1]. Use of biogas-derived fuels in the transportation sector can substantially reduce GHG emissions and can serve to promote effective organic waste management, as well as efficient biogas production, recovery, and utilization. There are many existing treatment processes for converting municipal sludge into biosolids, which are most landfilled but can be applied to land typically as a soil amendment (US EPA, 40 C.F.R. § 503. 10, 2008). According to EPA 2008 data and the EPA Clean Watersheds Needs Survey [2, 3], 16,675 WWTPs are treating 34,211 million gallons per day (MGD) of wastewater. Economic viability and environmental sustainability, however, have yet to be addressed for widespread deployment of WTE pathways for better utilization of biosolids. Thus, for economic viability and environmental sustainability, the techno-economic analysis (TEA) of WTE pathways are performed for converting biosolids anaerobic digestion then upgrading the biogas to various energy products, such as power and liquid fuel.

This work is to determine whether waste resources are large enough to justify the economies of scale and whether WTE pathways can increase the value proposition of the waste feedstocks. The criteria are whether waste resources within a practical transportation radius are large enough for, at a minimum, a small-scale biorefinery (e.g., 10–100 million gallons per year) and whether the investigated WTE pathways have the potential to increase the value proposition. Several alternative process concepts are defined and studied for TEA. Due to uncertainties in the data available for WTE, it is difficult to support a single economic target/potential. Reasonable cost ranges are established for base case as well as process alternatives. In addition to further investigate and update key TEA assumptions and to examine process sensitivity and alternatives to increase waste value proposition. Other more advanced technologies should be considered in the future to understand the economic viability of the conversion technologies using biogas from biosolids of WWTPs, particularly for pathways to make transportation fuels. For instance, methanol-to-gasoline, methanol-to-olefins, syngas catalytic upgrading pathways (syngas-to-oxygenates, syngas-to-mixed alcohols, syngas to triptyls), and methane biological pathways could have promising potential to meet cost and sustainability criteria set by this study. To meet production criteria, more feedstocks would be needed, either by increasing biosolids using centralized options or by blending biosolids with other types of waste, such as food waste.

References:

1. EPA Regulatory Announcement. EPA Issues Final Rule for Renewable Fuel Standard (RFS) Pathways II and Modifications to the RFS Program, Ultra Low Sulfur Diesel Requirements, and E15 Misfueling Mitigation Requirements. 2014.

2. EPA. Clean Watersheds Needs Survey (CWNS) 2008 Report to Congress. 2008.

3. USEPA. Clean Watersheds Needs Survey 2008 - Report to Congress. Washington, DC: US Environmental Protection Agency, 2010.

434905 Optimization of Two-Dimensional Hoist Scheduling and Production Line Design for Multi-Stage Material Handling

Abstract:

Multi-stage material handling (MSMH) process is widely applied in industries to manufacture large amount of various workpieces and products (jobs). Generally, a hoist will handle all jobs by following a preset movement schedule, including job lifting, releasing and moving along a production line. Different scheduling can cause huge difference in production period. Therefore, lots of engineers have studied the feasibility and optimization of hoist movement schedules. However, the design of production line, such as the spatial allocation of every processing unit, could also affect the production rate. During the production, hoist will travel forth and back to pick up job from its current unit and release it into the next one, following its processing requirement. In this way, a proper arrangement of unit position would dramatically shorten the total moving time and improve the processing capability.

In this paper, a new two-dimensional (2D) matrix arrangement of processing unit has been proposed to replace original one-dimensional (1D) line arrangement for the optimization of productivity, and a mixed integer linear programming (MILP) model has been developed to solve this problem. The major advantage of this study is that the compacted allocation provides shorter paths for hoist to move, so the productivity could be improved significantly. The efficacy of the proposed methodology is demonstrated by a case study. The comparisons between 1D and 2D production line arrangements have also been conducted.

430715 Tracking of Flare Streams Emissions and Tax Implication during Process Abnormal Situations

Abstract:

Upset flaring is the controlled burning of gas stream that cannot be processed due to a disturbance within the steady state process which leads to abnormal situation. The main concern with flaring is its contribution to global climate change due to GHG formation during the combustion process. Global Gas Flaring Reduction (GGFR) public-private partnership reported that flaring gas emits about 390 million tons of CO2 annually. Countries who have signed on with GGFR and other environmental regulatory authorities are concerned with reducing the GHG emissions footprint that results from flaring. Hence there is a need for, a standard, reliable and consistent approach to quantify GHG emissions. These values will be used to report to regulatory agencies; and furthermore such estimation methods will be very useful in assessing potential flare reduction mitigation methods. The aim of this work is to illustrate systematic procedure for estimating Carbon dioxide (CO2) and Carbon dioxide equivalent (CO2-e) of process upset flares based on standard published Global Warming Potential (GWP) values. The merit of the CO2 calculator is that it has the ability to estimate CO2 emissions for different process upset scenarios while each scenario has multiple known flare sources. As a part of carbon emissions reduction motivation, this CO2 calculator is able to estimate the carbon tax and credit using the latest carbon tax rates at various focal countries around the globe. The calculator is developed as part of an ongoing project that studies and develops mitigation methods for flare reduction during process upsets. In this paper the base case ethylene process is studied and simulated. The process prevalent flare sources during normal and abnormal (upset) operations are highlighted. The calculator is used to assess the process emissions during normal and during abnormal situations; in addition, the calculator was integrated with an optimization framework used to asses the impact of two proposed mitigation techniques for flare utilization. The calculator was useful in assessing both the environmental and economic implications of the studies cases.

436589 Mass Transfer Scale-up for Semi-Batch Polymer Production

Abstract:

For many polymerization processes it is of funamental importance to know the concentration of monomer at the polymerization site to determine, and tailor, the final properties of the polymer. This is particularly true is such processes in which three phases are involved, and the monomer needs to transfer from a gas phase to a liquid phase to the solid phase.

In this work, different approaches have been used to determine the concentration of the monomer in the diluent phase of the reaction, as a consequence of the existing mass transfer in the particular batch reactor of use. Namely, laboratory and pilot equipment of different sizes have been used to experimentally determine the gas-liquid mass transfer coefficient, and its dependency on different process parameters and geometry configurations. Based on literature relationships, the same mass transfer has been extrapolated to plant production conditions, and verified as well by meas of CFD simulations.

The results of the measurement and simulations at different scales have been put together to understand the governing factors in the reactor, and to be used for process and product development purposed.

It has been seen that the geometrical configuration of the different scale reactors if of fundamental importance to generate the right monomer concentration conditions, as well as to define and optimize the batch plant productivity.

423516 Design of a Large-Scale Micro-Channel Fischer-Tropsch Reactor: Model Development for Barrels-per-Day-Scale Reactor

Abstract:

In this study, a large-scale micro-channel Fischer-Tropsch (FT) reactor was designed. Based on 1-barrel per day of wax production rate, a model that solves momentum, material, and heat balance equations simultaneously was developed. Approximately 5,000 channels-1,000 for process and 4,000 for cooling channel-are decomposed into each unit cell, and every neighboring cells are then coupled together to solve the balance equations. The realistic flow distribution effect was incorporated into the model by using results obtained from computational fluid dynamics (CFD). The effect of several important variables, i.e., catalyst loading, coolant flow rate, channel layout, flow configurations, flow distribution, are clarified through the relevent case studies. The results of the model validation work, and some realistic issues that arises in the fabrication of the large scale reactor are presented as well.

435539 Effect of Friction Factor Correlations and Propagation Errors on Differential Pressure in a Crude OIL Fouling Measuring RIG

Abstract:

EFFECT OF FRICTION FACTOR CORRELATIONS AND PROPAGATION ERRORS ON DIFFERENTIAL PRESSURE IN A CRUDE OIL FOULING MEASURING RIG

Z. Tajudin[+], *E.Diaz-Bejarano*[+], *F. Coletti**, *S. Macchietto*[+]* *and G. F. Hewitt*[+]*

+Department of Chemical Engineering, Imperial College London, South Kensington Campus, London SW7 2AZ, UK

** Hexxcell Ltd, Imperial College Incubator, Bessemer Building Level 2, Imperial College London, London SW7 2AZ, UK*

Corresponding author: s.macchietto@imperial.ac.uk

Keywords: *pilot scale, error of propagations, experiment, validation*

Abstract

In order to detect crude oil fouling experimentally, primary measurements of differential pressure and temperatures must be obtained with high fidelity, accuracy and reproducibility at (or close to) industrial conditions. Information of the thermal and hydraulic effects of fouling can be studied by using robust models to decouple the various phenomena involved. To start with, it is important to have a reliable set of primary measurements in which the robust model could be validated against the experiment data.

A novel pilot scale experimental facility for study of crude oil fouling, High Pressure Oil Rig (HIPOR), was developed and tested with non-fouling oil and crude oil at different operating conditions. The experiments have been carefully designed in order to develop comprehensive and reliable baseline validation for both oils without any fouling deposition. The model, an extension of Hexxcell's model for HiPOR, for the tubular test section with vertical geometries, considers the two dimensional (axial and radial) heat transfer in different domains, boundary conditions and other aspects (for example, heat losses) which are carefully determined.

The objectives of this work are divided into two. First, a predictive model for differential pressure of the HIPOR test section is validated against experimental results for both oils. The model consists three friction factors correlation was used to assist in the analysis and interpretation of the experimental data collected, compared with the variation of flowrate as well as different heat inputs. Second, the errors of propagation which are observed from the validation process have been identified and its sensitivity towards inlet bulk temperature increments quantified.

The validation process shows that both oils show excellent agreement with model predictions for at all ranges of operating conditions and capturing all responses including the transient processes. The best agreement is achieved for different friction factor correlations for each oil. The reported results also show that the deviation between simulation data and measured data due to the errors of propagation starts to increase when the inlet bulk temperature of oil is increased from 80 °C onwards. Furthermore, the errors of propagation have a linear correlation with the temperature increment.

It is concluded that the model of the HIPOR rig for non-fouling and crude oil conditions for differential pressure were successfully validated and error of propagations with temperature effects have been quantified.

Acknowledgments

This research was partially performed under the UNIHEAT project. The authors wish to acknowledge the Skolkovo Foundation and BP for financial support. ZT wishes to also acknowledge MARA and UniKL for their support and funding. The support of Hexxcell Ltd, through provision of Hexxcell Studio™, is also acknowledged.

References

S. Macchietto, G.F. Hewitt, F. Coletti, B.D. Crittenden, D.R. Dugwell, A. Galindo, G. Jackson, R. Kandiyoti, S.G. Kazarian, P.F. Luckham, O.K. Matar, M. Millan-Agorio, E.A. Müller, W. Paterson, S.J. Pugh, S.M. Richardson, D.I. Wilson, "Fouling in crude oil preheat trains: a systematic solution to an old problem", *Heat Transfer Engineering*, **32**, 3, 197-215 (2011).

F. Coletti and S. Macchietto, A dynamic, distributed model of shell-and-tube heat exchangers undergoing crude oil fouling. *Ind. Eng. Chem. Res*. **50** (8), pp 4515–4533 (2011).

Z. Tajudin, Experiments, Modelling and Validation of Crude Oil Fouling on Large Scale Rig. PhD, *Imperial College London*, UK – submitted for PhD awards (2015).

Hexxcell Ltd., 2015. Hexxcell Studio. http://www.hexxcell.com.

411010 Redox Properties of Food-Derived Polyphenols

Abstract:

Polyflavanols (procyanidins), anthocyanins, and phenolic acids are known naturally-occurring antioxidants. Polyphenolic structures of these antioxidants enable one-electron transfer reactions to quench radicals; the reaction is reversible if the antioxidant forms stable semiquinone radicals. Because of its aromatic, polyphenolic structures, procyanidin extracted from sorghum reportedly have several-fold greater peroxyl radical-quenching ability than phenols and Trolox. Limited information is available on the chemical constituents responsible for the antioxidant activity of sugarcane extracts. For example, kokuto, "black sugar" in Japanese, is produced by evaporating water from sugarcane juice at elevated temperature. Like sugarcane molasses, kokuto contains phenolic acids, phenyl glycosides, phenyl propanoid glycosides, and phenyl propanoids. The objective of this study was to understand the relationships between the structure (by solid-state NMR and fluorescence/UV-visible spectrophotometry) and redox property of naturally occurring antioxidants: grape seed, evaporated sugarcane juice, and sorghum grain. Fully oxidized and reduced forms of sugarcane, sorghum, and grape seed extracts were prepared using Fremy's salt. Solid and liquid-state NMR techniques were then utilized to investigate the structural components responsible for the redox reactivity. Iron (III) reduction kinetics was examined, and compared with DPPH, acid-butanol, and vanillin assays.

429113 Hybrid Cell Culture Models for Efficient Process Development

Abstract:

Cell culture processes are dynamic systems governed by a large number of influential and highly interacting process variables. Despite the stringent control of several process variables, considerable variation in the process outcome is often observed. Moreover, generalized deterministic models for dynamic process prediction based on the impact of the process variables on the product quantity and quality attributes do not exist. According to the Process Analytical Technology (PAT) initiative, multivariate data acquisition and analysis should be implemented to achieve advanced process knowledge and to gain effective control on risk in manufacturing. While several studies have shown that such multivariate statistical models provide promising possibilities for process monitoring and prediction based on the process history, they are very limited regarding their capability for the support of experimental design.

This work demonstrates how the flexibility of statistical models combined with the formalized bioengineering knowledge of simplified deterministic models can improve the efficiency in cell culture process development. Based on a rigorous case study, the results obtained with the hybrid approach will be compared to the ones from the purely statistical and deterministic models to visualize its promising capability of providing significant process information using the least number of experiments. Different kinds of hybrid cell culture models will be discussed and the possibility to incorporate such models into a sequential experimental design methodology to define the process design space will be presented.

420830 The Impact of Glycation on Protein Color

Abstract:

Health Authorities expect that color of liquid formulations of monoclonal antibodies be monitored and controlled. Subcutaneous drug formulations often require high drug substance concentrations which may result in more intense product color. A potential mechanism of coloration is through the Maillard reaction, where reducing sugars in the cell culture media react with the secreted protein product in a series of non-enzymatic reactions that result in formation of glycated protein and advanced glycation end products (AGEs). We varied the glucose feeding strategy and cell culture media across multiple cell lines to study and delineate the effects. Presence of AGEs was also measured using mass spectrometry. We observed good correlation between the glucose fed to the culture and the level of the glycated protein. However, the glucose feeding strategy tested did not significantly impact color, and we observed no strong correlation between glycated protein and color. Changes in the levels of the AGEs for different culture conditions will also be discussed.

433606 Increase Cell Culture Development Value Using the Modular Automated Sampling Technology (MASTTM) Platform in an Integrated Bioprocess Lab Environment

Abstract:

Biopharmaceutical companies are continually striving to fundamentally understand and optimize cellular performance within their bioreactors. While on-line process analytical technology (PAT) tools like dielectric and Raman spectroscopy are helping to provide insight into cell processes that impact titer and antibody quality, the integration of these data with off-line measurements such as cell density, viability, and titer remains an elusive goal.

Bend Research Inc., in collaboration with Pfizer Inc. and other major biopharmaceutical companies, is advancing the Modular Automated Sampling Technology (MAST^{TM}) platform and integrating this system with automated analytical and data retrieval systems. This approach allows application of novel experimental methods providing greater insight into the bioreactor environment. In this process, aseptically collected bioreactor samples are delivered to multiple analytical devices with resulting data being automatically retrieved and curated. This data is presented at a graphical user interface (GUI) for real-time data analysis and predictive model development. Early studies with the MAST sample delivery system coupled with our data-processing software increased system observability and real-time process understanding by the end user.

This presentation describes how Bend Research has used the MAST platform as a cornerstone of an integrated lab environment. We will describe how the MAST system works and the features that set it apart from other autosampling solutions. Applications will be presented where MAST facilitated the use novel development approaches yielding enhance process understanding

From this new level of insight into the bioreactor, scientists can obtain enhanced data-driven guidance for key activities like optimizing process operation. With this new technology, the bioprocess industry can make major advances toward advanced real-time testing, predictive control, and overall enhanced bioprocess design and operation.

428887 Application of Raman Spectroscopy in Cell Culture Process Development

Abstract:

In recent years, the European Medicines Agency (EMA) and the United States Food and Drug Administration (FDA) have stressed the need of greater process understanding and control in the area of pharmaceutical development, manufacturing and quality assurance, all which is summarized in the Process Analytical Technology (PAT) initiative. Therefore, in bioprocesses, and particularly in cell cultures, multivariate data acquisition and analysis are likely to provide versatile and significant real-time information. Lately, Raman spectroscopy has been characterized as a promising tool to detect multiple process variables simultaneously.

This work presents several cases studies, where Raman spectroscopy was successfully applied for the quantification of significant variables in monoclonal antibody (mAb) cell culture processes. These studies include online monitoring of fed-batch cultures, offline analysis of (partially spiked) cell culture samples as well as the utilization in perfusion cell culture processes. Besides the evaluation of those applications of Raman spectroscopy, the central role of multivariate spectra processing techniques including pretreatment, spectral range selection and data reduction will be discussed with regards to robust calibration and prediction of the components of interest. Finally, the role of those spectral data with regards to overall process prediction, monitoring and control will be highlighted within the scope of all acquired online and offline data in bioprocesses.

411442 Density Surrogate for Fine Tuning Dissolution of an Extended Release Tablet to Increase Manufacturing Process Robustness

Abstract:

Tablet dissolution can be affected by several factors including raw materials, formulation and processing conditions. For a tablet with a fixed formulation, such as that of a commercial drug product, there remain few levers by which to improve tablet critical quality attributes (CQA). Tablet dissolution is one such known critical quality attribute to be affected by tablet hardness, which can be controlled by varying the compression force, typically in a directly proportional relationship. Beyond a certain threshold of compression force value, however, hardness reaches a maximum value and plateaus. Continuing to increase compression force after this threshold value could lead to loss in tablet cohesion due to high internal stresses or eventually tooling and equipment damage. Near the plateau threshold, there is a space over which tablet hardness may not adequately capture the tablet properties controlling its dissolution. This work explores this space for a polymer matrix extended release tablet where improved dissolution release robustness is desired. A parameter of weight/thickness ratio is suggested as a density surrogate to fine tune dissolution performance near and within the hardness plateau region.

429396 Greenscope Tool for Process Sustainability Assessment and Life Cycle Inventory Generation

Abstract:

The development of sustainable solutions for addressing the satisfaction of the human population, its material and energy needs, and simultaneously preserving the Earth systems that make human life possible has been an objective of researchers working in academia, government, and industry for decades. Commonly, new chemicals and their manufacturing processes are developed, however, some evaluation and analyses are needed for supporting the classification of these new contributions as sustainable alternatives. This contribution demonstrates the application of the U.S. Environmental Protection Agency's software tool, GREENSCOPE, for the assessment of process sustainability and life cycle inventory generation. GREENSCOPE is a web-based tool for evaluating sustainability of existing and new technologies by employing performance indicators for objectives in four main areas: environment, economics, energy, and efficiency. A case study demonstration using GREENSCOPE will be performed to provide users guidance on using this assessment and design tool, from data entries to sustainability evaluation outcomes and the related life cycle inventory generation. The indicator scores describe product and process performance in a sustainability measurement scale, determining whether chemical products and processes are more or less sustainable.

412272 LCA Needs to Move Toward Sustainability to Remain Relevant-

Abstract:

LCA, a methodology amply utilized to environmentally assess product systems under the requirements of ISO LCA standards, has been used for the last decade or so with important constraints or limitations. Among the most important of such limitations are specifically its use in comparing product systems and in its use as a predictor of sustainability. Albeit the issues around its capacity to accurately and fairly compare product systems have been amply discussed and the original standard revised or modified in those sections having to do with comparative assertions, etc. still. limitations as a predictive tool for environmental preference among product systems persist. However, the LCA standard capability to address sustainability issues has not been examined in the same depth and extent. Nevertheless, the demand of society for a proper indicator of sustainability of different product systems has resulted in the improper suggestion, and even use, of inadequate methodologies. LCA, because its popularity, has been considered one such a metric. Another, purely arithmetic reduction efficiency, has and its being used as a pseudo sustainability metric. In both cited approaches, the lack of context and thresholds invalidate its valid use as sustainability metrics. The paper analyzes the fundamental characteristics of LCA methodology in ISO standards vis a vis the needed characteristics for a proper sustainability metric for product systems. The results of the analysis clearly indicate that LCA study results are not proper metrics to evaluate sustainability of products systems. A more proper and effective use of LCA would be as a preliminary analysis of the product system from which to focus in more important elements or indicators of the LCA study analysis. A concerted effort should be initiated by academy and practice to incorporate needed thresholds or pseudo-thresholds, in the methodology and practice of LCA. As now, in general, the results of a LCA study, by themselves, are improper metrics for sustainability.

436855 Sustainability Metrics for Charging Infrastructure for Electric Vehicles

Abstract:

Sustainability metrics have value in life cycle analysis and the evaluation of alternatives. One approach to expanding the infrastructure for electric vehicles (EVs) is to install solar powered charging stations with solar panels above parking spaces to generate electricity and provide shade for cars parked under the panels. Grid connected systems with charging stations enable electricity to flow into the grid or into the batteries of the EVs. Social, environmental, and economic metrics are included in the analysis of benefits to society of an infrastructure of solar powered charge stations. Social metrics include the social value of shade, convenience, and availability, and better air quality. Environmental metrics include greenhouse gas emissions, air quality, water use, and land requirements. Economic metrics include capital costs, EV charging cost, and cost of transportation. Several alternatives will be compared.

NREL
NATIONAL RENEWABLE ENERGY LABORATORY

Combined Sustainability Assessment and Techno-Economic Analysis for the Production of Biomass-Derived High-Octane Gasoline Blendstock

Eric C.D. Tan

National Renewable Energy Laboratory (NREL) Golden, CO

November 12, 2015

Session: Sustainability Metrics at the Process and Product Level

2015 AIChE Annual Meeting, Salt Lake City, UT

NREL is a national laboratory of the U.S. Department of Energy, Office of Energy Efficiency and Renewable Energy, operated by the Alliance for Sustainable Energy, LLC.

Publications

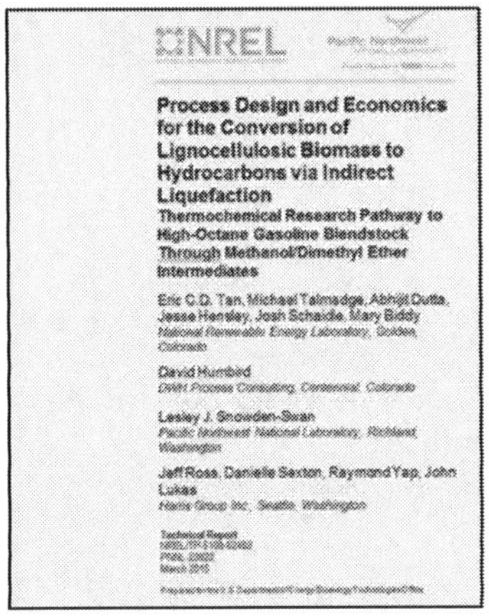

http://www.nrel.gov/docs/fy15osti/62402.pdf

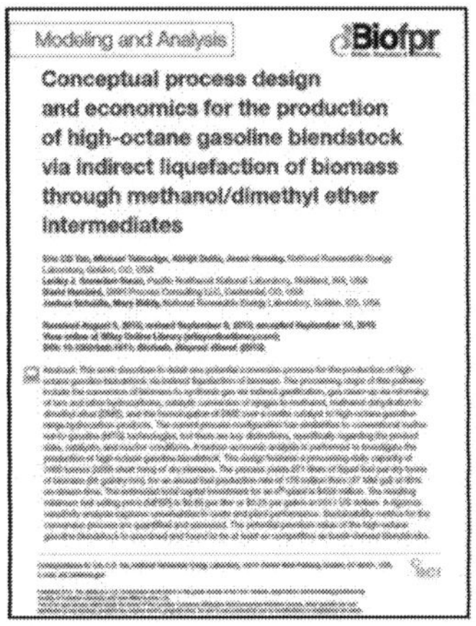

http://onlinelibrary.wiley.com/doi/10.1002/bbb.1611/abstract

http://www.energy.gov/sites/prod/files/2015/03/f20/mypp_beto_march2015.pdf

Highlights of this IDL conversion pathway

- Produces a high-octane, gasoline-range (C5+), hydrocarbon product.
- Leverages advances in syngas production and clean-up technologies developed for the thermochemical pathway from biomass to mixed alcohols, demonstrated in 2012.
- Operates at lower severity conditions (lower operating temperature and pressure) relative to traditional methanol-to-gasoline (MTG) processes.
- Combines homologation (-CH2- addition) and conversion of recycled C4-olefins and iso-paraffins to larger products in a single reactor system and does not require a separate alkylation step.
- Possesses high selectivity to C7 (mostly 2,2,3-trimethylbutane and 2,2,3-trimethyl-but-1-ene) and C4 paraffins and olefins, which can be recycled to maximize yield of gasoline blendstock product.
- Produces low aromatic-content product and has a low coke formation rate.

High-Octane Gasoline Pathway vs. MTG

Process Attribute	High-Octane Gasoline Pathway Target	Methanol to Gasoline (MTG) Pathway	Impact on Techno-Economic Analysis
Molecular structures favored in synthesis reactions	Branched paraffins 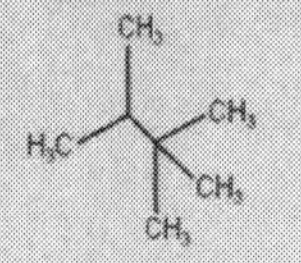	Aromatics 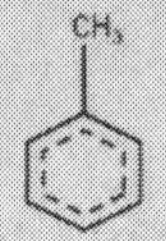	High octane product rich in branched paraffins, similar to a refinery alkylate. H-saturation decreases density, increasing product volume.
Example Compound Specific Gravity	Triptane 0.70	Toluene 0.87	
Hydrocarbon synthesis catalyst	Beta-Zeolite (12-membered rings)	ZSM-5 (10-membered rings)	Different pore sizes and structures result in different compound selectivities.
Octane number of gasoline-range product	RON: 95+ MON: 90+	RON: 92 MON: 83	Octane number increases value of product as a finished fuel blendstock.
Selectivity of C_5+product	C_5+ product only (~65 Gal / Ton)	~ 85% C_5+ (~55 Gal / Ton)	High selectivity to primary (premium quality) product maximizes overall product value.
Severity of synthesis operating conditions	350 – 450 Deg. F 130 PSIA	650 – 950 Deg. F 315 PSIA	The lower severity operating conditions result in lower capital and operating costs relative to MTG.
Coke formation	Coke formation is minimized by hydrogen addition and selectivity to branched paraffins rather than aromatics.	High propensity for coke formation due to aromatic coke pre-cursors.	Minimizing coke formation helps to maximize product yield / carbon efficiency and maximizes catalyst regeneration and replacement cycles.

Sources: (1) 2011 NREL MTG Design Report (Phillips et al), (2) Methanol to Gasoline (MTG) Production of Clean Gasoline from Coal (ExxonMobil Research and Engineering)

Simplified Process Flow Diagram

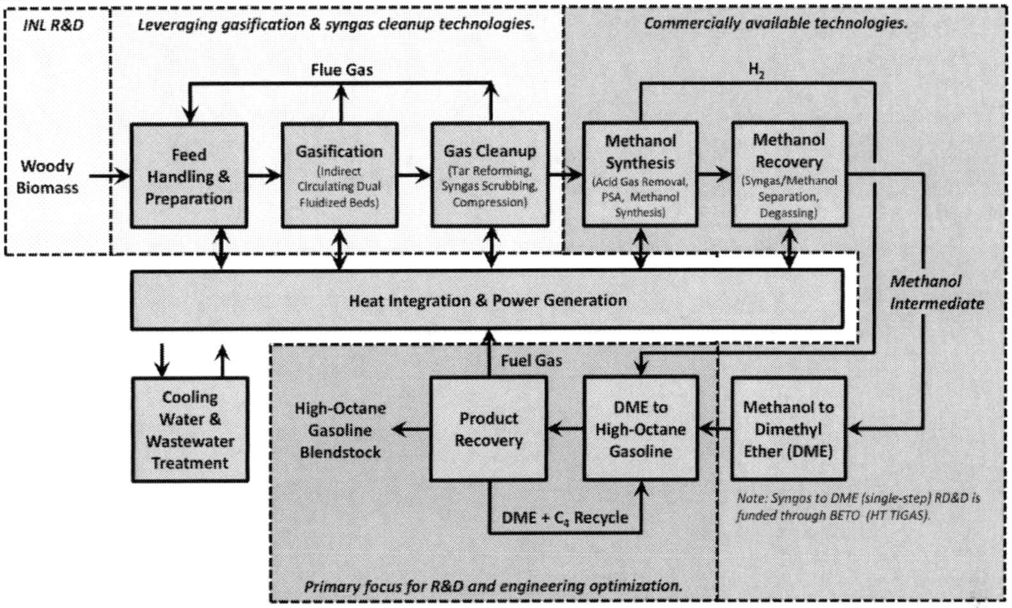

Gasification and Gas Clean-up

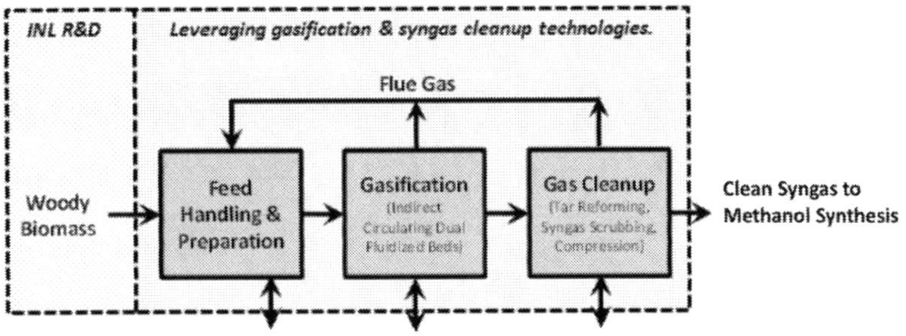

Indirect Gasifier

- INL woody feedstock blend (<1% ash, <10% moisture). Waste heat required only to preheat the feedstock
- Consistent with 2012 mixed alcohol design / demonstration

  Dutta, A., et al. "Technoeconomic Analysis for the Production of Mixed Alcohols via Indirect Gasification of Biomass Based on Demonstration Experiments," Industrial & Engineering Chemistry Research, 53(30), 12149-12159 (2014).

Gas Cleanup

- Tar reforming (based on 2012 mixed alcohol design / demonstration)
- Scrubbing to remove residual fines

Acid Gas Removal & Methanol Synthesis

Acid Gas Removal

- Combined AGR and LO-CAT systems
- Remove CO2 and H2S
- ZnO sulfur guard bed is required

Methanol Synthesis

- Syngas further conditioned to meet CO_2 (5 mol%) and H_2S (0 ppm) specifications
- Methanol conversion based on commercial technologies
- Fixed-bed shell & tube reactor containing a copper/zinc oxide/alumina catalyst
- Operating conditions: 500 F, 730 psia
- 95 wt% of the unconverted syngas is recycled back to the synthesis reactor
- Portion of syngas is sent through PSA to recover hydrogen for use downstream

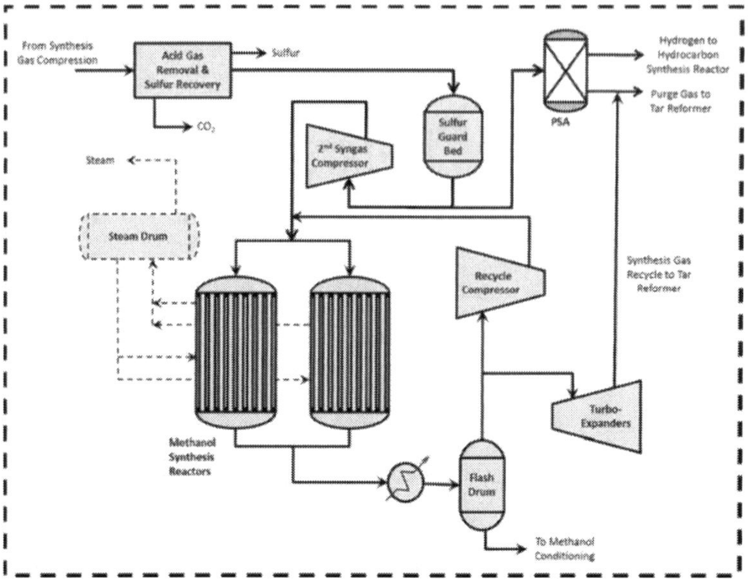

DME to High-Octane Gasoline

DME Synthesis

- Vapor-phase dehydration of methanol to DME on gamma alumina catalyst at 480 F and 140 psia.

Hydrocarbon Synthesis

- DME is converted to gasoline-range hydrocarbons over 3 parallel fixed bed reactor trains (with inter-bed cooling) containing metal-modified beta-zeolite catalyst.
- Operating conditions: 438 F, 130 psia.

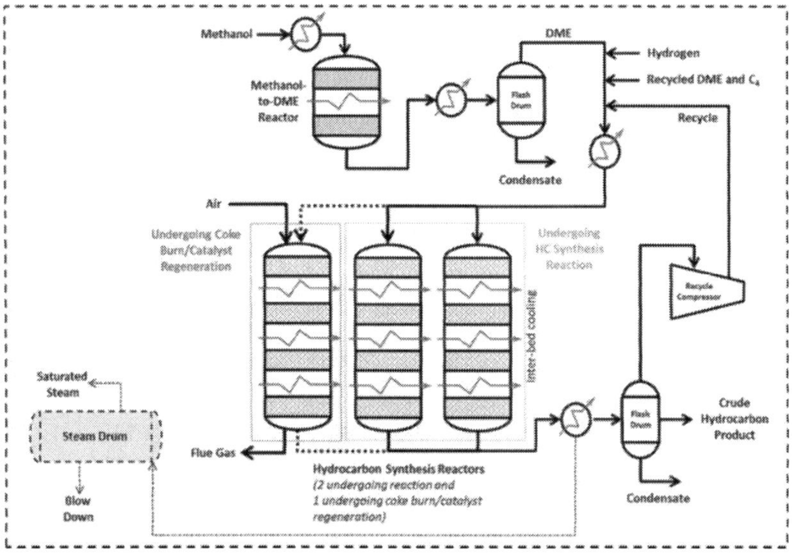

- DME conversion: 40% (single-pass); 93% (overall).
- H2 is introduced to reduce aromatics formation and C4s are recycled to improve yields.
- Potential for dimerization step to produce jet-range molecules.

Product Recovery

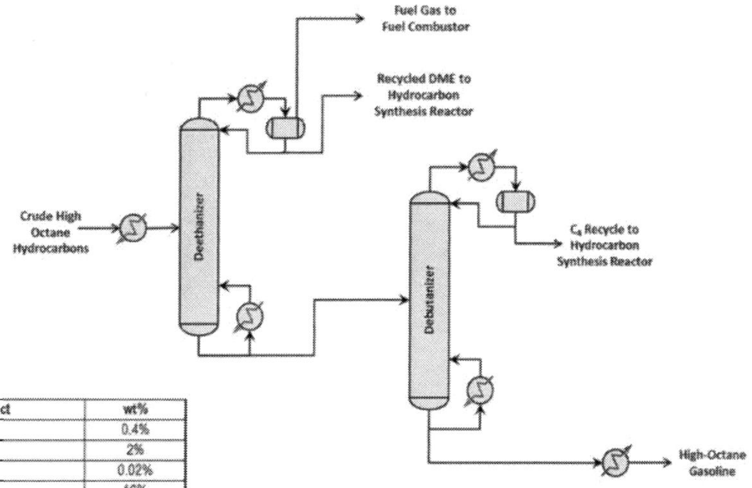

Crude Hydrocarbon Product	wt%
Light gas	0.4%
Propane	2%
Propylene	0.02%
Isobutane	16%
n-Butane	0.3%
Butenes	0.4%
C5+ gasoline	81%
Total	100%

Finished Fuel Products	wt%
Gasoline	95%
LPG	0%
Fuel gas	5%
Total	100%

Typical Crude Gasoline Separation

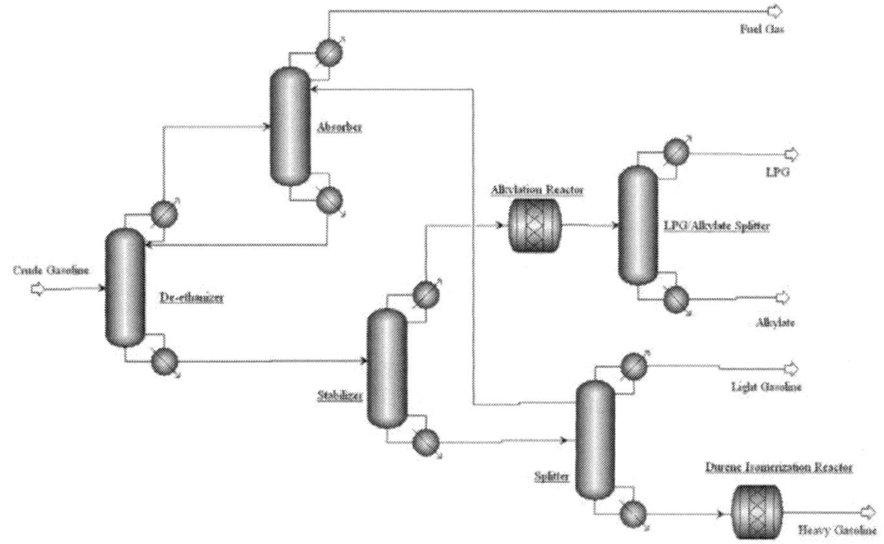

Source: Phillips, S.D.; Tarud, J.K.; Biddy, M.J.; Dutta, A. Gasoline from Wood via Integrated Gasification, Synthesis, and Methanol-to-Gasoline Technologies. NREL/TP-5100-47594. Golden, CO: National Renewable Energy Laboratory, 2011.

Comparison of gasoline composition

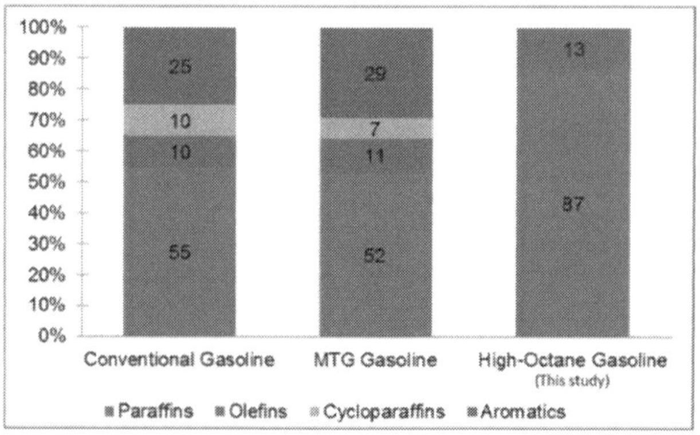

- Gasoline produced from this high-octane gasoline pathway contains only paraffins and olefins.
- High octane rating is largely attributed to its high concentration of iso-paraffins (98%).
- As opposed to MTG, no durene (1,2,4,5-tetramethylbenzene) is formed in this high-octane gasoline process.

Operating and Financing Assumptions

Description of Economic Parameter	Analysis Value / Basis
Delivered Feedstock Cost	$80.00 / US Dry Ton
Internal Rate of Return (IRR)	10.0 %
Plant Financing by Equity / Debt	40% / 60% of Total Capital Investment
Plant Life	30 Years
Income Tax Rate	35.0 %
Interest Rate for Debt Financing	8.0 % Annually
Term for Debt Financing	10 Years
Working Capital Cost	5.0% of Fixed Capital Investment
Depreciation Schedule	Total Plant: 7-Year MACRS
Construction Period (Spending Schedule)	3 Years (8% Y1, 60% Y2, 32% Y3)
Start-Up Time	6 Months
Revenue and Costs During Start-Up	Revenue = 50% of Normal Operation Variable Costs = 75% of Normal Operation Fixed Costs = 100% of Normal Operation
On-Stream Percentage After Start-Up	90% (7,884 Operating Hours per Year)

Design Case Economic Summary

Description of Economic Parameter	Current Value
Minimum Fuel Selling Price	$3.41 / Gal GE
Minimum Fuel Selling Price	$3.25 / Gallon
Feedstock Costs	$1.23 / Gallon
Operating Costs & Credits	$0.55 / Gallon
Capital Charges & Taxes	$1.47 / Gallon
Gasoline Blendstock Production	47.0 MM Gallons / Year
Gasoline Blendstock Product Yield	64.9 Gallons / US Dry Ton
Delivered Feedstock Cost	$80.00 / US dry ton
Total Purchase Equipment Cost (TPEC)	$122.2 MM
Installation Factor / Total Installed Cost (TIC)	x 2.06 Factor = $251.2 MM
Total Fixed Capital	$415.2 MM
Working Capital / Land Costs	$20.1 MM / $1.6 MM
Total Capital Investment	$437.5 MM
Lang Factor (TCI / TPEC)	3.58
TCI Lang Factor (TCI / ISBL TPEC)	4.54

NATIONAL RENEWABLE ENERGY LABORATORY 13

Design Report Cost Distribution Details

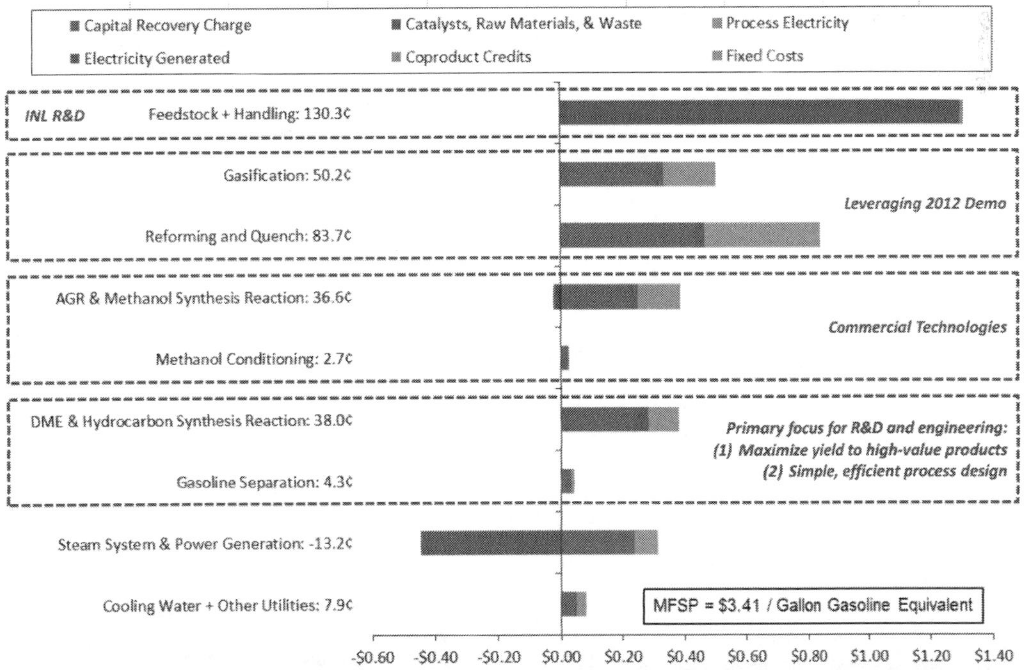

NATIONAL RENEWABLE ENERGY LABORATORY 14

Conversion Stage Sustainability Metrics

GHGs (g CO_2-e/MJ fuel) – (fossil emissions; biogenic emissions)	0.6; 153
Fossil energy consumption (MJ fossil energy/MJ fuel product)	6.0E-03
Total fuel yield (gal/dry ton feedstock)	64.9
Biomass Carbon-to-Fuel Efficiency (C in fuel/C in biomass)	31%
Water Consumption (m^3/day; gal/GGE)[1]	911; 1.76
Wastewater Generation (m^3/day; gal/GGE)	123; 0.24

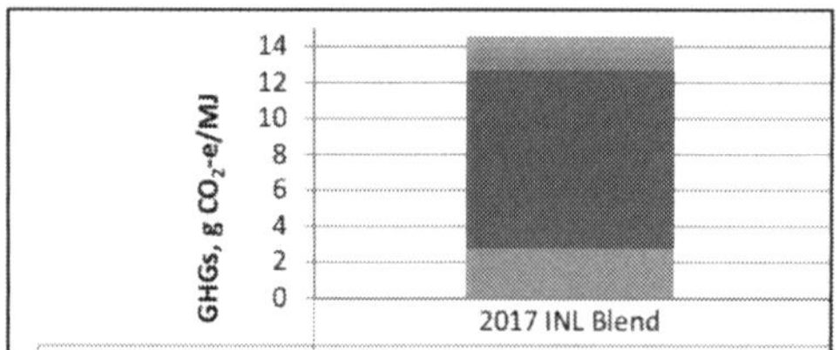

	2017 INL Blend
Fuel Use	0.49
Fuel Distribution	0.72
Conversion	0.57
Harvest, Collection and Preprocessing	10.0
Crop Establishment and Growing	2.8
Total	14.6

The overall life-cycle GHG emissions (14.6 g CO2-e/MJ from feedstock production to vehicle operation) reduction exceeds the 60% reduction criteria relative to the 2005 petroleum gasoline baseline (93.08 g CO2-e/MJ).

Single-Point Sensitivity Analyses

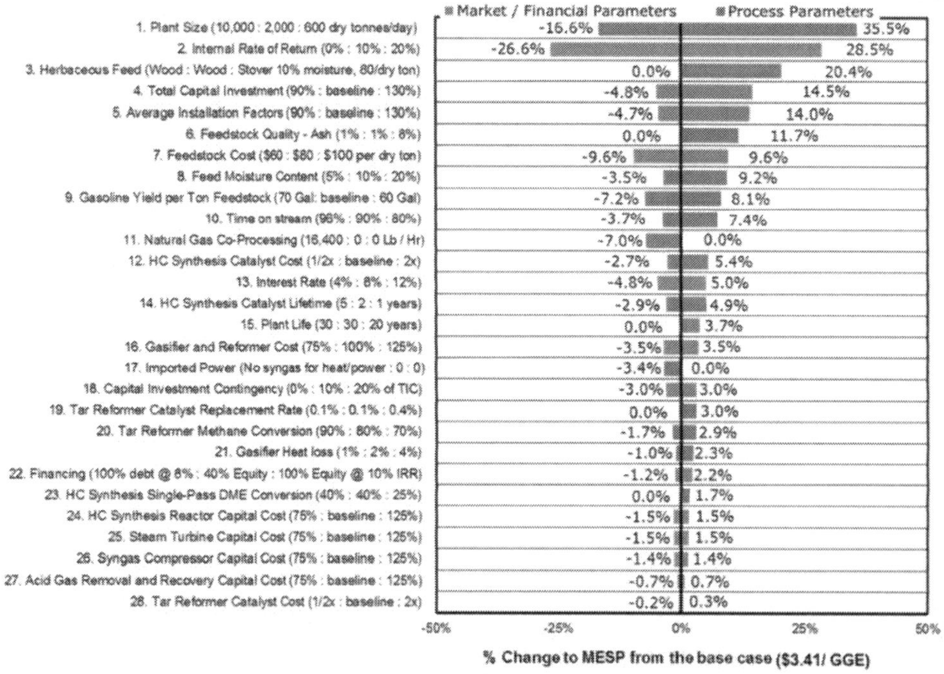

GHG/Cost Trade-Offs & Synergies: Gasification
Co-conversion with natural gas

- **Thermochemical conversion pathway to high-octane gasoline blendstock through methanol/DME intermediates**
  - Co-conversion of biomass with natural gas can simultaneously increase fuel yields and reduce fuel production costs provided that the life-cycle GHG thresholds specified in EISA is not violated.
  - Co-processing natural gas at the life-cycle GHG threshold limit (i.e., 60% GHG emissions reduction relative to the petroleum baseline) decreases the MFSP by 7%, from $3.41/GGE to $3.17/GGE.

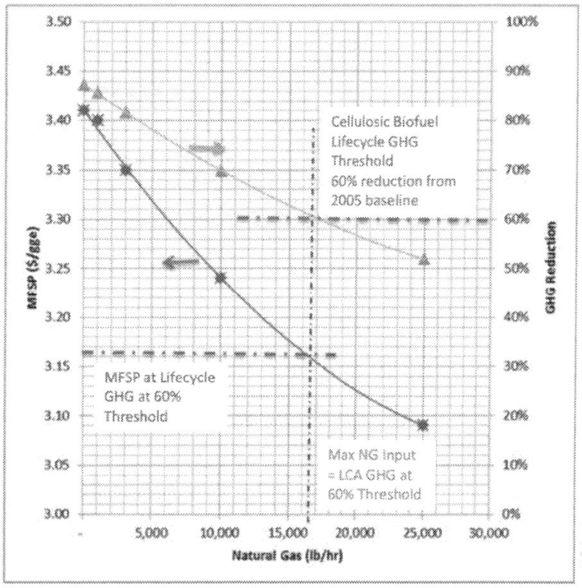

Potential Premium Value of the Design Case Modeled High-Octane Gasoline Blendstock

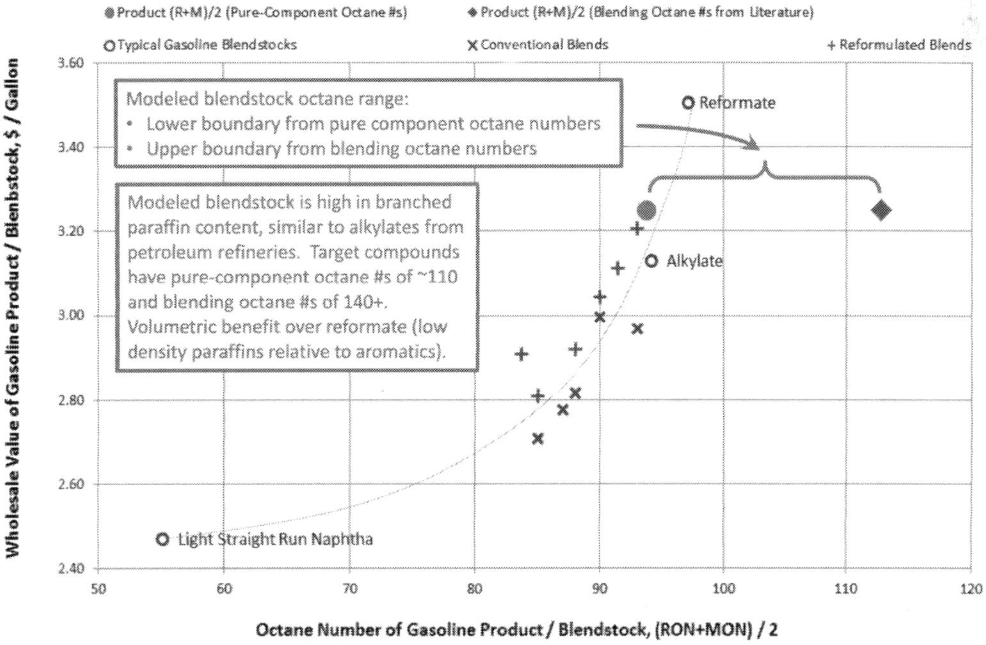

Sources: (1) May 2014 Spot Prices from EIA, (2) Platts Oilgram Price Report - July 11, 2013, (3) Modern Petroleum Technology; 5th Edition Part II; Edited by G.D Hobson, Wiley 1984.

Major Technical Targets for High-Octane Gasoline Pathway

Process Parameter	2014 SOT	2015 Target	2016 Target	2017 Target	2022 Target / Design Case
Hydrocarbon Synthesis Catalyst	Commercially available beta-zeolite		Beta-zeolite modified with copper (Cu) and gallium (Ga) for performance improvement		
H_2 Addition to HC Synthesis	No	Yes			
C_4 Co-Product (Function of Recycle)	Yes (No Recycle of C_4s)		No (Recycle C_4s)		
Single-Pass DME conversion	15%	15%	20%	30%	40%
Productivity of Hydrocarbon Synthesis Catalyst (kg/kg-cat/h)	0.02	0.03	0.04	0.05	0.10
Carbon Selectivity to C_5+ Product	46.2%	50.8%	86.1%	89.9%	93.1%
Carbon Selectivity to Aromatics (HMB represents coke / pre-cursers)	25% Aromatics (10% HMB)	15% Aromatics (7% HMB)	8% Aromatics (4% HMB)	4% Aromatics (2% HMB)	0.5% as HMB
Dimerization of C_4-C_8 Olefins to Jet	No	Consider in SOT cases as sensitivity or modify target case			

SOT / Target Waterfall for High-Octane Gasoline Pathway

Minimum Fuel Selling Price (MFSP) Breakdown ($ / Gallon of Gasoline Equivalent)	2014 State of Technology	2015 Target	2016 Target	2017 Target	2018 Target (Interpolated)	2019 Target (Interpolated)	2020 Target (Interpolated)	2021 Target (Interpolated)	2022 Target / Design Case
Feedstock	$ 1.99	$ 1.82	$ 1.48	$ 1.32	$ 1.31	$ 1.31	$ 1.31	$ 1.31	$ 1.30
Gasification	$ 0.70	$ 0.67	$ 0.56	$ 0.54	$ 0.53	$ 0.52	$ 0.52	$ 0.51	$ 0.50
Synthesis Gas Clean-up (Reforming and Quench)	$ 1.06	$ 1.00	$ 0.84	$ 0.84	$ 0.84	$ 0.84	$ 0.84	$ 0.84	$ 0.84
Acid Gas Removal, Methanol Synthesis and Methanol Conditioning	$ 0.59	$ 0.55	$ 0.44	$ 0.42	$ 0.42	$ 0.41	$ 0.41	$ 0.40	$ 0.39
Hydrocarbon Synthesis	$ 1.01	$ 1.01	$ 0.67	$ 0.60	$ 0.55	$ 0.51	$ 0.47	$ 0.42	$ 0.38
Hydrocarbon Product Separation	$ 0.05	$ 0.05	$ 0.05	$ 0.04	$ 0.04	$ 0.04	$ 0.04	$ 0.04	$ 0.04
Balance of Plant	$ 0.04	$ (0.00)	$ (0.00)	$ (0.04)	$ (0.04)	$ (0.04)	$ (0.05)	$ (0.05)	$ (0.05)
Minimum Fuel Selling Price (MFSP)	**$ 5.45**	**$ 5.09**	**$ 4.04**	**$ 3.72**	**$ 3.66**	**$ 3.59**	**$ 3.53**	**$ 3.47**	**$ 3.41**

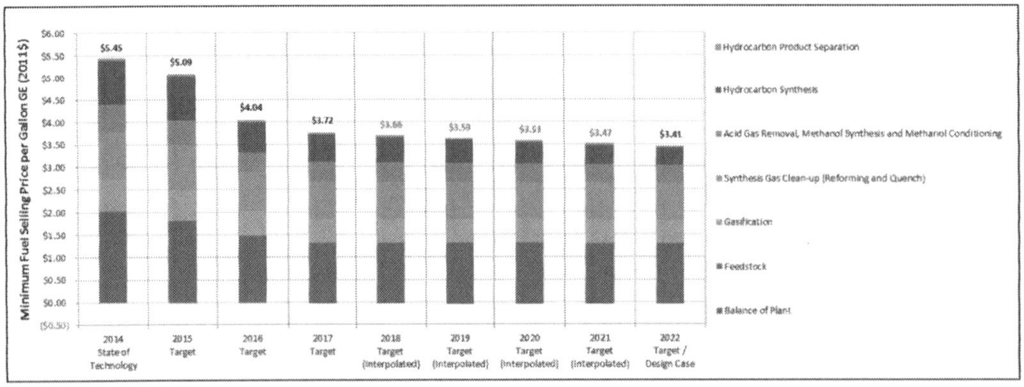

Product Quality Outlook for High-Octane Gasoline Pathway

Modeled product quality trends (1 of 3)

Product Quality Trends	2014 State of Technology	2015 Target	2016 Target	2017 Target	2018 Target (Interpolated)	2019 Target (Interpolated)	2020 Target (Interpolated)	2021 Target (Interpolated)	2022 Target / Design Case
Aromatics Content of C5+ Product (Wt%)	25.3%	15.0%	8.0%	4.0%	3.3%	2.6%	1.9%	1.2%	0.5%
C5+ Product Specific Gravity	0.75	0.73	0.71	0.70	0.70	0.70	0.70	0.70	0.70
C5+ Product API Gravity	58.17	62.41	68.57	69.37	69.52	69.68	69.83	69.98	70.14
Mixed C4 Product (Wt% Total Products)	27.2%	27.4%	0.0%	0.0%	0.0%	0.0%	0.0%	0.0%	0.0%
C5+ Aromatics (Wt% Total Products)	18.5%	10.9%	8.0%	4.0%	3.3%	2.6%	1.9%	1.2%	0.5%
C5+ Non-Aromatics (Wt% Total Products)	54.4%	61.7%	92.0%	96.0%	96.7%	97.4%	98.1%	98.8%	99.5%

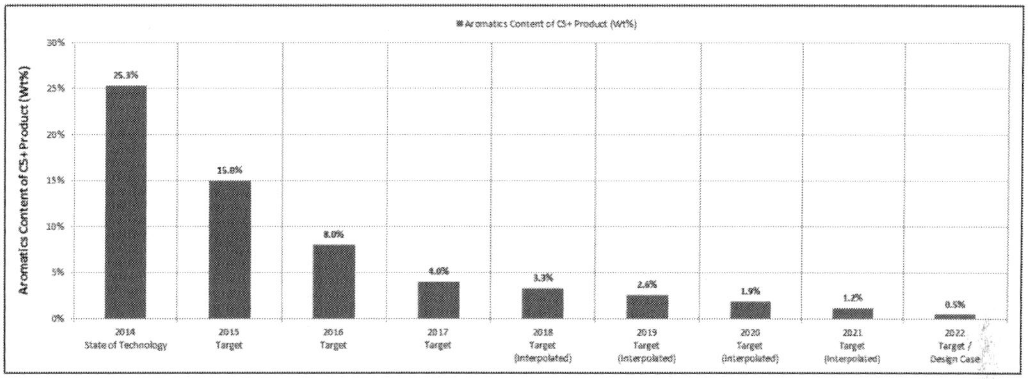

Sustainability Metrics for High-Octane Gasoline Pathway

Modeled trends in sustainability metrics (1 of 2)

Sustainability Metric Trends	2014 State of Technology	2015 Target	2016 Target	2017 Target	2018 Target (Interpolated)	2019 Target (Interpolated)	2020 Target (Interpolated)	2021 Target (Interpolated)	2022 Target / Design Case
Fossil GHG Emissions (g CO2-e / MJ Fuel)	1.64	1.42	1.19	0.96	0.88	0.81	0.74	0.67	0.60
Fossil Enegy Consumption (MJ Fossil Energy / MJ Fuel)	0.023	0.019	0.011	0.013	0.011	0.010	0.009	0.007	0.006
Total Fuel Yield (Gallons / Dry Ton)	57.5	58.5	61.8	64.2	64.4	64.5	64.6	64.8	64.9
Total Fuel Yield (Gallons GE / Dry Ton)	51.3	52.2	59.1	61.4	61.5	61.6	61.7	61.8	61.9
Overall Carbon Efficiency to Hydrocarbon Products (% C in Feedstock)	28.2%	28.7%	29.9%	31.0%	31.0%	31.0%	31.1%	31.1%	31.2%
Water Consumption (Gal H2O / Gal C5+)	12.4	9.3	5.6	5.2	4.5	3.8	3.1	2.4	1.7
Electricity Export (kWh / GGE)	0	0	0	0	0	0	0	0	0

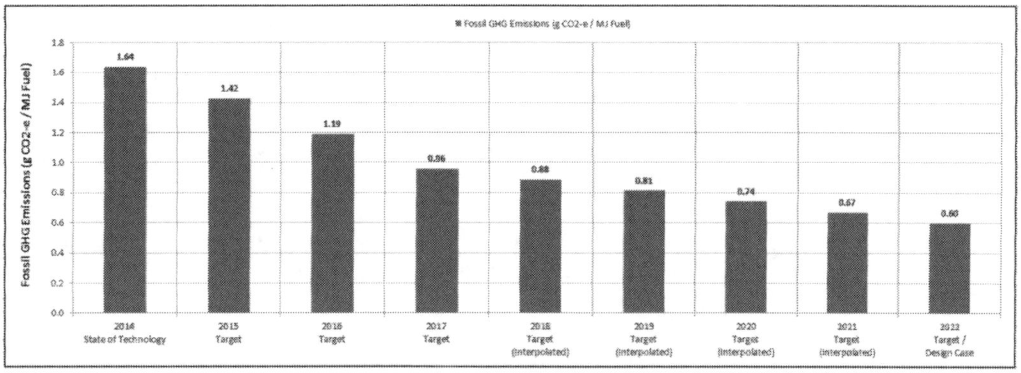

Sustainability Metrics for High-Octane Gasoline Pathway

Modeled trends in sustainability metrics (2 of 2)

Sustainability Metric Trends	2014 State of Technology	2015 Target	2016 Target	2017 Target	2018 Target (Interpolated)	2019 Target (Interpolated)	2020 Target (Interpolated)	2021 Target (Interpolated)	2022 Target / Design Case
Fossil GHG Emissions (g CO2-e / MJ Fuel)	1.64	1.42	1.19	0.96	0.88	0.81	0.74	0.67	0.60
Fossil Enegy Consumption (MJ Fossil Energy / MJ Fuel)	0.023	0.019	0.011	0.013	0.011	0.010	0.009	0.007	0.006
Total Fuel Yield (Gallons / Dry Ton)	57.6	58.5	61.8	64.2	84.4	64.5	64.6	64.8	84.9
Total Fuel Yield (Gallons GE / Dry Ton)	51.3	52.2	59.1	61.4	81.5	61.6	81.7	61.9	81.9
Overall Carbon Efficiency to Hydrocarbon Products (% C in Feedstock)	28.2%	28.7%	29.9%	31.0%	31.0%	31.0%	31.1%	31.1%	31.2%
Water Consumption (Gal H2O / Gal C5+)	12.4	9.3	5.8	5.2	4.5	3.8	3.1	2.4	1.7
Electricity Export (kWh / GGE)	0	0	0	0	0	0	0	0	0

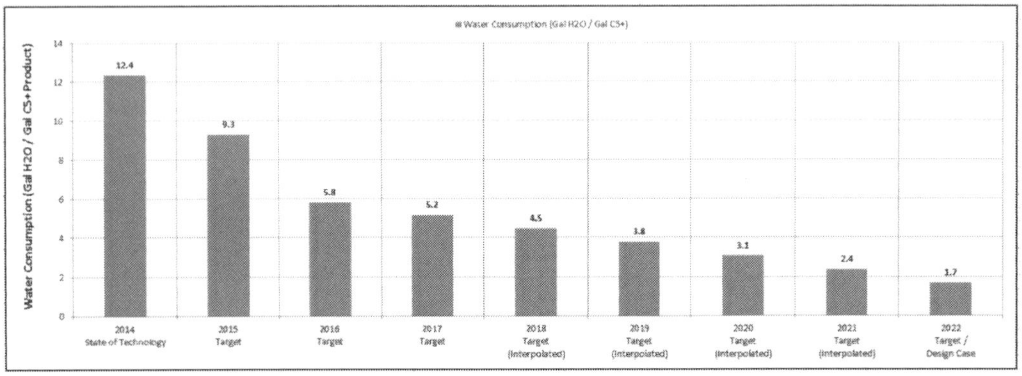

Product Quality Outlook for High-Octane Gasoline Pathway

Modeled product quality trends (3 of 3)

Product Quality Trends	2014 State of Technology	2015 Target	2016 Target	2017 Target	2018 Target (Interpolated)	2019 Target (Interpolated)	2020 Target (Interpolated)	2021 Target (Interpolated)	2022 Target / Design Case
Aromatics Content of C5+ Product (Wt%)	25.3%	15.0%	8.0%	4.0%	3.3%	2.6%	1.9%	1.2%	0.5%
C5+ Product Specific Gravity	0.75	0.73	0.71	0.70	0.70	0.70	0.70	0.70	0.70
C5+ Product API Gravity	58.17	62.41	66.57	69.37	69.52	69.66	69.83	69.98	70.14
Mixed C4 Product (Wt% Total Products)	27.2%	27.4%	0.0%	0.0%	0.0%	0.0%	0.0%	0.0%	0.0%
C5+ Aromatics (Wt% Total Products)	18.5%	10.9%	8.0%	4.0%	3.3%	2.6%	1.9%	1.2%	0.5%
C5+ Non-Aromatics (Wt% Total Products)	54.4%	61.7%	92.0%	96.0%	96.7%	97.4%	98.1%	98.8%	99.5%

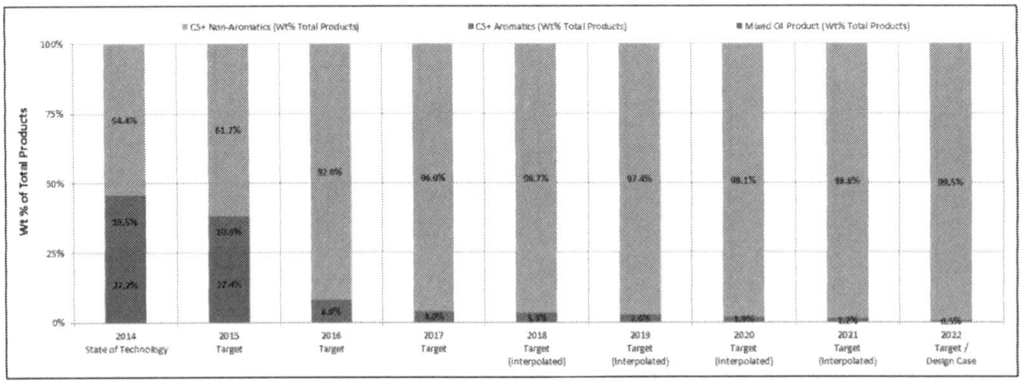

Acknowledgements

Michael Talmadge (NREL, co-author)
Abhijit Duta (NREL, co-author)
Jesse Hensley (NREL)
Jose Schaidle (NREL)
Mary Biddy (NREL)
Lesley Snowden-Swan (PNNL)

DOE's Bioenergy Technologies Office (BETO)
http://www.eere.energy.gov/biomass

Speaker information:
Eric.Tan@nrel.gov
http://www.nrel.gov/biomass

Assessing the Recovery of N-methyl-2-pyrrilodone from Waste Streams in the Specialty Chemical Industry

C. Stewart Slater, Mariano J. Savelski, Brigitte Pastore
Rowan University, Glassboro, NJ, USA

Frank A. Richetti, Nancy A. Uff
DuPont, Parlin, NJ, USA

Paper 752f
2015 Annual Meeting American Institute of Chemical Engineers
Salt Lake City, UT
November 2015

Rowan University

1

Introduction

- Case study is underway on waste reduction in specialty chemical manufacture
- EPA P2 Program – Region 2
- Partnership with DuPont plant in Parlin, NJ
  - Resins and resin precursors
  - Cyrel® printing products
  - ChromaPro® colorants
  - Teflon® and Autograph® coatings
- Case study focuses on resin precursor manufacture
- Batch-based process with relatively high solvent use
  - Focus on solvent reduction – parallels to Pharma industry

Rowan University

2

Background

- Polyimide (PI) and Polybenzoxazole (PBO) resin precursors
  - Applied and cured by DuPont's customers to form PI or PBO resins
- Resins essential to make
  - Microelectronic devices
  - Microelectromechanical systems
  - Aerospace components
- Synthesis requires solvents

A. J. Kirby, *Polyimides - Materials, Processing, and Applications*, Tarrytown, NY: Pergamon Press Inc., 1992.
TCS Electronics, Inc. *Assembly of Turn Flex Printed Circuit Board used within the Medical Industry* [Online]. Available: www.tcs-electronics.com
S. Leonard. (2010, May 10). *Making Sense of Medical MEMS Technologies* [Online]. Available: www.qmed.com

Rowan University

3

Background

- Solvent of interest in this study
  - N-Methyl-2-pyrrolidone (NMP)
- Issues
  - Suspected reproductive and developmental toxin
  - Significant life cycle emissions from manufacture
  - Expense for purchase and disposal
- Beneficial properties
  - Dipolar aprotic solvent
  - High thermal stability
  - Produces high molecular weight resin precursor

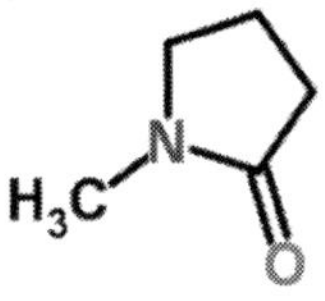

Royal Society of Chemistry (2015). *N-Methylpyrrolidone* [Online]. Available: www.chemspider,com
M. Reisch, "Solvent Users Look To Replace NMP," *Chemical & Engineering News Archive*, p. 32, 2008.
Scientific Committee on Consumer Safety, "Opinion on N-Methyl-2-pyrrolidone (NMP)," European Union, 2011.

Rowan University

Data generated from SimaPro®

4

Background: Toxic Release Inventory

U.S. NMP Waste in 2013 by Sector

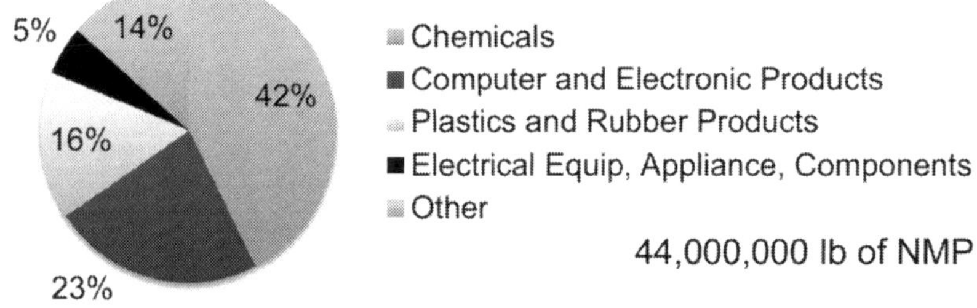

44,000,000 lb of NMP waste in 2013

Historical Trends in U.S. NMP Use

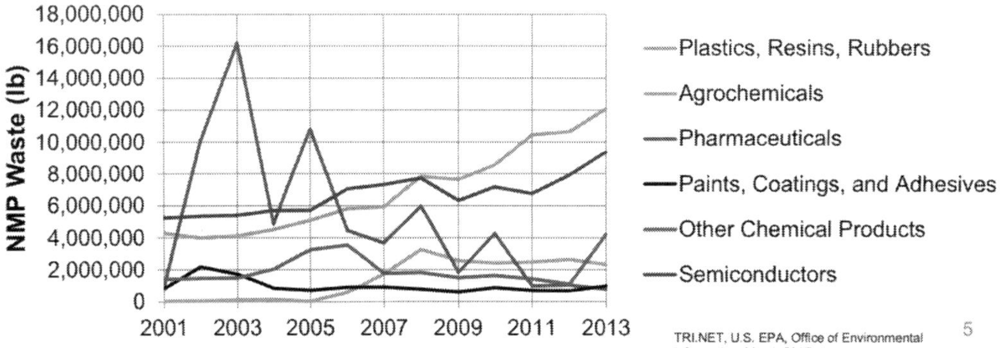

TRI.NET, U.S. EPA, Office of Environmental Information, March 2015.

5

Case Study Approach

- Evaluate current process
- Explore green engineering alternatives
  - Recover NMP from "hazardous" solvent waste
    - Reduce virgin NMP use and disposal
    - Design and analyze recovery system
  - Substitution of NMP with more benign solvents
  - Purify and reuse water from washwater waste
    - Design and analyze recovery system
- Analyze green engineering alternatives
  - Environmental and financial

Rowan University

6

Current Case

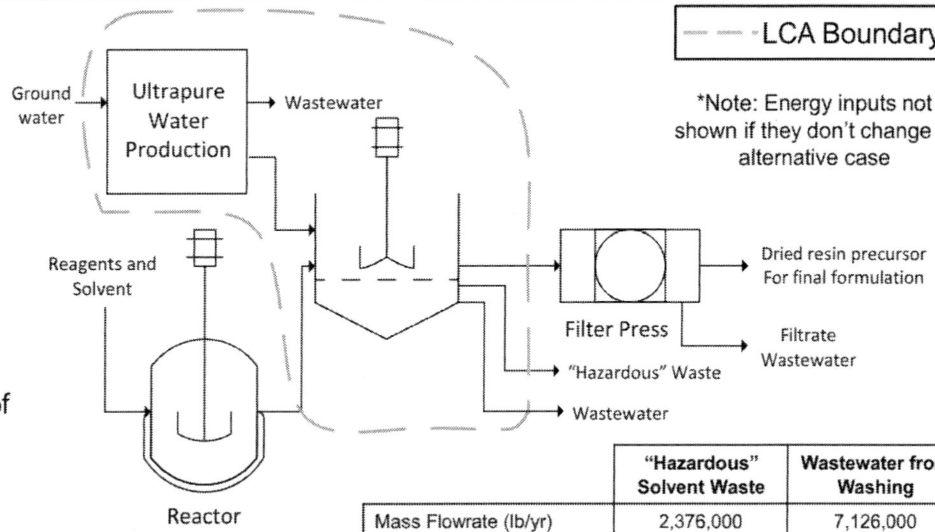

*Note: Energy inputs not shown if they don't change in alternative case

Basis: Annual amount of solvent used

Virgin NMP Flowrate (lb/yr)	404,000
Ground Water Flowrate (lb/yr)	15,535,000
Ultrapure Water Flowrate (lb/yr)	9,062,000
Ultrapure Water Production Wastewater Flowrate (lb/yr)	6,473,000

	"Hazardous" Solvent Waste	Wastewater from Washing
Mass Flowrate (lb/yr)	2,376,000	7,126,000
Water Composition (wt%)	81.5	99.85
NMP Composition (wt.%)	17	0.15
HEMA Composition (wt.%)	0.5	~50 ppm
HCl Composition (wt.%)	0.5	~50 ppm
TFA Composition (wt.%)	0.5	~50 ppm

Preliminary Analysis

- "First pass" approach examines separation of binary system of major components: NMP / H_2O
- Evaluate thermodynamics, concentration ranges, etc
- Evaluate different separation processes: distillation, pervaporation, etc.
- Screen alternative solvents
- Results: Distillation preferred method (non-azeotropic)

Rowan University

8

Proposed NMP Recovery

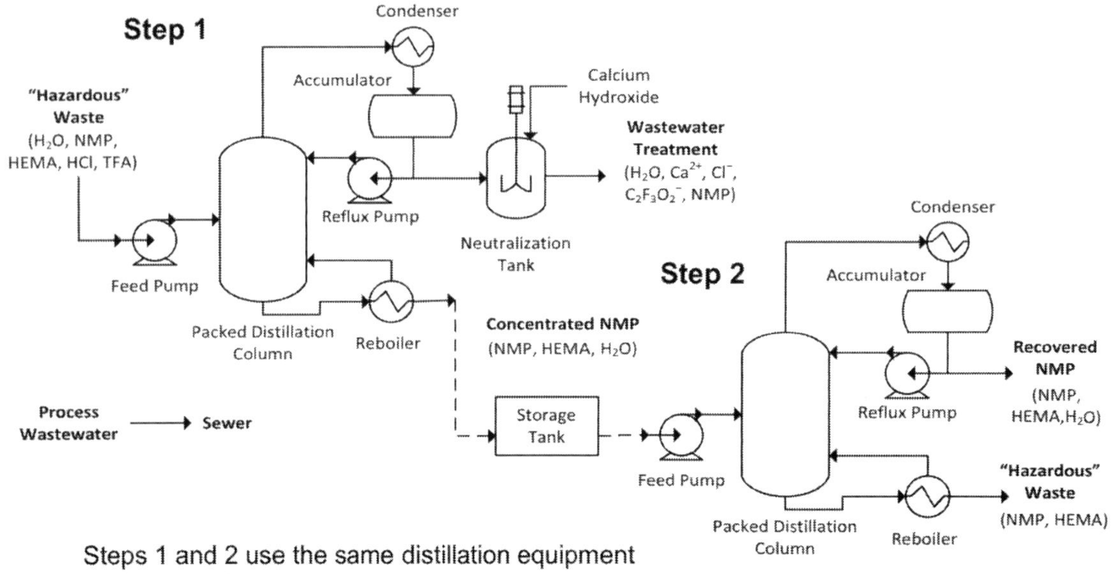

Steps 1 and 2 use the same distillation equipment operated in a sequential batch operation

Rowan University

9

Proposed NMP Recovery: Step 1

Step 1 operates 832 hr/yr

Calcium Hydroxide Slurry

	wt.%	Flow (lb/yr)
Water	85.0	90,000
Calcium Hydroxide	15	16,000
Total	100	106,000

Distillate — neutralized and sent to wastewater treatment

	wt.%	Flow (lb/yr)
Water	98.36	2,034,000
NMP	0.05	1,000
HEMA	trace	trace
Ca^{2+}	0.44	9,000
Cl^-	0.58	12,000
$C_2F_3O_2^-$	0.58	12,000
Total	100	2,068,000

"Hazardous" waste — from resin process

	wt.%	Flow (lb/yr)
Water	81.5	1,936,000
NMP	17	404,000
TFA	0.5	12,000
HCl	0.5	12,000
HEMA	0.5	12,000
Total	100	2,376,000

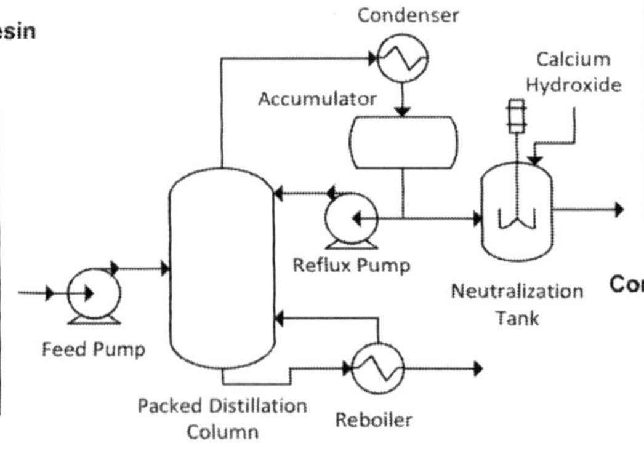

Concentrated NMP — sent to Step 2

	wt.%	Flow (lb/yr)
Water	0.03	100
NMP	97.11	402,000
TFA	trace	trace
HCl	trace	trace
HEMA	2.86	12,000
Total	100	414,000

Rowan University

10

Proposed NMP Recovery: Step 2

Step 2 operates 145 hr/yr

Concentrated NMP — from Step 1

	wt.%	Flow (lb/yr)
Water	0.03	100
NMP	97.11	402,000
HEMA	2.86	12,000
Total	100	414,000

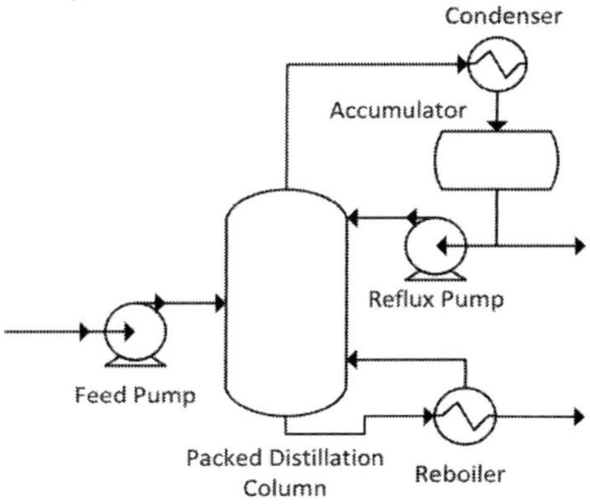

Recovered NMP

	wt.%	Flow (lb/yr)
Water	0.03	100
NMP	99.97	382,000
HEMA	trace	trace
Total	100	382,000

Bottoms — Disposed of as "Hazardous" Waste

	wt.%	Flow (lb/yr)
Water	trace	trace
NMP	62.96	20,000
HEMA	37.04	12,000
Total	100	32,000

This process recovers 95% NMP at 99.97% purity

Rowan University

11

Method of Analysis

- SimaPro®
  - Life Cycle Assessment
    - Total Emissions
    - ReCiPe
- AspenTech®
  - Process Design
  - Economic Evaluation
- R.SWEET
  - Software tool developed by Rowan

aspentech

Rowan University

E. J. Cavanagh, M. J. Savelski and C. Slater, "Optimization of environmental impact reduction and economic feasibility of solvent waste recovery using a new software tool," *Chemical Engineering Research and Design*, vol. 92, no. 10, pp. 1942-1954, 2014.

12

ReCiPe Damage Assessment

- Transforms individual life cycle category results into limited number of individual scores
- ReCiPe (H/A) — hierarchist approach, most common
- Scores express relative severity on an impact category: 18 Midpoints are combined into 3 Endpoints
  - Human Health, Ecosystems, Resources
- Endpoints are normalized and weighted to produce a single score
  - Human Health and Ecosystems weighted by 400 Pt
  - Resources weighted by 200 Pt

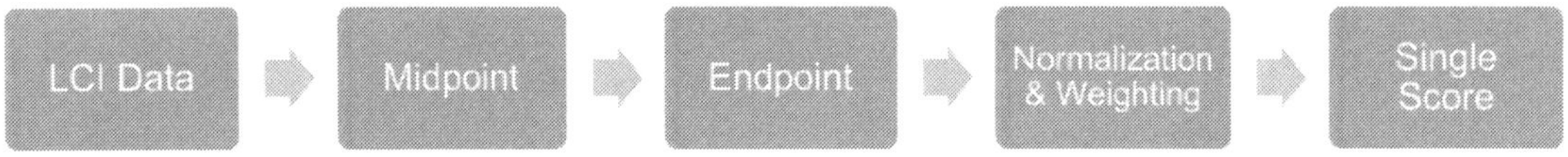

Rowan University

www.icia-recipe.net

M. Goedkoop, R. Heijungs, M. Huijbregts, A. De Schryver, J. Struijs and R. van Zelm, ReCiPe 2008, 2009.

13

Life Cycle Inventories and Costs

Manufacturing ("Cradle") LCIs, Damage factors, and Costs for Raw Materials
(1 lb basis)

	NMP	Ultrapure Water	HEMA*	HCl	TFA*	$Ca(OH)_2$
Total Emissions (lb)	4.22	0.0115	5.78	1.50	4.80	0.751
CO_2 (lb)	3.72	0.00913	5.59	1.27	4.16	0.740
Human Health (mPt)	66.2	0.143	105	31.9	112	10.3
Ecosystems (mPt)	34.9	0.0781	55.4	11.9	40.0	6.28
Resources (mPt)	97.7	0.104	148	22.2	92.6	4.98
Purchasing Cost ($)	1.95[a]	0.00613[a]	1.06[b]	0.126[b]	0.126[b]	0.113[c]

*Required chemical tree or substitution technique to estimate LCI

[a]Provided by DuPont
[b]ICIS (2008). *Indicative Chemical Prices A-Z* [Online]. Available: http://www.icis.com/chemicals/channel-info-chemicals-a-z/
[c]M.C. Schrock and J.K. Mercer, "Types of Calcium Reagents and Their Application in Acid Mine Drainage and Abondoned Mine Land Reclamation," *National Meeting of the American Society of Mining and Reclamation*, Pittsburgh, PA, 2010.

Data generated from SimaPro®

Rowan University

14

Life Cycle Inventories and Costs

LCIs, Damage factors, and Costs for Utility Production and Waste Disposal ("Grave")

	Utilities (1 MJ basis)		Waste Disposal (1 lb basis)	
	Electricity	HP Steam	"Hazardous" Waste	Wastewater
Total Emissions (lb)	0.245	0.148	0.217	0.0280
CO_2 (lb)	0.211	0.147	0.193	0.0275
Human Health (mPt)	4.26	2.01	2.52	0.404
Ecosystems (mPt)	1.91	1.26	1.62	0.235
Resources (mPt)	4.78	3.38	-3.95	0.103
Cost ($)	0.0314[a]	0.00682[b]	0.241[c]	0.00475[c]

Hazardous solvent waste sent to off-site incinerator
Wastewater from water washing and recovery distillation step 1 is sent to publically–owned treatment works (POTW)

[a]US Energy Information Administration (2015). *Electricity Data Browser* [Online]. Available: http://www.eia.gov/electricity/
[b]G. Towler and R. Sinnott, *Chemical Engineering Design*, New York: Butterworth-Heinemann, 2013.
[c]Provided by DuPont

Rowan University

Data generated from SimaPro®

15

Life Cycle Emissions

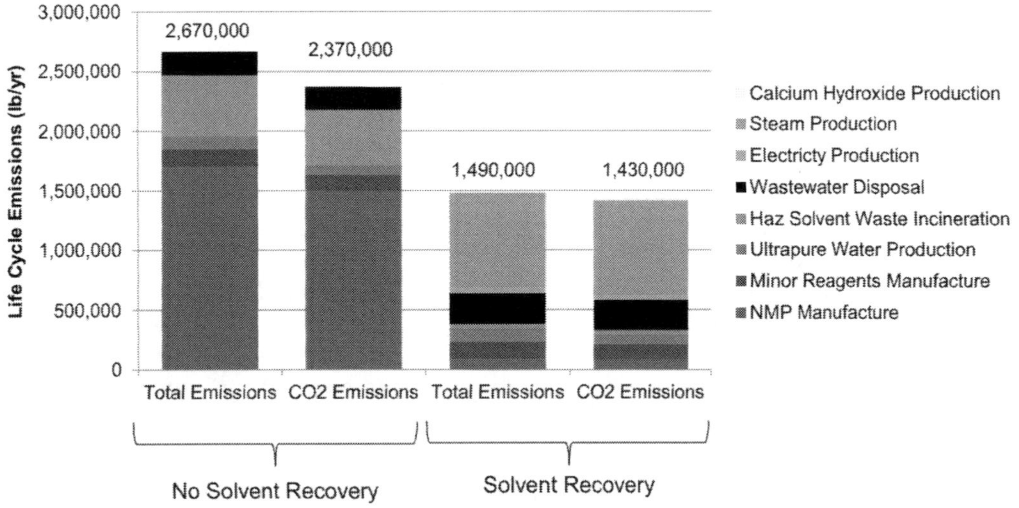

44% Reduction in Total Emissions
40% Reduction in CO_2 Emissions

Rowan University

16

ReCiPe Damage Assessment

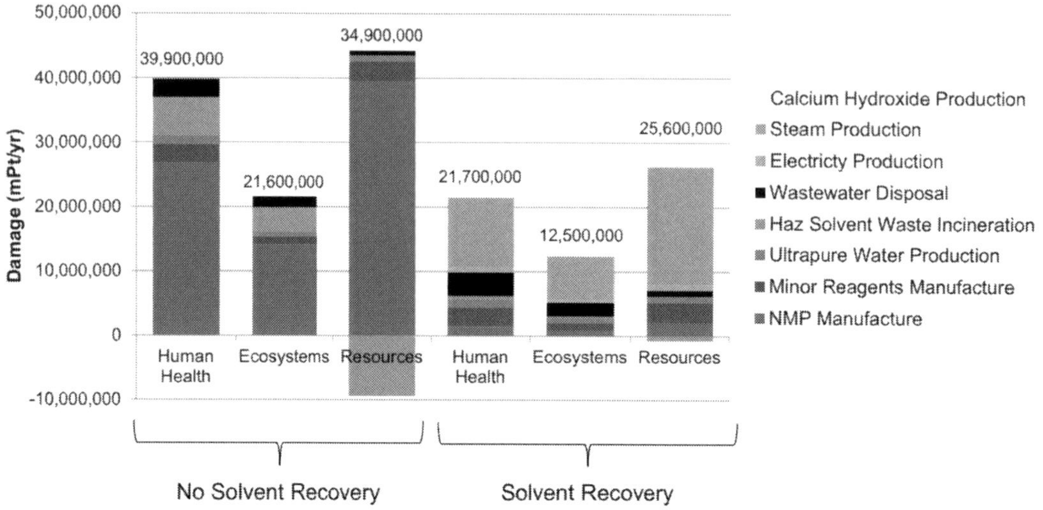

46% Reduction in Human Health Damage
42% Reduction in Ecosystems Damage
27% Reduction in Resources Damage
38% Reduction in Total Damage

Rowan University

17

Economic Life Cycle Assessment

Operating Costs

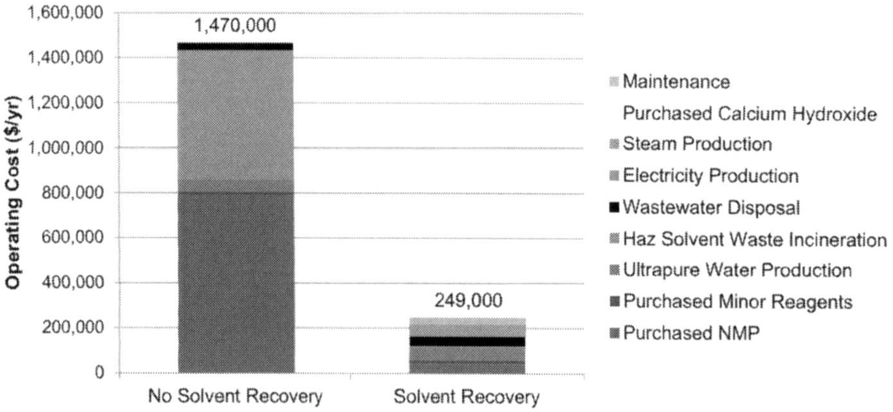

83% Annual Cost Savings

Rowan University

18

Alternative Solvents

- **DMSO** and **sulfolane** are favorable substitutes
  - Less hazardous
  - Reduced environmental impact
- Solvent substitution analysis
  - Substituted at a 1:1 ratio
  - Distillation recovery design varies for each solvent

Manufacturing LCIs and Costs for Solvents (1 lb basis)

	NMP	DMSO	Sulfolane*
Total Emissions (lb)	4.22	1.31	1.10
CO_2 (lb)	3.72	1.18	0.749
Human Health (mPt)	66.2	21.8	29.3
Ecosystems (mPt)	34.9	11.8	7.2
Resources (mPt)	97.7	54.9	44.3
Purchasing Cost ($)	1.95[a]	2.10[b]	3.09[b]

*Required chemical tree technique to estimate LCI

Data generated from SimaPro®

[a]Provided by DuPont
[b]"ACS Grade Solvents," Spectrum Chemical MGF Corp. 2015. [Online]. Available: https://www.spectrumchemical.com/

Rowan University

U. Tilstam "Sulfolane: A Versitle Dipolar Aprotic Solvent," *Org. Process Res. Dev.*, vol. 16, pp. 1273-1278, 2012.
A. Figoli *et al.* "Towards non-toxic solvents for membrane preparation: a review," *Green Chemistry*, vol. 16, pp. 4034-4059, 2014.

19

Life Cycle Emissions

- NMP, DMSO, and sulfolane are high boiling solvents – high pressure steam is needed for recovery
- DMSO and sulfolane recovery designs require vacuum distillation
- Sulfolane is recovered in 1 step - b.p. of sulfolane > b.p. of HEMA

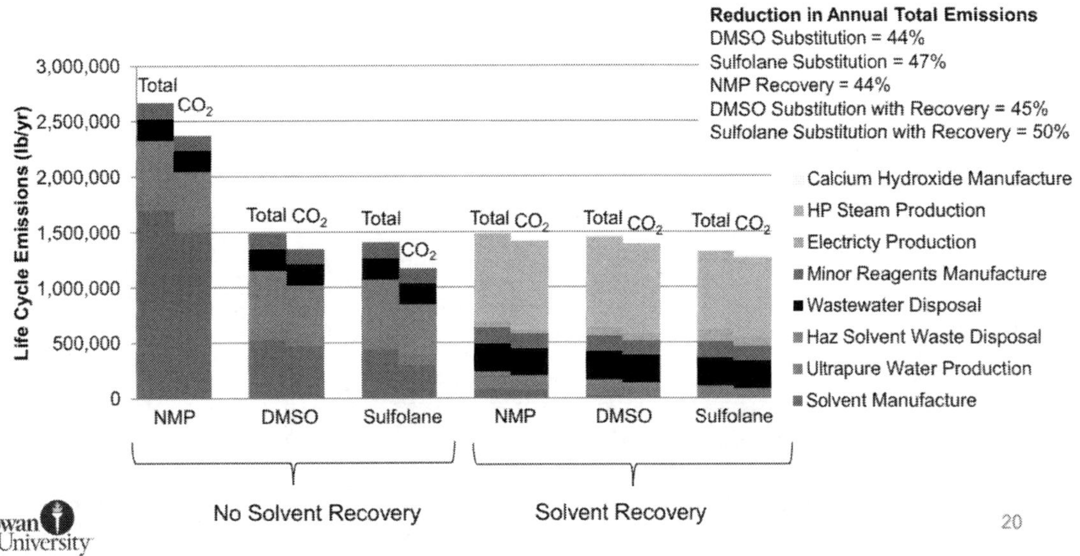

Rowan University

20

Economic Life Cycle Assessment

Operating Costs

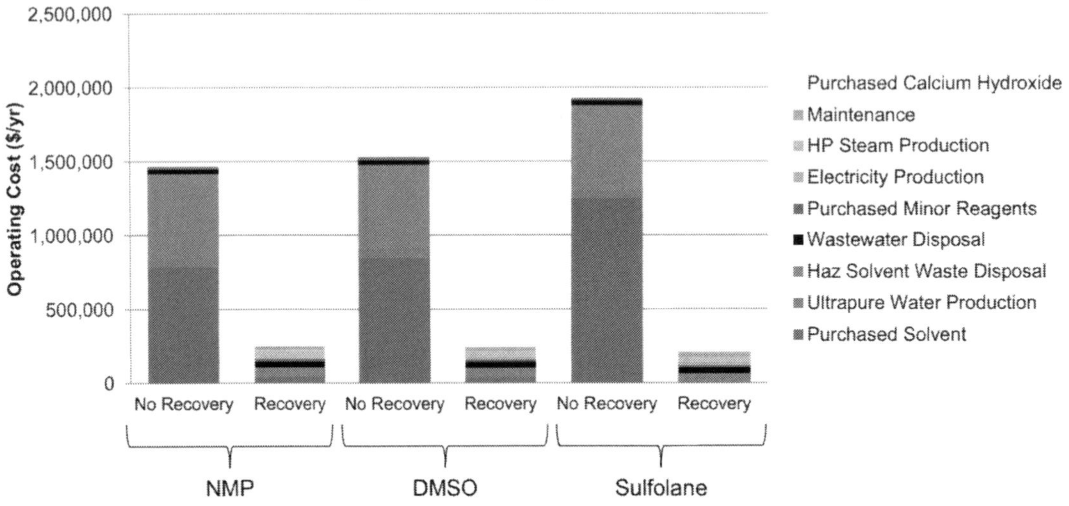

Reduction in Annual Operating Costs
DMSO Substitution = -4%
Sulfolane Substitution = -31%
NMP Recovery = 83%
DMSO Substitution with Recovery = 83%
Sulfolane Substitution with Recovery = 86%

Rowan University

Solvent Recovery Economics

Installed Capital Cost of Solvent Recovery Equipment (Distillation)

Solvent	Capital Cost ($)
NMP	1,497,000
DMSO	1,480,000
Sulfolane	1,566,000

Projected Economic Results for Solvent Recovery

Solvent	ROI	IRR	Payback Time (after tax)	NPV at 5 yr	NPV at 10 yr
NMP	59%	67%	2.2 yr	$1,750,000	$3,128,000
DMSO	60%	69%	2.2 yr	$1,781,000	$3,165,000
Sulfolane	58%	66%	2.3 yr	$1,790,000	$3,214,000

Notes: 7 year MACRS depreciation method
35% Tax Rate
15% Interest Rate

Rowan University

22

Overall Analysis of Alternatives

Scenario	Total Emission Reduction	CO_2 Emission Reduction	Damage Assessment Reduction	NPV at 10 yr
DMSO substitution without recovery	44%	43%	46%	$-298,000
Sulfolane substitution without recovery	47%	51%	50%	$-2,306,000
▶ NMP recovery	44%	40%	38%	$3,128,000
DMSO substitution with recovery	45%	41%	39%	$3,165,000
Sulfolane substitution with recovery	50%	46%	44%	$3,214,000

- Solvent substitutes appear slightly better in environmental categories, but don't yield significant economic improvement
- Recommended approach — NMP with recovery
  - If purity is sufficient, avoids validating alternative solvents

Rowan University

23

Other Processing Options

- NMP recovery without capital expenditure – assuming distillation equipment exists elsewhere
  - Transportation cost to ship NMP waste to alternative plant or Solvent Recovery facility
  - "Toll" cost to reprocess – will be higher if sent to outside Solvent Recovery facility
  - Transportation cost to ship purified NMP back to plant or for 3rd party use

- Water recovery from wastewater - production of purified water
  - Limited potential impact on LCA and cost benefit

Rowan University

www.cleanharbors.com

24

Summary

- Solvent recovery is a viable option for manufacturing platforms in the specialty chemical industry
- Case study shows potential benefits in $ saved and emissions reduced if a recovery scheme can be implemented
- Much more attractive if recovery equipment (distillation) capacity is available
- Future work
  - Low-energy separations
  - Alternative solvents
  - Reach out to other plant locations
  - Water reuse evaluation/design and LCA
  - Reverse osmosis / Pervaporation / Adsorption / IX for washwater recovery

Rowan University

www.ddpsltd.co.uk/distillation-and-evaporation,49910/

25

Acknowledgements

- Support for this research is provided by a grant from the U.S. Environmental Protection Agency – Region 2 - NP-96291314-0
- We acknowledge the support of DuPont Corporation and Michael Hatton Jr, Charles Richwine, Jeffrey Staniszewski (ret) of DuPont
- We acknowledge assistance of the following students: Matt Buchanan, Ben Ficara, Joe Garrett, Steve Hertzog, Ryan Liberty, Kyle Surette, Christian Wisniewski

Rowan University

26

THANK YOU
FOR YOUR INTEREST
AND BEING HERE
THURSDAY PM

THANK YOU TO THE
SESSION CHAIRS

Rowan University

27

431372 Looking into the Future of the Ethylene Industry: A Generic Assessment Model for Emerging Technologies

Abstract:

Ethylene is the largest volume chemical derived from natural gas liquids (NGL)[1]. Recent shale gas boom leads to decreasing price of NGL and natural gas, and as a result boosts its production and exports[2]. At the same time, ethylene is one of the largest energy consumer and GHG emissions resources in the chemical industry. Based on IEA's analysis, ethylene production accounts for 13% of energy use and 15% of GHG emissions for global chemical industry[3, 4]. Thus the rapid growth of ethylene production adds considerable amount of energy use and GHG emissions to the chemical industry. A lot of efforts have been taken to develop energy efficient technologies for the ethylene production, and it is critical to assess the potential life cycle impacts of them on energy saving and GHG emissions mitigation over temporal and geographic scale. However, due to lack of inventory data, it is difficult to conduct such an assessment for new technologies at a economy-wide scale, especially for those in their early stages.

In this work, the impacts of several emerging technologies that aim to improve energy efficiency of ethylene production are evaluated by a generic modeling framework. This model integrates chemical process design and modeling, life cycle assessment, techno-economic analysis and scenario analysis. A brief introduction of this model will be given in the beginning of the presentation. Then a case study of ethylene industry will be presented in the details. First, the conventional technology module in the modeling framework is used to model all direct and indirect unit processes involved in current ethylene production life cycle. Second, new technology module is activated to assess the net impacts of emerging technologies on energy use and GHG emissions. These impacts are compared with the convention technologies to shed light on improving opportunities in a product level. Last, the results of both conventional and emerging technologies are integrated with scenario analysis module to reflect the energy, emissions and economic implications of technology changes for the national-level ethylene production system in future decades. This model provides a reliable methodology to overcome knowledge barriers and deliver robust insights of potential influences that emerging technologies can have at a macro-level from life cycle perspective. The results can provide policy makers with good references on decision making for technology investment and promotion, offer manufactures with good understanding of key drivers of energy, emissions and resources use for a certain production system, provide researchers with insightful information on the possible bottleneck and future direction of R&D.

References

[1] B. Fallas and P. Pavlov, "Platts special report: petrochemicals, time to get cracking " McGraw-Hill Companies, Inc. , New York, NY, USA2013.

[2] ACC, "Year-end 2013 chemical industry situation and outlook," American Chemistry Council Washington, DC, USA2013.

[3] Y. Yao, D. Graziano, M. Riddle, J. Cresko, and E. Masanet, "Greener pathways for energy-intensive commodity chemicals: opportunities and challenges," *Current Opinion in Chemical Engineering,* vol. 6, pp. 90-98, 11// 2014.

[4] IEA, ICCA, and DECHEMA, "Technology Roadmap, Energy and GHG Reductions in the Chemical Industry via Catalytic Process," IEA, ICCA, DECHEMA, France2013.

AUTHOR INDEX

AUTHOR INDEX

AUTHOR INDEX

AUTHOR INDEX

AUTHOR INDEX

AUTHOR INDEX